LECTURES ON ELECTROCHEMICAL CORROSION

A publication of CEBELCOR, Brussels

LECTURES ON ELECTROCHEMICAL CORROSION

Marcel Pourbaix
Professor, Université Libre de Bruxelles
Manager, Centre Belge d'Etude de la Corrosion
CEBELCOR
Brussels, Belgium

Translated by J. A. S. Green
Research Institute for Advanced Studies (RIAS)
Martin Marietta Corporation
Baltimore, Maryland

Translation Edited by Roger W. Staehle
Corrosion Center
Department of Metallurgical Engineering
The Ohio State University
Columbus, Ohio

With a Foreword by Jerome Kruger
National Bureau of Standards
Washington, D.C.

PLENUM PRESS · NEW YORK-LONDON · 1973

Library of Congress Catalog Card Number 69-12537
ISBN 0-306-30449-X

A Division of Plenum Publishing Corporation
227 West 17th Street, New York, N. Y. 10011

United Kingdom edition published by Plenum Press, London
A Division of Plenum Publishing Company, Ltd.
Davis House (4th Floor), 8 Scrubs Lane, Harlesden, London, NW10 6 SE, England

Printed in the United States of America

To
Jean Baugniet
André Jaumotte
As a Testimony of Gratitude

And with affection
to my co-workers

Jean Van Muylder
Antoine Pourbaix
Leo Braekman

- Truth will sooner come out from error than from confusion.

Francis Bacon, 16th Century

- Pour atteindre la vérité, il faut une fois dans sa vie se défaire de toutes les opinions que l'on a reçues et reconstruire de nouveau, et dès le fondement, tous les systèmes de ses connaissances.

  To arrive at the truth it is necessary, at least once in life, to rid oneself of all the opinions one has received, and to construct anew, and from the fundamentals, all the systems of one's knowledge.

René Descartes, 1637

- On a opposé à Henry Le Chatelier une tradition qui ignorait tout de la thermodynamique. On lui a reproché de favoriser l'envahissement de la physique et des mathématiques dans une science qui pouvait à la rigueur les ignorer. Mais, en matière scientifique, il est toujours scabreux de faire de l'ignorance une arme de combat.

  There was opposed to Henry Le Chatelier a tradition which knew nothing of thermodynamics. He was reproached for encouraging the encroachment of physics and mathematics into a science which could perhaps ignore them. But in scientific matters, it is always a ticklish matter to make ignorance a weapon in your arsenal.

Georges Urbain, 1925

FOREWORD

Workers in the field of corrosion and their students are most fortunate that a happy set of circumstances brought Dr. Marcel Pourbaix into their field in 1949. First, he was invited, while in the USA, to demonstrate at a two-week visit to the National Bureau of Standards the usefulness of his electrochemical concepts to the study of corrosion. Secondly, also around the same time, Prof. H. H. Uhlig made a speech before the United Nations which pointed out the tremendous economic consequences of corrosion. Because of these circumstances, Dr. Pourbaix has reminisced, he chose to devote most of his efforts to corrosion rather than to electrolysis, batteries, geology, or any of the other fields where, one might add, they were equally valuable. This decision resulted in his establishing CEBELCOR (Centre Belge d'Étude de la Corrosion) and in his development of a course at the Free University of Brussels entitled "Lectures on Electrochemical Corrosion." This book is the collection of these lectures translated into English.

English-speaking students will now be able to study the fundamentals of the electrochemical aspects of corrosion from the special point of view of Dr. Pourbaix. While a special point of view in itself is not necessarily a virtue, in this case it is decidedly so. These lectures place a great deal of emphasis on the potential–pH equilibrium diagrams devised by their author and known the world over as "Pourbaix diagrams." Invaluable tools in understanding complex corrosion phenomena, these diagrams enable the relevant thermodynamic parameters in a corroding system to be evaluated by looking at the "whole picture." As useful as this is in studying corrosion phenomena, it is still more valuable in corrosion education. This book clearly and elegantly develops the fundamentals and thereby makes clear to the student the thermodynamic basis of corrosion, deriving in the end the potential–pH diagrams.

Corrosion, being of the real world, must be connected to real situations. To bolster this feeling for the subject the lectures relate many of the principles to laboratory demonstrations. After an understanding is carefully built, it is shown how the fundamental concepts can be used in the protection of metals against corrosion, as by cathodic protection, corrosion inhibition, coatings, etc. Therefore, there is no chance that the student will question the relevance of idealized thermodynamic concepts to the real world. Moreover, while Dr. Pourbaix places much greater emphasis in this book on thermodynamics than on kinetics, the latter is not neglected. His approach to electrochemical kinetics is a pragmatic one that lends itself to attacking problems rather to the consideration of detailed mechanisms. This conforms with Dr. Pourbaix's approach—one foot firmly planted in fundamentals, the other in the world of problems that these principles have to attack. The result, this book, gives the student the sort of understanding that will allow him either to go into the more advanced fundamental aspects of electrochemistry or to attack applied problems by an intelligent use of scientific principles.

The English painter John Constable has said, "We see nothing truly until we understand it." Dr. Pourbaix has provided a most valuable aid for the understanding to help us see corrosion more truly.

Jerome Kruger

National Bureau of Standards
Washington, D.C. 20234

PREFACE

This book is based on a course given in the School of Applied Sciences of the University of Brussels. The aims of this course are to elucidate fundamental principles which govern corrosion in aqueous solutions and to apply these principles to the resolution of industrial corrosion problems.

The subject matter of this book is intimately related to the work of the *Centre Belge d'Étude de la Corrosion, CEBELCOR*, and of the *Comité International de Thermodynamique et de Cinétique Electrochimiques, CITCE.*[1]

These organizations publish extensively in areas directly related to fundamental and practical aspects of corrosion:

Publications of CEBELCOR

Recueils de Mémoires (RM) *Corrosion et Protection des Matériaux*
Rapports Techniques (RT) (in French and in English)
Bulletin d'Information (BI)
Review *Corrosion Science*
Various publications in French (F), German (D), English (E), Italian (I), Japanese (J), Dutch (N), Portuguese (P) and Spanish (S).

Publications of CITCE

Proceedings of annual meetings
- 2nd meeting (Milan 1950)
- 3rd meeting (Bern 1951)
- 6th meeting (Poitiers 1954)
- 7th meeting (Lindau 1955)
- 8th meeting (Madrid 1956)
- 9th meeting (Paris 1957)

Review *Electrochimica Acta*

[1] In 1970 *CITCE* became the *International Society of Electrochemistry* (*I.S.E.*).

This book has its origin in research which began in 1937 and was subsequently presented as a thesis to the University of Brussels on May 3, 1940. The University was closed during World War II. Afterwards, one of the reports from this research was, at the kind invitation of F.E.C. Scheffer, defended as a thesis at the Technical University in Delft on October 31, 1945 (Refs. 2, 22).[2] The other reports were defended as a thesis at the University of Brussels on March 16, 1945 (Ref. 3). See also CEBELCOR publications in French (F) 4, 5, 21, 22, 24.

A general bibliography is given at the end of the book. This bibliography includes books in which the reader may find more extensive discussions of the scientific and applied aspects alluded to in this book. Special attention should be given to the publications of U. R. Evans because of his enthusiasm and competence in the area of corrosion as well as the high quality of his work. We believe that he should be considered the father of the scientific study of aqueous corrosion.

Laboratory experiments related to these lectures are described in CEBELCOR's Publications RT. 92 (in French) and E. 65 (in English).

I take this opportunity to express my gratitude to those who, too numerous to name individually, have helped me pursue the studies upon which this book is based. I especially remember three departed friends: F. E. C. Scheffer, Professor at the Technical University, Delft, without whose encouragement this task would not have been undertaken; Louis Baes, Vice President of CEBELCOR, whose integrity, energy, and generosity has protected this institution on many occasions; Leo Cavallaro, Scientific Advisor of the Commission of Fundamental Studies and Applications CEFA and National Secretary of CITCE for Italy whose enthusiasm, kindness, and support was forthcoming as soon as he knew of us. Particularly I thank Messrs. Jean Baugniet, Willy G. Burgers, Gaston Charlot, George Chaudron, Jacques Errera, Ulick R. Evans, Edward C. Greco, Thomas P. Hoar, André Juliard, Félix Leblanc, Roberto Piontelli, Gabriel Valensi, and Pierre van Rysselberghe for the support they have constantly lavished, and our other friends concerned with the origination and development of CITCE. I thank the faithful team of researchers which has carried out much of the work of CEBELCOR discussed in this book: Jean van Muylder, Pierre van Laer, Antoine Pourbaix, Jean Meunier, and my old friend Leo Braekman.

The work of CEBELCOR has been supported by the Institute for the Encouragement of Scientific Research in Industry and Agriculture IRSIA,

[2] The numbers refer to references in the general bibliography at the end of the book.

by the Collective Fundamental Research Foundation FRFC, by the Free University of Brussels, by the US Air Force System Command (Material Laboratory Research and Technology Division), and by members of the Commission of Fundamental Studies and Applications CEFA.

Also my thanks go to those who have presented this work in the English language: to the names of U. R. Evans, T. P. Hoar, and P. van Rysselberghe, I add those of J. N. Agar, P. Delahay, J. A. S. Green, A. G. Guy, and F. N. Rhines, as well as those of M. C. Bloom, B. F. Brown, H. W. Paxton, and E. D. Verink, and of the Advanced Research Projects Agency (ARPA) of the USA. I should also like to thank particularly warmly R. W. Staehle, who suggested the publication of the lectures upon which this book is based and edited the translation with greatest care and competence. Finally, I would like to express my appreciation to G. J. Theus, who assisted R. W. Staehle with the editing of the book and helped coordinate the final preparation of the manuscript.

M. Pourbaix

CONTENTS

NOTATION

a	activity
ⓐ	hydrogen-equilibrium line, for 1 atm (in diagrams of electrochemical equilibria)
A	hydrogen-equilibrium electrode potential, for 1 atm
A	affinity
ⓑ	oxygen-equilibrium line, for 1 atm (in diagrams of electrochemical equilibria)
B	oxygen-equilibrium electrode potential, for 1 atm
c	concentration; molarity (moles $liter^{-1}$)
e^-	electron
E	internal energy
E	electrode potential
E_0	equilibrium electrode potential
$E_0{}^0$	standard equilibrium electrode potential
f	activity coefficient
F	Helmholtz free energy ($F = E - TS$)
g	specific molar free enthalpy ($g = h - Ts$)
G	free enthalpy (or Gibbs free energy) ($G = H - TS$)
h	specific molar enthalpy
H	enthalpy ($H = E + pV$)
i	reaction current (positive for oxidations, negative for reductions)
i_0	exchange current
I	electrolysis current (positive for oxidations, negative for reductions)
K	equilibrium constant
ln	napierian logarithm (natural logarithm)
log	decimal logarithm

M	symbol of a chemical species; the following typographic convention is employed:
	M for liquid water and for species dissolved in water
	M for solid species
	M for liquid species other than water
	M for gaseous species
$[M]$	activity of dissolved species, and fugacity of gaseous species
p	partial pressure
P	total pressure
R	gas constant (1.985 cal/°K)
s	specific molar entropy
S	entropy
SCE	saturated calomel electrode
SHE	standard hydrogen electrode
SOE	standard oxygen electrode
t	centigrade temperature
T	absolute temperature
V	volume
Z	valency
α	Tafel constant
β	Tafel slope
∂	partial differential
ε	electromotive force
φ	fugacity
γ	constituent of a chemical system
μ	chemical potential
μ^0	standard chemical potential
ν	stoichiometric reaction coefficient of a chemical species
ξ	advancement of a reaction
$\sum$	sum
ψ	electric potential

Chapter 1

INTRODUCTION TO CORROSION

The word *corrosion* means the destruction of a material under the chemical or electrochemical action of the surrounding environment. For this reason corrosion is a subject of technical and economic interest as well as of scientific interest. This chapter gives examples of the variety of corrosion phenomena, traces historical developments, describes the economic significance, and presents the results of illustrative experiments.

1.1. ECONOMIC AND TECHNICAL SIGNIFICANCE

It is well known that iron and ordinary unalloyed steels unless adequately protected, for example, by painting, corrode easily and are transformed with varying rapidity into rust. While the rapid rusting of many commercial steel products is commonplace, examples exist of very slow rusting of articles exposed to the atmosphere. For example, in Delhi there is a very old iron column which, although unprotected, has scarcely deteriorated. However, this case is exceptional and is probably due to the particular composition of the metal and to the dry and often hot climate. Atmospheric corrosion of some low-alloy steels known as "weathering steels" may sometimes lead to the formation of a protective patina which prevents further corrosion of the steel.

For most metal users, corrosion is a common source of anxiety. A person is said to be "rusting" when he begins to lose his faculties! Among the consequences of corrosion occurring in daily life, one may mention:

- the necessity of painting and repainting structures fabricated of ordinary steels;

- the premature destruction of automobiles due to the corrosive action of deicing salts on the lower part of the vehicle, to the corrosive action of cooling water or antifreeze in radiators, and to the corrosive action of combustion gases on mufflers;
- damage in zinc and lead roofing.

The economic significance of corrosion is illustrated by the following:

(a) The quantity of iron destroyed annually by corrosion has been estimated as 1/4 or 1/3 of the annual production. At an exhibition dealing with corrosion organized in Brussels in 1937 by the Shell Company, the following poster was displayed:

While you read this 760 kg of iron has been corroded

(b) Each year, 750,000 tons of sulfuric acid from the combustion of coal falls on London. This quantity is capable of completely dissolving more than 400,000 tons of iron.

(c) In 1949, H. H. Uhlig[1] estimated the cost of the corrosion of metals in the United States to be about 5.5 billion dollars a year. The breakdown is shown in Table I. Estimates made in 1966 by Stanley Liechtenstein of National Bureau of Standards[2,3] run to 10 billion dollars (Table II).

These sums of about 34 or 50 dollars per inhabitant per year correspond for Belgium to about 16 or 25 billion francs per year, equaling the total amount of all public investments in Belgium.

These evaluations only represent the *direct loss*, the cost of the protection and replacement of the corroded material. The incalculable and often much greater *indirect* losses are not included. These result from the breakdown of installations (water distribution, electrical networks, electricity supply, factories, transport), from the loss of products (water, gas, gasoline), from diminished output, from explosions (gas), from contamination (foodstuffs in tins and packages; the presence of iron in chemical or other products), and even from the loss of human life.

[1] H. H. Uhlig, Cost of corrosion in the United States, *Corrosion* **6**, 29–33 (1950).
[2] Stanley Liechtenstein, The many faces of corrosion, *Technical News* STR-3454, National Bureau of Standards NBS, Washington, D.C. (1966).
[3] Corrosion costs $ 10 billions per year, *Materials Protection* **16**, 24 (1967).

TABLE I. Cost of Corrosion in the United States in 1949 According to Uhlig

1. Paint, varnish, and lacquers	$ 2,045,000,000
2. Phosphate coatings	$ 20,000,000
3. Galvanized sheet, pipe, and wire	$ 136,500,000
4. Tin and tin plate	$ 316,000,000
5. Cadmium electroplate	$ 20,100,000
6. Nickel and nickel alloys	$ 182,000,000
7. Copper and copper-base alloys	$ 50,000,000
8. Stainless Fe–Cr and Fe–Cr–Ni alloys	$ 620,400,000
9. Boiler and other water conditioning	$ 66,000,000
10. Underground pipe	$ 600,000,000
11. Oil refinery	$ 50,000,000
12. Domestic water heaters	$ 225,000,000
13. Internal combustion engines	$ 1,030,000,000
14. Mufflers	$ 66,000,000
Total	$ 5,427,000,000

It has been stated that corrosion in condenser tubes of warships at a crucial period of the First World War caused greater anxiety to the British Admiralty than all their purely operational problems. The "British Non-Ferrous Metals Research Association," alerted to this danger during the war, promptly introduced "Admiralty" brass (76Cu, 22Zn, 2Al).

The economic importance of corrosion has increased considerably in recent years, mostly due to the development of relatively new types of equipment whose satisfactory operation requires very low corrosion rates: cor-

TABLE II. Cost of Corrosion in the United States in 1966 According to Liechtenstein

1. Water heaters	$ 300,000,000
2. Tankers gasoline	$ 75,000,000
3. Petroleum refineries	$ 50,000,000
4. Automobile mufflers and tail pipes	$ 500,000,000
5. Petroleum structures underground	$ 365,000,000
6. Waterworks pumping, added capacity	$ 40,000,000
7. Miscellaneous costs	$ 8,670,000,000
Total	$ 10,000,000,000

rosion may cause, and actually already has caused, considerable disasters to nuclear reactors, airplanes, missiles, and automatic apparatus. Corrosion is a limiting problem presently in the nuclear, aerospace, and electronics industries.

In the United States, a 1% reduction of corrosion loss would produce, according to the evaluation made by Uhlig in 1949, a yearly savings of 55 million dollars. The situation outlined by Uhlig constitutes a problem of the greatest magnitude to engineers and scientists. As an incentive to encourage additional work in combating corrosion Uhlig suggested that a plan of 5 years in research and education could allow a savings of up to 20% of the actual loss, i.e., a 1000 million dollars yearly saving to the United States. Assuming an outlay of 5–10 million dollars for this effort, or 0.5–1% of the sum saved annually, the invested capital would yield a return of 10,000–20,000%. Furthermore, a thorough understanding of corrosion phenomena would often prevent large blunders. While such a prognostication may be optimistic, there is clearly a strong incentive for starting substantial investigations.

CEBELCOR was created and has operated with the main goal of applying the Uhlig plan to industrial corrosion problems: to fight against corrosion simultaneously on the four fronts of fundamental research, applied research, direct help to industry, and information and education. These are the four aspects of a polyvalent and combined technical and scientific action: to know, to know making, to make, and to make knowing. Besides CITCE, which is an international forum for scientific discussion in the field of fundamental electrochemistry, CEBELCOR is mainly an international organization for applying electrochemistry to corrosion. Both of these organizations have been created on the basis of the doctrine which underlies the theme of this book.

To illustrate the complexity and often unpredictable nature of corrosion, the following examples of major corrosion "blunders" are cited:

(a) To wash linen, a Danish laundry had to use hard water which required an excessive amount of soap and consequently the linen was prematurely worn. So the laundry decided to soften the water. Unfortunately, this corroded the water supply system and led to the formation of rust spots on the linen.

(b) Similarly, a hospital had its water softened and then distributed by lead piping; the lead, which had resisted hard water, corroded in contact with the softened water, and this produced symptoms of lead poisoning among the hospital patients.

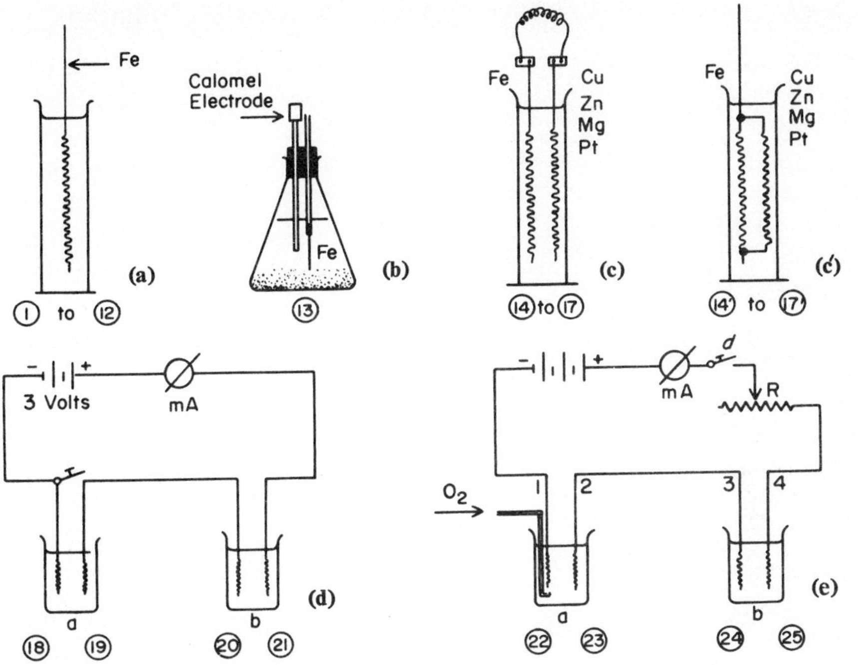

Figure 1. Demonstration experiments.

(c) Before moving some lead storage batteries an organization decided, because of the danger of transporting sulfuric acid, to replace the acid in the batteries by distilled water; this resulted in extensive corrosion which completely destroyed the battery plates in a few hours.

The explanation for the above will become apparent during the reading of subsequent chapters.

1.2. COMPLEXITY OF CORROSION PHENOMENA

Corrosion phenomena are very complex. Their causes and solutions are often obscure. Often, a normally resistant metal corrodes; conversely, but unfortunately less often, a metal thought to be nonresistant behaves reasonably well. Figure 1 shows a number of experimental arrangements for observations which illustrate the apparently capricious nature of corrosion.[4]

[4] Applications of potential–pH diagrams for iron and hydrogen peroxide, *Rapports Techniques CEBELCOR* **2**, RT. 2 (1954) (in French).

1.2.1. Iron in the Presence of Different Aqueous Solutions

Lengths of piano wire are introduced into tubes containing various solutions shown in Figure 1a:

1. Distilled water
2. Solution of sodium chloride (1 g/liter)
3. Solution of sulfuric acid (2 g/liter)
4. Solution of sodium bisulfite (2 g/liter)
5. Solution of caustic soda (1 g/liter)
6. Solution of potassium chromate (1 g/liter)
7. Solution of potassium chromate (1 g/liter) and sodium chloride (1 g/liter)
8. Dilute solution of potassium permanganate (0.2 g/liter)
9. Concentrated solution of potassium permanganate (1 g/liter)
10. Water of relatively low hydrogen peroxide content (0.3 g/liter of H_2O_2)
11. Water of relatively high hydrogen peroxide content (3.0 g/liter of H_2O_2)
12. Tap water from the city of Brussels

It is observed that iron corrodes in distilled water (1) with formation of brown rust; the corrosion is enhanced by the presence of chloride (2), which causes a portion of the corrosion products to turn greenish in color. Similarly, corrosion occurs in the presence of sulfuric acid (3) together with evolution of hydrogen gas. The corrosion is dimished by the addition of a reducing agent, such as bisulfite (4), in which case the metal is covered with a black layer of magnetite, $\mathbf{Fe_3O_4}$;[5] the corrosion may be suppressed by the addition of chromate (6); however, if the solution contains chloride (7) it will result in a severe localized corrosion at certain points of the surface. Permanganate and hydrogen peroxide greatly increase the corrosion if present in small concentrations (8, 10) and suppress it entirely when present in large concentrations (9, 11). This protective action of large concentrations, permanent in the case of permanganate, is only transitory for hydrogen peroxide. Stagnant tap water corrodes iron (12).

[5] As a general rule, for differentiating chemical species belonging to different phases, we shall use the following typographic conventions:

M will be used for liquid water and for species dissolved in water,
M will be used for solid species,
M will be used for liquid species other than water,
M will be used for gaseous species.

1.2.2. Iron Filings in a Solution of NaOH

Experiment 5 above shows that iron is not generally corroded in the presence of caustic soda. However, if, shown as in Figure 1b, iron filings are etched and placed in a degassed caustic solution (40 g/liter) contained in a closed flask, the iron corrodes with the evolution of hydrogen gas, which may persist for several months. If the flask is opened for a short time corrosion may cease immediately.

1.2.3. Iron in Nitric Acid Solutions

A piece of piano wire is introduced into a series of test tubes containing aqueous nitric acid solutions of different concentrations (between 396 g/liter, density 1.2, and 939 g/liter, density 1.4). For low concentrations (below about 494 g/liter) considerable corrosion of the metal occurs with continuous evolution of gas (nitrogen and oxides of nitrogen); with relatively high concentrations (above about 494 g/liter) corrosion occurs with evolution of gas which lasts for only a short period, after which the metal is no longer attacked. For intermediate concentrations it is possible to produce alternate periods of corrosion with gas evolution and periods of no corrosion.

1.2.4. Iron as an Anode or Cathode in an Electrolytic Cell

Figure 1d represents an experimental arrangement where four identical lengths of piano wire are etched and immersed in identical concentrations of sodium bicarbonate (8.4 g/liter) and through which an electrolysis current is passed. It is observed that both samples used as negative electrodes are free from corrosion; one of the two samples used as the positive electrode (cell b) does not corrode, and the other positive electrode (cell a) does corrode. The only difference between the two electrolysis cells is that cell a, unlike cell b, incorporates a switch, which was closed only for a few seconds when the experiment was started. There is then, for the same conditions of metal, solution, and current density, a case of *anodic corrosion* and a case of *anodic protection.*

Figure 1e represents an analogous experiment where four identical piano wires are immersed in identical solutions of sodium bicarbonate (8.4 g/liter) and the same electrolysis current is passed. A stream of oxygen is bubbled around the negative electrode of cell a. It is observed that the samples used as positive electrodes are both free from corrosion. One of the two

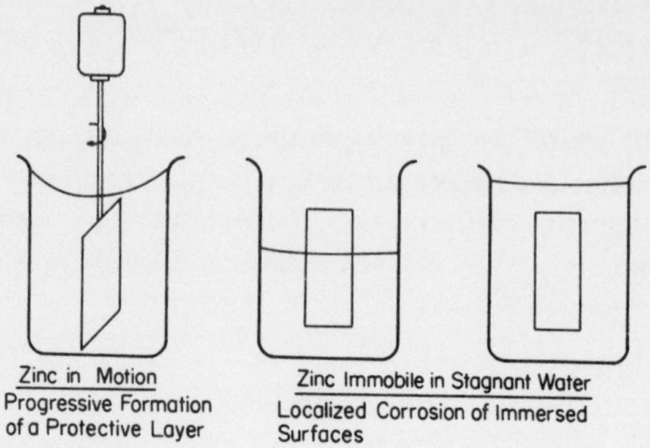

Figure 2. Action of distilled water on zinc. Influence of relative movement zinc–water.

samples used as a negative electrode (cell b) does not corrode while the other negative electrode (cell a) does corrode. There is, then, for the same conditions of metal, solution, and current density, a case of *cathodic corrosion* and a case of *cathodic protection.*

1.2.5. Iron in Contact with Different Metals

In figures 1c and 1c′ a piece of piano wire electrically connected to a piece of another metal (either copper, platinum, zinc, or magnesium) is placed in a series of cylinders containing tap water. It is observed that coupling the iron with copper or platinum accelerates corrosion, while coupling with zinc or magnesium generally suppresses corrosion. Sometimes this suppression of corrosion is permanent, sometimes transitory; the permanent character of the protection is more likely if the coupling is external (Figure 1c) rather than internal (Figure 1c′).

1.2.6. Zinc in Contact with Distilled Water

Sheets of zinc are placed into three cylindrical flasks containing distilled water in the following conditions: agitated while totally immersed, stationary while totally immersed, and stationary while partially immersed (see Figure 2).

The two stationary sheets are observed to corrode rapidly with local formation of basic white zinc carbonate while the agitated sheet is progressively covered with an appreciable protective layer and then ceases to corrode.[6]

[6] M. Pourbaix, Corrosion and protection of zinc roofing, CEBELCOR Publication F. 11 (1951) (in French).

1.3. INFLUENCE OF OXIDANTS, ELECTRIC CURRENTS, MOTION OF SOLUTIONS

Eventually we will return to each of the above experiments seeking an interpretation. For the present we wish to point out particularly the following facts relating to the influences of oxidants, electric currents, and motion of the solution:

1.3.1. Oxidants

Normally, corrosion of iron consists of the transformation of metal into an oxide or hydroxide, for example, according to the overall reaction,

$$4Fe + 6H_2O + 3O_2 \rightarrow 4Fe(OH)_3$$

the approximately 7 mg of oxygen contained in 1 liter of water saturated with air under atmospheric pressure are capable of transforming 16 mg of iron into 31 mg of $Fe(OH)_3$. Then, in attempting to combat corrosion, it is reasonable to eliminate oxygen from the water. In practice one attempts to eliminate as much oxygen as possible by physical degassing or by the addition of reducing agents such as sulfite or hydrazine. These precautions diminish considerably the corrosion of boilers.

It is observed that if dissolved oxygen is eliminated from urban water supplies corrosion is greatly diminished; on the other hand, it is sometimes possible to decrease corrosion even further or to suppress it entirely if, instead of eliminating oxygen, oxygen or another oxidant (chromate, nitrite, hydrogen peroxide) is *added.*

1.3.2. Influence of Electric Current

It has been known for a long time that it is possible to protect iron from corrosion by applying a current source of negative polarity; in fact this procedure, termed *cathodic protection,* is generally employed for underground piping. Conversely, the corrosion of iron generally is enhanced by the application of positive polarity.

But the results can sometimes be the inverse, as shown by the experiments of Figures 1d and e; sometimes a positive current will protect the iron; sometimes a negative current will cause the corrosion of protected iron.

1.3.3. Influence of Motion of Solution

Generally, tap water does not corrode iron; but the water may become corrosive for installations which remain out of service for some time. When a supply network that has remained idle for several weeks is reused, the first water that flows is often laden with rust.

The above examples appear very complicated and paradoxial and very often corrosion was considered as an unavoidable hazard, uncontrollable and incomprehensible. U. R. Evans[7] suggested that "possibly it is really the strangeness of corrosion reactions which causes the orthodox physical chemist to regard the whole subject with suspicion." Corrosion reactions did not seem to obey well-defined laws; and due to the lack of scientific research, the study of corrosion and methods of protection against corrosion has remained empirical for a long time.

1.4. APPLICATION OF CHEMICAL THERMODYNAMICS

All experimental science has had, as a first stage, a period of fragmentary discovery, during which progress was haphazard and dependent on the intuition and ingenuity of a few experimenters. Their pioneering work, however, is no less admirable. For example, *physics*, *electricity*, and *chemistry* have all known this romantic period; the development of these sciences was laborious and shaky until the time when *energetics* and *mathematics* established the essential basis. Since then, facts have been well ordered, the research has lost glamor but gained logic, and considerable progress has been accomplished in a relatively short time: it is possible to *understand* and *predict*. The impetus given to chemistry by *chemical thermodynamics* is well recognized. This is the science of the application of energetics to chemistry and depends on the concept of *chemical equilibrium*.

Chemical thermodynamics, whose fundamentals were established around 1876 by an American, Gibbs, and whose applications have developed particularly since 1913, has helped the comprehension of many obscure chemical phenomena. It enabled the prediction of facts before they were experimentally verified and helped to explain many topics in chemistry and metallurgy. Of the work notably connected with the application of thermodynamics are the names of Frenchmen Sainte Claire Deville, Boudouard,

[7] U. R. Evans, *An Introduction to Metallic Corrosion*, Arnold, London (1948), p. xx.

Le Chatelier, and Chaudron who elucidated the function of the blast furnace and showed that it was useless to construct ever larger blast furnaces at vast expense in the hope of transforming into CO_2 a greater proportion of CO contained in the outlet gases. The laws of chemical equilibrium oppose this conversion; at temperatures for which the reaction rate is appreciable, the content of CO falls below a certain limit. As further example, because of chemical thermodynamics, the German Haber in 1913 invented the synthesis of ammonia, which enabled Germany to manufacture nitric acid and explosives without recourse to the nitrates of Chile. As is well known, this was exploited in 1914.

Like all other phenomena, corrosion obeys the thermodynamic laws; metals tend to return to the state in which they are found in nature. With the exception of gold and platinum for example, which are "noble" and are found on the earth in an uncombined state, metals are gradually destroyed under the action of water, air, and other atmospheric agents. They *corrode* and transform themselves into substances similar to the mineral ores from which they were originally extracted: e.g., oxides, carbonates, sulfates. Iron changes to rust similar to limonite; copper becomes covered with a green patina like brochantite, basic copper sulfate. P. Erculisse (1) has remarked that one of the commentators at the corrosion exhibition organized by the Shell Company, to which we have previously made reference, visualized an application from a verse of Genesis: "till thou return unto the ground; for out of it wast thou taken; for dust thou art, and unto dust shalt thou return" (Genesis 3:19). This appears to be the "first statement of the second law of thermodynamics."

Unfortunately, the aid that chemical thermodynamics can give to the study of metallic corrosion by aqueous solutions is extremely meager and insufficient. For example, when concerned with the transformation of iron into ferric hydroxide in the presence of oxygen according to the equation

$$4\mathrm{Fe} + 6\mathrm{H_2O} + 3O_2 \rightarrow 4\mathrm{Fe(OH)_3}$$

chemical thermodynamics expresses the condition of equilibrium at 25°C as the following:

$$P_{\mathrm{O_2}} = 10^{-81} \text{ atm}$$

This means that the transformation of iron into ferric hydroxide tends to be produced until the partial pressure of oxygen in the atmosphere becomes virtually nothing. This result confirms that minute quantities of air suffice to transform iron into rust. However this statement is sometimes wrong, for oxygen is able, on the contrary, to suppress any corrosion of iron.

Chemical thermodynamics states practically nothing about measures to avoid corrosion of iron. As U. R. Evans (see footnote 7) said, those who wished to apply chemical thermodynamics to the study of corrosion phenomena found themselves in "quicksands." The basis of their study was real but insufficient, and the researcher was engulfed without the ability to progress.

1.5. APPLICATION OF ELECTROCHEMICAL THERMODYNAMICS

The lack of success in applying chemical thermodynamics to corrosion is principally due to the fact that the phenomena of metallic corrosion in aqueous solutions are not only *chemical* but also *electrochemical*, as has been illustrated by the classical experiments of U. R. Evans related to differential aeration (Figure 3). In 1923 Evans[8] observed that, if two samples of iron connected by a galvanometer are immersed in two solutions of potassium chloride separated by a porous membrane and if a stream of air is bubbled through one of these solutions, then an electric current circulates between the aerated sample (positive pole) and the nonaerated sample (negative pole) which corrodes. On the other hand, if after slightly modifying the experimental arrangement illustrated in Figure 1c, a sample of iron and a sample of another metal (copper, platinum, zinc, or magnesium) connected through a galvanometer are immersed in water then the passage of electric current is also observed. For copper or platinum, iron becomes the negative pole and corrodes; for zinc or magnesium, iron is the positive pole and does not corrode.

It becomes clear then that *chemical thermodynamics* is not sufficient for a study of corrosion phenomena: it is necessary to employ *electrochemical thermodynamics*. The laws of chemical equilibrium, such as the law of mass action of Guldberg and Waage, and the law of solubility product—involving partial pressures, p (or fugacities, f) and concentrations, c (or activities, a)—do not enable one to study all the equilibria concerned with corrosion. It is necessary to employ the laws of electrochemical equilibrium which include besides pressure, and concentration, an *electrode potential.*

The *scale of dissolution potentials* emphasized by W. Nernst[9] has been applied to many corrosion studies. Values of potential in volts, measured

[8] U. R. Evans, *J. Inst. Metals* **30**, 263, 267 (1923).

[9] W. Nernst, *Z. Phys. Chemie* **4**, 129 (1889).

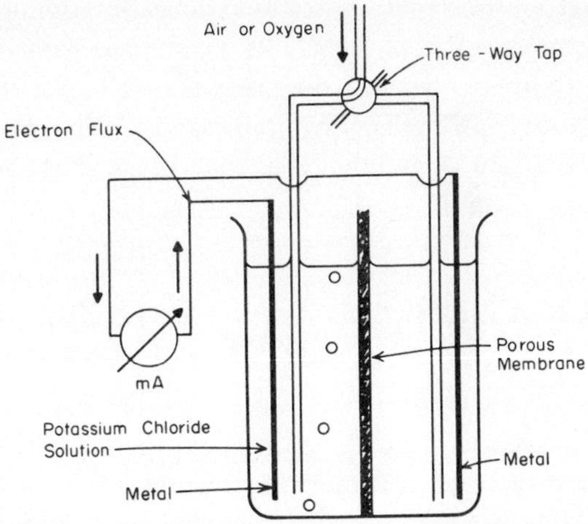

Figure 3. Production of a current by differential aeration (Evans experiment).

with respect to a hydrogen reference electrode conforming to the pattern illustrated in Figure 4, are given in Figure 27. Under certain conditions which will be examined later, in the case of metals, these values relate to the case where the metal is taken to be in equilibrium with a solution containing 1 gram ion of the metal per liter.

As a general rule,[10] metals with a dissolution potential less than hydrogen ($H_2 \rightleftarrows 2H^+ + 2e^-$)[11] are corroded with evolution of hydrogen by aqueous solutions. The lower the value of the dissolution potential, the greater the corrosion tends to be. Examples of such metals are lead, tin, nickel, cobalt, thallium, cadmium, iron, chromium, zinc, manganese, aluminum, magnesium, sodium, cerium, potassium, strontium, beryllium, and lithium.

Metals with values of dissolution potential greater than the dissolution potentials of oxygen (+1.23 V), for example gold, are generally not corroded in aqueous solutions.

These trends led Vernon,[12] in 1949, to remark that the decreasing order of the values of dissolution potential is only the order in which these metals were discovered and used by man. The use of gold and silver, noble metals which are found in the earth in the uncombined state, is known from the remote past; then copper and its alloys and then lead appeared, these being

[10] There are exceptions to this which are examined later.

[11] The dissolution potential of this reaction (when occurring under 1 atm and at pH 0) is taken as zero by convention at all temperatures and is thus a standard.

[12] W. H. J. Vernon, The corrosion of metals, *J. Roy. Soc. of Arts* **97**, 579 (1949).

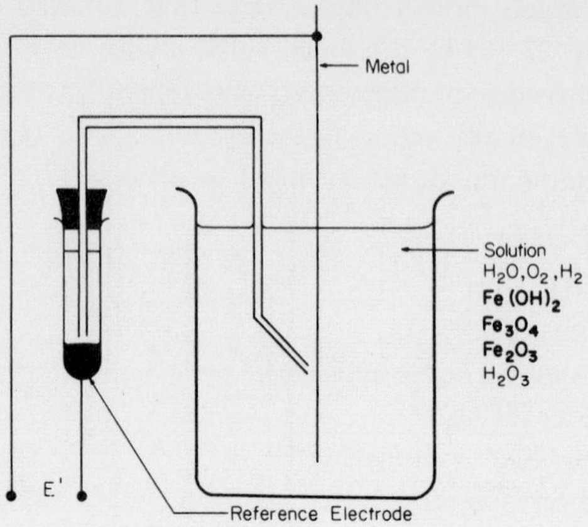

Figure 4. Measurement of the electrode potential.

relatively easy to extract from the ores; then iron and zinc. The alkali and alkaline earth metals were not obtained until relatively recently.

Metals with a dissolution potential greater (more positive) than the dissolution potential of hydrogen (zero volt) and lower (more negative) than the dissolution potential of oxygen (+1.23 V) are not corroded with the evolution of hydrogen, but they may be corroded by solutions containing dissolved oxygen. Such metals are antimony, bismuth, copper, and silver.

However, often metals which should corrode remain practically uncorroded; and conversely, metals not generally susceptible are corroded.

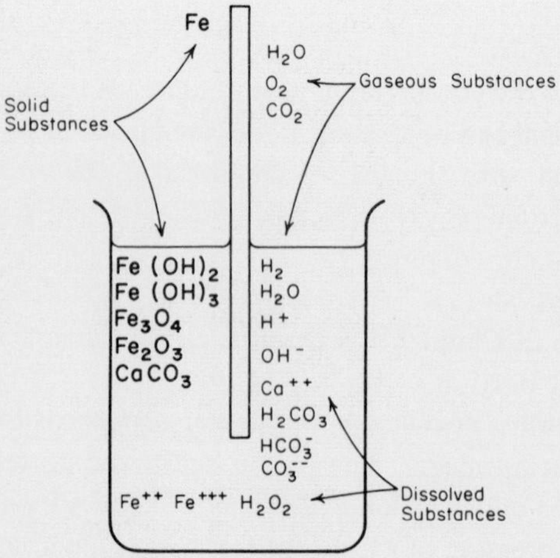

Figure 5. Iron in the presence of city water.

The problem is much more complex than that outlined by the simplified scheme of Figure 27 (to be discussed subsequently in Chapter 4) because when a metal corrodes multiple reactions generally occur. For example, when iron corrodes in city waters in the presence of air (Figure 5) the reacting medium contains numerous chemical species:

Water and dissolved species

- H^+ and OH^- ions
- calcium ions, Ca^{++}
- undissociated carbonic acid, H_2CO_3
- bicarbonate, HCO_3^-
- carbonate, CO_3^{--}
- ferric and ferrous ions, Fe^{+++}, Fe^{++}
- hydrogen peroxide, H_2O_2 from the reduction of oxygen

Solid species

- white ferrous hydroxide, $\mathbf{Fe(OH)_2}$
- magnetite, $\mathbf{Fe_3O_4}$, black if anhydrous and green if hydrated
- ferric hydroxide $\mathbf{Fe(OH)_3}$ and ferric oxide $\mathbf{Fe_2O_3}$, brown rust
- white calcium carbonate, $\mathbf{CaCO_3}$

Gaseous species

- oxygen, O_2, from the air
- carbon dioxide, CO_2, from the air
- hydrogen formed by the reaction, $2H^+ + 2e^- \rightarrow H_2$

These different species reacting together chemically and electrochemically make it impossible to separately examine any reaction in an engineering situation. It is convenient to use graphical means to study such complex phenomena. Then *all* the competing and simultaneous reactions, both chemical and electrochemical can be studied. It has been possible to establish one such graphical method with the use of *diagrams of electrochemical equilibria*, drawn as a function of pH (abscissa) and electrode potential (ordinate). Figure 6 shows such a diagram for the case of iron in the presence of dilute aqueous solutions.

As is shown in Chapter 4, where the establishment of such diagrams is outlined, the left portion of these diagrams represents acid media and the right portion alkaline media; the top portion represents oxidizing media and the bottom reducing media. The region below the dotted line *a* represents the circumstances under which water may be reduced with the evolution of hydrogen under pressure of 1 atmosphere; the region above the dotted line *b* represents the circumstances under which water may be oxidized with evo-

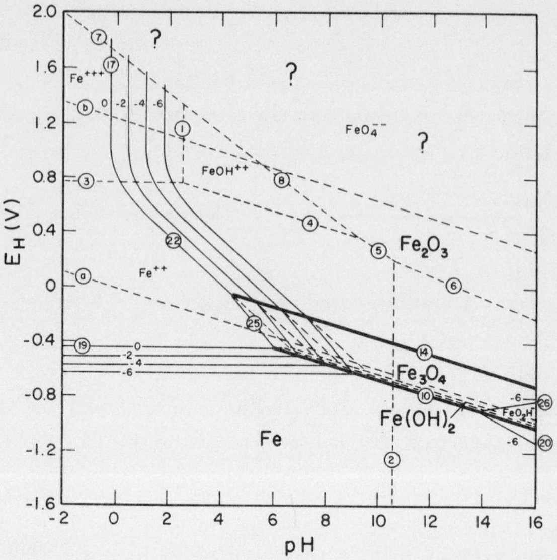

Figure 6. Electrochemical equilibria of the iron–water system.

lution of oxygen; and the region between the two lines *a* and *b* represents the circumstances in which both this reduction and oxidation are impossible. Water then is thermodynamically stable and this region represents the *region of thermodynamic stability* of water under 1 atm pressure.

Depending on the actual conditions of pH and electrode potential, Figure 6 shows that the oxidation of iron may give rise to soluble products —green ferrous ions, Fe^{++}, yellow ferric ions, Fe^{+++}, and green dihypoferrite ions, FeO_2H^-—or to insoluble products—white ferrous hydroxide $\mathbf{Fe(OH)_2}$ (unstable relative to black magnetite, $\mathbf{Fe_3O_4}$) and brown ferric oxide, $\mathbf{Fe_2O_3}$, which may be variously hydrated and is the main constituent of rust. For the sake of definition, iron is said to be *corroding* in the presence of an iron-free solution when the quantity of iron that this solution may dissolve is greater than a given low value (e.g., 10^{-6} g-atoms/liter or 0.056 ppm), and conversely iron may be rendered passive if it becomes covered by a protective insoluble oxide or hydroxide (e.g., $\mathbf{Fe_2O_3}$). Then the lines which are drawn in Figure 6 corresponding to a solubility of metal and its oxide equal to 10^{-6} delineate various regions or areas. There are two areas where corrosion is possible (areas of corrosion), an area where corrosion is impossible (area of immunity[13] or cathodic protection), and an area where pas-

[13] This definition of "immunity" has been suggested by J. N. Agar.

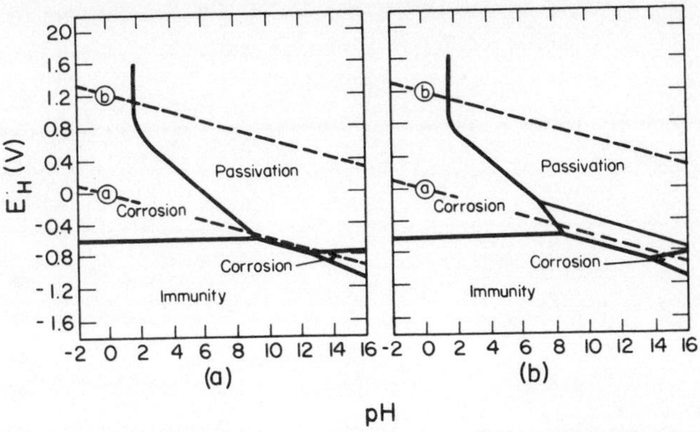

Figure 7. Theoretical conditions of corrosion, immunity, and passivation of iron: (a) assuming passivation by a film of Fe_2O_3; and (b) assuming passivation by films of Fe_2O_3 and Fe_3O_4.

sivation is possible (area of passivation). Figures 7a and b represent the theoretical conditions of corrosion, immunity, and passivation. This assumes that the insoluble products, Fe_2O_3 and Fe_3O_4, are sufficiently adherent and impermeable that corrosion of the underlying metal is essentially stifled and the metal is then "passive."

The existence of the various regions of corrosion, immunity, and passivation as a function of potential and pH suggests that some definite experimental correlation is possible. If the pH and potential of the experiments of Figure 1 are measured (see Table III) and a notation is made of general corrosion (●), local corrosion (◐), or absence of corrosion (○), then these results can be correlated with the *theoretical predictions*. This correlation is shown in Figure 8.[14,15] This figure shows that the conditions under which there is effective corrosion or absence of corrosion are in good agreement with the theoretical predictions. Especially, this figure shows that an oxidizing action either protects iron, or conversely increases the corrosion, depending on whether the particular value of electrode potential of the metal falls within the area of passivation or not. Also the figure shows that the corrosion of iron in contact with a degassed solution of caustic soda is due to the existence of an "area of corrosion" in alkaline solutions free from oxidants; an area which becomes more important at higher temperatures and

[14] Applications of potential–pH diagrams relating to iron and hydrogen peroxide. Demonstration-experiments, *Rapports Techniques CEBELCOR* **2**, RT. 2 (1954) (in French).
[15] Laboratory experiments for lectures on electrochemical corrosion, CEBELCOR Publication E. 65 (1967).

TABLE III. State of Environments and Metals for Experiments in Figure 1 (see also Figure 8, where values of potential and pH in this table are plotted graphically)

Expt.	Sample No.	Solution	pH	E_H (V)	State of metal	Gas
a	1	H_2O distilled	8.1	−0.486	●	—
	2	NaCl 1 g/liter	6.9	−0.445	●	—
	3	H_2SO_4 1 g/liter	2.3	−0.351	●	H_2
	4	$NaHSO_4$ 1 g/liter	6.4	−0.372	●	—
	5	NaOH 1 g/liter	11.2	+0.026	○	—
	6	K_2CrO_4 1 g/liter	8.5	+0.235	○	—
	7	K_2CrO_4 + NaCl 1 g/liter	8.6	−0.200	◐	—
	8	$KMnO_4$ 0.2 g/liter	6.7	−0.460	●	—
	9	$KMnO_4$ 1 g/liter	7.1	+0.900	○	—
	10	H_2O_2 0.3 g/liter	5.7	−0.200	●	—
	11	H_2O_2 3.0 g/liter	3.4	+0.720	○	O_2
	12	Brussels city water	7.0	−0.450	●	—
b	13	NaOH 40 g/liter degassed	13.7	−0.810	●	H_2
c	14	city water–iron–copper	7.5	−0.445	●	—
	15	city water–iron–zinc	7.5	−0.690	○	H_2
	16	city water–iron–magnesium	7.5	−0.910	○	H_2
	17	city water–iron–platinum	7.5	−0.444	●	—
c′	14′	city water–iron–copper	7.8	−0.385	●	—
	15′	city water–iron–zinc	7.7	−0.690	○	H_2
	16′	city water–iron–magnesium	8.7	−0.495	○	H_2
	17′	city water–iron–platinum	—	—	●	—
d	18	$NaHCO_3$ 0.1M Pole −	8.4	−0.860	○	H_2
	19	$NaHCO_3$ 0.1M Pole +	8.4	−0.350	●	—
	20	$NaHCO_3$ 0.1M Pole −	8.4	−0.885	○	H_2
	21	$NaHCO_3$ 0.1M Pole −	8.4	+1.380	○	O_2
e	22	$NaHCO_3$ 0.1M Pole −	8.4	−0.500	●	—
	23	$NaHCO_3$ 0.1M Pole +	8.4	+1.550	○	O_2
	24	$NaHCO_3$ 0.1M Pole −	8.4	−1.000	○	H_2
	25	$NaHCO_3$ 0.1M Pole +	8.4	+1.550	○	O_2

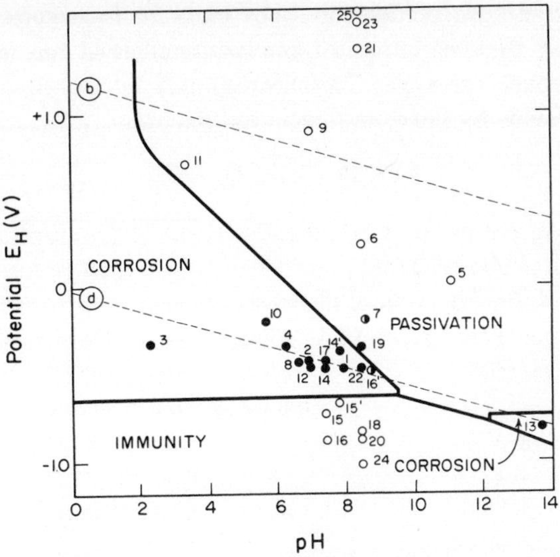

Figure 8. Theoretical and experimental conditions of corrosion, ●, and noncorrosion, ○, of iron at 25°C.

which causes the form of corrosion known in boiler technology as "caustic cracking" when stresses are present.

We have just used the potential–pH equilibrium diagram relating to the system Fe–H_2O for studying the behavior of iron in the experiments outlined in Figure 1. Similarly, equilibrium diagrams relating to other substances involved in these experiments may yield useful information on the behavior of these substances: we shall consider now the behavior of hydrogen peroxide in experiments (10) and (11), and the behavior of permanganate in experiments (8) and (9) (see Section 1.2.1).

Figure 9[16] represents the conditions of equilibrium for the reduction of hydrogen peroxide (and its ion HO_2^-) to water and for the oxidation of hydrogen peroxide to oxygen according to the reactions

$$H_2O_2 + 2H^+ + 2e^- = 2H_2O \tag{2}$$

and

$$H_2O_2 = O_2 + 2H^+ + 2e^- \tag{4}$$

In the area below the family of lines (2, 3), hydrogen peroxide may be reduced to water according to reaction (2); in the area above the family of

[16] *Proceedings of the 2nd meeting of CITCE—Milan 1950*, CEBELCOR Publication F. 16 (1954), p. 42.

lines (4, 5), hydrogen peroxide may be oxidized to oxygen according to reaction (4); in the area between these two families of lines both reactions are simultaneously possible. This corresponds to the decomposition of hydrogen peroxide by the overall chemical reaction

$$2H_2O_2 \rightarrow 2H_2O + O_2$$

Comparing points 10 (dilute solution) and 11 (concentrated solution) in this figure, it is observed that point 10, for which there is corrosion of iron, is in the area of the reduction of hydrogen peroxide to water. Then the reaction of hydrogen peroxide in dilute solution results from a combination of the corrosion reaction

$$Fe \rightarrow Fe^{++} + 2e^-$$

with the reaction

$$H_2O_2 + 2H^+ + 2e^- \rightarrow 2H_2O$$

corresponding to the overall reaction

$$Fe + H_2O_2 + 2H^+ \rightarrow Fe^{++} + 2H_2O$$

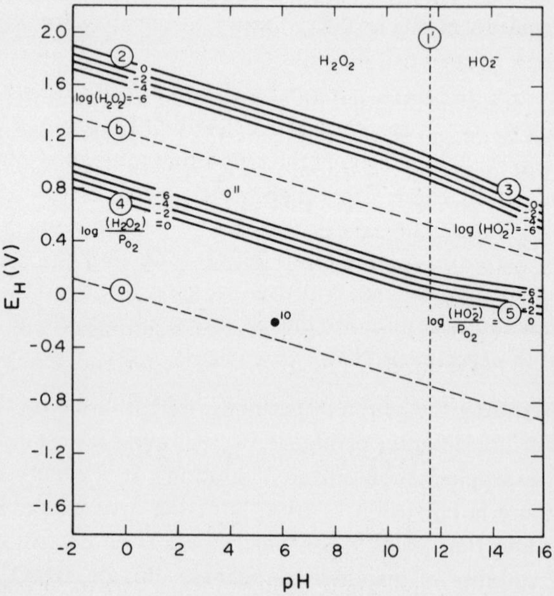

Figure 9. Electrochemical equilibria of the hydrogen peroxide–water system. Theoretical and experimental conditions of reduction and catalytic decomposition of hydrogen peroxide in the presence of iron (25°C). From *Proc. 2nd Meeting of CITCE–Milan, 1950*, p. 42, CEBELCOR Publication F. 16 (1954).

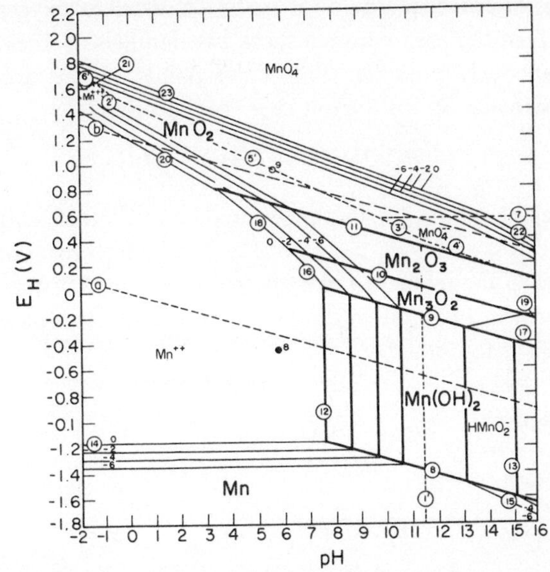

Figure 10. Electrochemical equilibria of the manganese–water system. Theoretical and experimental conditions of the reduction of permanganate into manganous ions and into $\mathbf{MnO_2}$ in the presence of iron at 25°C.

Point 11, for which there is no corrosion of iron, is in the area of double instability (or of chemical decomposition) of hydrogen peroxide; iron passivated by the formation of a film of $\mathbf{Fe_2O_3}$ then catalyses the decomposition of hydrogen peroxide according to the reaction

$$2H_2O_2 \rightarrow 2H_2O + \boldsymbol{O_2}$$

This reaction accounts for the evolution of oxygen on the surface of passive iron in experiment (11).

Understanding the action of permanganate is based on the manganese–water equilibrium diagram of Figure 10.[17] Comparing example 8 (dilute solution) and example 9 (concentrated solution), it is observed that point 8, for which there is corrosion of iron, is in the area of stability of soluble Mn^{++} ions; and that point 9, for which there is no corrosion of iron, is in the area of stability of insoluble manganese dioxide, $\mathbf{MnO_2}$. Thus the activating action of dilute permanganate additions results from the combina-

[17] Electrochemical behavior of manganese, *Rapports Techniques CEBELCOR* **18**, RT. 18 (1954) (in French).

tion of the corrosion reaction

$$\mathbf{Fe} \rightarrow Fe^{++} + 2e^-$$

with the reaction

$$MnO_4^- + 8H^+ + 5e^- \rightarrow Mn^{++} + 4H_2O$$

corresponding to the overall reaction

$$2MnO_4^- + 5\mathbf{Fe} + 16H^+ \rightarrow 2Mn^{++} + 5Fe^{++} + 8H_2O$$

The passivating action of permanganate in concentrated solution is due to a combination of the passivation reaction

$$2\mathbf{Fe} + 3H_2O \rightarrow \mathbf{Fe_2O_3} + 6H^+ + 6e^-$$

with the reaction

$$MnO_4^- + 4H^+ + 3e^- \rightarrow \mathbf{MnO_2} + 2H_2O$$

giving the overall reaction

$$2MnO_4^- + 2\mathbf{Fe} + 2H^+ \rightarrow 2\mathbf{MnO_2} + \mathbf{Fe_2O_3} + H_2O$$

Hence, the passive film is made up not only of iron oxide as is generally the case, but of a mixture of iron oxide and manganese dioxide.

The following chapters are devoted essentially to the establishment of the diagrams of electrochemical equilibria, of which Figures 6, 9, and 10 are three examples, and to the use of these diagrams for the theoretical and experimental study of electrochemical phenomena, particularly those concerning the corrosion and protection of metals.

Chapter 2

CHEMICAL AND ELECTROCHEMICAL REACTIONS

A *chemical reaction* is a reaction in which only chemical bodies participate (neutral molecules or positively or negatively charged ions). For example[1]:

- dissociation of water vapor

$$2\boldsymbol{H_2O} = 2\boldsymbol{H_2} + \boldsymbol{O_2}$$

- electrolytic dissociation of liquid water

$$H_2O = H^+ + OH^-$$

- precipitation of ferrous hydroxide

$$Fe^{++} + 2OH^- = \mathbf{Fe(OH)_2}$$

- dissolution of gaseous CO_2

$$\boldsymbol{CO_2} + H_2O = H_2CO_3$$

- corrosion of iron with evolution of hydrogen

$$\mathbf{Fe} + 2H^+ = Fe^{++} + \boldsymbol{H_2}$$

- oxidation of ferrous ions by permanganate

$$MnO_4^- + 8H^+ + 5Fe^{++} = Mn^{++} + 4H_2O + 5Fe^{+++}$$

[1] The significance of the different typefaces has been explained in footnote 5 of Chapter 1. This series of symbols is used for plotting diagrams of electrochemical equilibria, of which Figure 6 and Figure 10 are examples.

According to de Donder such reactions may generally be written as

$$|\nu_1'|\,M_1' + |\nu_2'|\,M_2' + \ldots = |\nu_1''|\,M_1'' + |\nu_2''|\,M_2'' + \ldots$$

where the stoichiometric coefficients, ν, are considered as positive. The reaction equation may also be written in the abbreviated form

$$\sum_{\gamma} \nu_\gamma M_\gamma = 0 \tag{2.1}$$

where the coefficients ν_γ are positive or negative depending on whether the corresponding substances M_γ are on the right- or left-hand side of the equilibrium equation. For example, for the reaction

$$MnO_4^- + 8H^+ + 5Fe^{++} = Mn^{++} + 4H_2O + 5Fe^{+++}$$

already discussed, the stoichiometric coefficients have the values

$$\nu_{MnO_4^-} = +1 \qquad \nu_{H^+} = +8 \qquad \nu_{Fe^{++}} = +5 \qquad \nu_{Mn^{++}} = -1$$
$$\nu_{H_2O} = -4 \qquad \nu_{Fe^{+++}} = -5$$

in which case the reaction may be written

$$MnO_4^- + 8H^+ + 5Fe^{++} - Mn^{++} - 4H_2O - 5Fe^{+++} = O$$

Conversely, the sign may be changed:

$$\nu_{Mn^{++}} = +1 \qquad \nu_{H_2O} = +4 \qquad \nu_{Fe^{+++}} = +5 \qquad \nu_{MnO_4^-} = -1$$
$$\nu_{H^+} = -8 \qquad \nu_{Fe^{++}} = -5$$

in which case the reaction may be written

$$Mn^{++} + 4H_2O + 5Fe^{+++} - MnO_4^- - 8H^+ - 5Fe^{++} = O$$

An *electrochemical reaction* is a reaction in which both chemical species and free electric charges take part (e.g., negative electrons dissolved in a metallic electrode). Such reactions are *oxidations* if they proceed in the direction which corresponds to the liberation of negative charge; they are *reductions* if they proceed in the direction corresponding to the absorption of negative charge. Examples are as follows:

- oxidation of hydrogen gas into hydrogen ions

$$H_2 \rightarrow 2H^+ + 2e^-$$

- reduction of oxygen gas into hydroxide ions

$$O_2 + 4H_2O + 4e^- \rightarrow 4OH^-$$

- oxidation of iron to ferrous ions

$$Fe \rightarrow Fe^{++} + 2e^-$$

- oxidation of ferrous ions to ferric ions

$$Fe^{++} \rightarrow Fe^{+++} + e^-$$

- reduction of permanganate to manganous ions

$$MnO_4^- + 8H^+ + 5e^- \rightarrow Mn^{++} + 4H_2O$$

These reactions may generally be written

$$|\nu_1'|\,M_1' + |\nu_2'|\,M_2' + \ldots + ne^- = |\nu_1''|\,M_1'' + |\nu_2''|\,M_2'' + \ldots$$

or again

$$\boxed{\sum_{\gamma} \nu_\gamma M_\gamma + ne^- = 0}$$

It is well known that, for all *chemical reactions* taking place between gases and/or dissolved species, there is an *equilibrium constant* whose value is constant for a given temperature. The pressures (or *fugacities*) of the reacting gases and the *concentrations* (or *activities*) of the reacting bodies in solution are related to each other at equilibrium through the equilibrium constant. For example:

- dissociation in the gas phase (constant of Guldberg and Waage)

$$2H_2O = 2H_2 + O_2 \qquad K = \frac{p^2_{H_2} \cdot p_{O_2}}{p^2_{H_2O}}$$

- dissociation in solution (Ostwald's constant)

$$H_2O = H^+ + OH^- \qquad K = [H^+] \cdot [OH^-]$$

- dissolution of slightly soluble bodies (solubility product)

$$Fe(OH)_2 = Fe^{++} + 2OH^- \qquad K = [Fe^{++}] \cdot [OH^-]^2$$

- dissolution of gaseous bodies (Henry's constant)

$$CO_2 + H_2O = H_2CO_3 \qquad K = \frac{[H_2CO_3]}{p_{CO_2}}$$

Likewise, it will be seen that *for all electrochemical reactions* taking place between gases and/or dissolved species, there is also an *equilibrium constant* whose value is constant for a given temperature. This value is a function not only of the partial pressures (or fugacities) of the reacting gases and the concentrations (or activities) of the reacting species in solution but also of a *difference of electrical potential* at the metal–solution interface (or electrode potential).

Thus, to establish equilibrium diagrams as a function of *pH* and *electrode potential* we are concerned with determining the influence of pH and electrode potential on the equilibrium characteristics of the various reactions. It will be useful to write these reactions in a given manner so that it is possible to define, in the reaction equation, the H^+ ions and the electric charges, e^-, involved in the reaction. For this reason, the following method is adopted for writing and balancing the equations. For example, to write the reaction of the transformation of permanganate ions, MnO_4^-, into manganous ions, Mn^{++} (reduction of permanganate in acid solution), the symbols of these two species are written on both sides of the equivalence sign:

$$MnO_4^- = Mn^{++}$$

the O is equated with H_2O

$$MnO_4^- = Mn^{++} + 4H_2O$$

the H is equated with H^+

$$MnO_4^- + 8H^+ = Mn^{++} + 4H_2O$$

the electric charge is balanced using e^-

$$MnO_4^- + 8H^+ + 5e^- = Mn^{++} + 4H_2O$$

Here are other examples, the reduction of permanganate in a higher pH solution and the reduction of dichromate in an acid solution:

$$\begin{aligned} MnO_4^- &= MnO_2 \\ MnO_4^- &= MnO_2 + 2H_2O \\ MnO_4^- + 4H^+ &= MnO_2 + 2H_2O \\ MnO_4^- + 4H^+ + 3e^- &= MnO_2 + 2H_2O \end{aligned}$$

$$\begin{aligned} Cr_2O_7^{--} &= 2Cr^{+++} \\ Cr_2O_7^{--} &= 2Cr^{+++} + 7H_2O \\ Cr_2O_7^{--} + 14H^+ &= 2Cr^{+++} + 7H_2O \\ Cr_2O_7^{--} + 14H^+ + 6e^- &= 2Cr^{+++} + 7H_2O \end{aligned}$$

TABLE IV. Examples of Classification of Reactions

	Chemical reactions		Electrochemical reactions	
With or without H^+	—	H^+	—	H^+
With or without e^-	—	—	e^-	e^-
Equilibria dependent on	Neither pH nor electrode potential	pH	Electrode potential	pH and electrode potential
Homogeneous reactions (solution)	$CO_2 + H_2O = H_2CO_3$	$H_2O = H^+ + OH^-$ $H_2CO_3 = HCO_3^- + H^+$	$Fe^{++} = Fe^{+++} + e^-$	$MnO_4^- + 8H^+ + 5e^-$ $= Mn^{++} + 4H_2O$
Heterogeneous reactions (solid–solution)	$\mathbf{H_2O} = H_2O$ $\mathbf{As_2O_3} + H_2O = 2HAsO_2$	$\mathbf{Fe(OH)_2} + 2H^+$ $= Fe^{++} + 2H_2O$	$\mathbf{Fe} = Fe^{++} + 2e^-$ $\mathbf{S} + 2e^- = S^{--}$	$\mathbf{Fe(OH)_3} + 3H^+ + e^-$ $= Fe^{++} + 3H_2O$ $MnO_4^- + 4H^+ + 3e^-$ $= \mathbf{MnO_2} + 2H_2O$
Heterogeneous reactions (gas–solution)	$\mathit{H_2O} = H_2O$ $\mathit{NH_3} + H_2O = NH_4OH$	$\mathit{CO_2} + H_2O = HCO_3^- + H^+$	$\mathit{Cl_2} + 2e^- = 2Cl^-$	$\mathit{H_2} = 2H^+ + 2e^-$ $\mathit{O_2} + 4H^+ + 4e^- = 2H_2O$

To fit the general scheme of emphasizing the dependence of pH and potential, it is useful to write the equilibrium of precipitation of ferrous hydroxide not as

$$Fe^{++} + 2OH^{-} = \mathbf{Fe(OH)_2}$$

but as

$$Fe^{++} + 2H_2O = \mathbf{Fe(OH)_2} + 2H^{+}$$

For maximum convenience in plotting graphically, the transformation of an oxidized body A to a reduced body B should be written as

$$a\mathrm{A} + c\mathrm{H_2O} + ne^{-} = b\mathrm{B} + m\mathrm{H}^{+}$$

When chemical equilibria are written in this form, the dependence on pH and potential is clear. To repeat:

pH measures the effect of H^{+} ions,
the electrode potential measures the effect of charge.

In accordance with these variables it is instructive to distinguish among reactions as to the dependence on pH, potential, both, or neither. Examples of these dependencies are given in Table IV. This table also emphasizes the fact that there are both chemical and electrochemical reactions.

Chapter 3

CHEMICAL EQUILIBRIA

The previous chapter categorized equilibria as "chemical" or "electrochemical," respectively, depending whether they do not or do involve electrons. This chapter will discuss exclusively *chemical equilibria* and therefore will not consider those equilibria involving oxidation–reduction processes. These electrochemical equilibria are discussed in Chapter 4. At the outset this chapter will discuss the important formalisms which underlie chemical equilibria; thereafter, numerous examples of various calculations will be discussed. Table V gives examples of various chemical equilibria and their conditions of equilibrium at 25°C.

During the discussion of this chapter, the following should become clear:

1. All the equilibrium relations of the particular cases in Table V constitute only one *general law of equilibrium.* Simply, this law says that the state of thermodynamic equilibrium of a transformation is attained when the *affinity* of the transformation (or the variation of free enthalpy relative to this transformation) is nil.
2. The use of *simple graphical methods* involving the *logarithmic scale* may often facilitate the practical application of these equilibrium relations.

3.1. GENERAL FORMULA OF CHEMICAL EQUILIBRIA

In order to introduce the usefulness of thermodynamic formalisms, two examples will be mentioned initially. For a transformation such as

$$\mathrm{H_2O} = \boldsymbol{H_2O}$$

TABLE V.[a] Physicochemical Transformations and Their Conditions of Equilibrium at 25°C

	Equilibrium constant K at 25°C $\log K = -\dfrac{\Sigma \nu \mu^0}{1363}$
1. Gaseous bodies	
(dissociation in the gas phase)	Law of Guldberg and Waage
$\boldsymbol{2H_2O = 2H_2 + O_2}$	$\dfrac{p_{H_2}{}^2 \cdot p_{O_2}}{p_{H_2O}{}^2} = 10^{-80.16}$ atm
2. Dissolved bodies	
(dissociation in aqueous solution)	Ostwald's law
$H_2O = H^+ + OH^-$	$\dfrac{[H^+] \cdot [OH^-]}{[H_2O]} = 10^{-14.00}$ (g-ion/liter)^2
$H_2CO_3 = HCO_3^- + H^+$	$\dfrac{[HCO_3^-] \cdot [H^+]}{[H_2CO_3]} = 10^{-6.37}$ (g-ion/liter)
$HCO_3^- = CO_3^{--} + H^+$	$\dfrac{[CO_3^{--}] \cdot [H^+]}{[HCO_3^-]} = 10^{-10.34}$ (g-ion/liter)
3. Gaseous and dissolved bodies	
(dissolution of gas)	Henry's law
a. Dissolved bodies not ionized	
$\boldsymbol{CO_2} + H_2O = H_2CO_3$	$\dfrac{p_{CO_2}}{[H_2CO_3]} = 10^{1.43}$ atm/mol/liter
b. Ionized dissolved bodies	
$\boldsymbol{CO_2} + H_2O = HCO_3^- + H^+$	$\dfrac{p_{CO_2}}{[HCO_3^-] \cdot [H^+]} = 10^{7.80}$ atm/(g-ion/liter)^2
$\boldsymbol{CO_2} + H_2O = CO_3^{--} + 2H^+$	$\dfrac{p_{CO_2}}{[CO_3^{--}] \cdot [H^+]^2} = 10^{18.14}$ atm/(g-ion/liter)^3
4. Gaseous and condensed bodies	
(vaporization)	Vapor pressure
$H_2O = \boldsymbol{H_2O}$	$p_{H_2O} = 10^{-1.507}$ atm $= 0.0313$ atm $= 23.7$ mm Hg
5. Condensed and dissolved bodies	
(dissolution of solid body or liquid)	
a. Dissolved body not ionized	Solubility
$\mathbf{As_2O_3} + H_2O = 2HAsO_2$	$[HAsO_2] = 10^{-0.68}$
b. Ionized dissolved bodies	Solubility product
$\mathbf{Ca(OH)_2} = Ca^{++} + 2OH^-$	$[Ca^{++}] \cdot [OH^-]^2 = 10^{-5.10}$ (g-ion/liter)^3
$\mathbf{CaCO_3} = Ca^{++} + CO_3^{--}$	$[Ca^{++}] \cdot [CO_3^{--}] = 10^{-8.35}$ (g-ion/liter)^2

[a] The *physical* transformation of the vaporization of water is included in Section 4 of this table.

(vaporization of water), the state of thermodynamic equilibrium is attained when

$$\mu_{H_2O_l} = \mu_{H_2O_g}$$

(where μ represents the molar chemical potentials of H_2O liquid and H_2O gas) or

$$\mu_{H_2O_l} - \mu_{H_2O_g} = 0$$

For the transformation,

$$2H_2O = 2H_2 + O_2$$

(dissociation of water vapor), the state of thermodynamic equilibrium is attained when

$$2\mu_{H_2O} = 2\mu_{H_2} + \mu_{O_2}$$

or

$$2\mu_{H_2O} - 2\mu_{H_2} - \mu_{O_2} = 0$$

Thus, for the general physicochemical transformation a generalized notation can be used:

$$|\nu_1'|\, M_1' + |\nu_2'|\, M_2' + \ldots = |\nu_1''|\, M_1'' + |\nu_2''|\, M_2'' + \ldots$$

or

$$\boxed{\sum_\gamma \nu_\gamma M_\gamma = 0} \qquad (3.1)$$

where M is the identity or index of the various chemical species.

According to the convention of De Donder, the condition of thermodynamic equilibrium is

$$\boxed{\sum_\gamma \nu_\gamma \mu_\gamma = 0} \qquad (3.2)$$

where μ is the "molar chemical potential" of the constituent γ.

It should be recalled that

T = absolute temperature

S = entropy

p = pressure

V = volume

Thermodynamicists make use of the following four "*thermodynamic potentials*":

Internal energy	$= E$
Helmholtz free energy	$= F = E - TS$
Enthalpy	$= H = E + pV$
Gibbs free energy (*or free enthalpy*)	$= G = H - TS$

It may also be recalled that the value of molar chemical potential μ_γ of the constituent γ of a chemical system is related to the values of these four thermodynamic potentials by the following equations[1]:

$$\mu_\gamma = \left(\frac{\partial E}{\partial n_\gamma}\right)_{S,V} = \left(\frac{\partial F}{\partial n_\gamma}\right)_{V,T} = \left(\frac{\partial H}{\partial n_\gamma}\right)_{S,p} = \left(\frac{\partial G}{\partial n_\gamma}\right)_{p,T} \tag{3.3}$$

As a result, for given values of pressures p and temperature T,

$$\mu_\gamma = \frac{\partial G}{\partial n_\gamma} = g_\gamma \text{ (specific molar Gibbs free energy)}$$

$$= h_\gamma - Ts_\gamma \text{ (specific molar free enthalpy)}$$

where h_γ is the specific molar enthalpy and s_γ is the specific molar entropy.

The numerical values of "free energy of formation," "heat of formation," and "affinity of formation" are found in the tables of thermodynamic constants, expressed in calories or kilocalories per gram molecule (or gram ion) of the body under consideration. The *molar chemical potential*, μ_γ of a constituent γ is equal *in magnitude and sign* to its specific molar *free energy of formation*, ΔG (of Gibbs), to its *free energy* ΔF (of Lewis), and to its *Bildungsarbeit* (of Ulich); it is *equal but opposite in sign* to its *affinity of formation* (Prigogine and Defay). Elsewhere,[2] we give a table of the values of standard chemical potentials of different metals and metalloids and of some of their compounds all calculated at a constant pressure.

It is possible to show the following:

1. *For a condensed pure body* (pure solid or liquid) μ_γ depends only on the pressure p and temperature T.

[1] See Eq. (6.11) of I. Prigogine and R. Defay, *Chemical Thermodynamics* (translated by D. H. Everett), Longmans–Green (1954).

[2] Standard free enthalpies of formation, at 25°C, *Rapports Techniques CEBELCOR* **72,** RT. 87 (1960) (in French).

2. *For a dissolved body*

$$\mu_\gamma = \mu_\gamma{}^0 + RT \ln a_\gamma$$

where a_γ is the activity of the species γ. This activity may be expressed in numerous ways.[3] In practice, for dilute solutions it is convenient to define the activity as a corrected concentration ($a_\gamma = c_\gamma f_\gamma$), where the concentration c_γ is expressed in gram-ions per liter or in *gram-molecules per liter*; f_γ is the coefficient of correction.

The particular value $\mu_\gamma = \mu_\gamma{}^0$, then, is related to the case where for the given temperature and pressure, the activity a_γ of species γ is equal to *1 gram-ion (or 1 gram-molecule) per liter*. The dissolved body is then in a reference state, called the "*standard*" *state for this dissolved body* at the given temperature.

3. *For a gaseous body*

$$\mu_\gamma = \mu_\gamma{}^0 + RT \ln \varphi_\gamma$$

where φ_γ is the *fugacity* of body γ, the fugacity being a *corrected partial pressure* ($\varphi_\gamma = p_\gamma \cdot f_\gamma$).

In practice it is convenient to express p_γ and φ_γ in atmospheres. Then the particular value $\mu_\gamma = \mu_\gamma{}^0$ is relative to the case where for the given temperature, the fugacity of the considered body is equal to 1 *atmosphere*. The gaseous body is then in a reference state, called the "*standard*" *state for this gaseous body*.

Thus, if we call $\mu_\gamma{}^0$ the "standard chemical potential"[4] of the constituent γ at a given temperature T and assume that for this temperature a condensed body is in the "standard" state if it is pure, a dissolved body is in the "standard" state if it has an activity (corrected concentration) of 1 gram-ion (or gram-molecule) per liter and a gaseous body is in the "standard" state if it has a fugacity (corrected partial pressure) of 1 atmosphere.

[3] See Prigogine and Defay, *loc. cit.*, footnote 2, Eqs. (20.4), (20.48), and (20.52).

[4] Standard chemical potentials, $\mu_\gamma{}^0$, considered here are related to the notations used by Prigogine and Defay as follows (pp. 111–112 and 181): For *pure condensed bodies*: $\mu_\gamma{}^0$ depends on p and T and is the function $\zeta(T, p)$ of Defay, corresponding to a *molar titer* equal to 1. For *dissolved bodies*: $\mu_\gamma{}^0$ depends on p and T and is the function $\zeta(T, p)$ of Defay, corresponding to a *corrected concentration* $c_\gamma f_\gamma$ equal to 1 gram equivalent per liter. For *gaseous bodies*: $\mu_\gamma{}^0$ depends on T and is the function $\eta(T)$ of Prigogine and Defay corresponding to a *fugacity* equal to 1 atm.

Then

- for pure condensed bodies $\mu_\gamma = \mu_\gamma{}^0$
- for dissolved bodies $\mu_\gamma = \mu_\gamma{}^0 + RT \ln a_\gamma$ (3.4)
- for gaseous bodies $\mu_\gamma = \mu_\gamma{}^0 + RT \ln \varphi_\gamma$ (3.5)

or, in general, for a temperature T

$$\boxed{\mu_\gamma = \mu_\gamma{}^0 + RT \ln [M_\gamma]} \tag{3.6}$$

where $[M_\gamma]$ is not considered for pure condensed bodies (solids and liquids), $[M_\gamma] = a_\gamma$ (activity or corrected concentration) for dissolved bodies, and $[M_\gamma] = \varphi_\gamma$ (fugacity or corrected partial pressure) for gaseous bodies.

The condition of equilibrium of the transformation is

$$\sum_\gamma \nu_\gamma M_\gamma = 0$$

which we have written in the form

$$\sum_\gamma \nu_\gamma \mu_\gamma = 0 \tag{3.2}$$

By virtue of Eq. (3.6), this may be written as

$$\sum_\gamma \nu_\gamma \mu_\gamma{}^0 + RT \sum_\gamma \nu_\gamma \ln [M_\gamma] = 0$$

or

$$\sum_\gamma \nu_\gamma \ln [M_\gamma] = -\frac{\sum \nu_\gamma \mu_\gamma{}^0}{RT}$$

or in decimal logarithms

$$\sum_\gamma \nu_\gamma \log [M_\gamma] = -\frac{\sum_\gamma \nu_\gamma \mu_\gamma{}^0}{2.3026\ RT}$$

or, substituting $R = 1.985$ cal/°K,

$$\boxed{\sum_\gamma \nu_\gamma \log [M_\gamma] = -\frac{\sum_\gamma \nu_\gamma \mu_\gamma{}^0}{4.575\ T}} \tag{3.7}$$

or

$$\sum_{\gamma} \nu_\gamma \log [M_\gamma] = \log K$$

$$\log K = -\frac{\sum_{\gamma} \nu_\gamma \mu_\gamma^0}{4.575\, T} \tag{3.8}$$

where K is the "equilibrium constant" of the reaction at temperature T. At the temperature of 25.0°C ($T = 298.1$°K) the value of the equilibrium constant is given by the relation

$$\log K = -\frac{\sum_{\gamma} \nu_\gamma \mu_\gamma^0}{1363} \tag{3.9}$$

Recall that μ_γ^0 represents the standard chemical potentials of the reacting bodies relative to the temperature of 25.0°C.

Relation (3.7) is the general law of *all* physicochemical equilibria at the temperature T. This law also may be written in the more usual forms

$$\log \frac{[M_1']^{\nu_1'} \times [M_2']^{\nu_2'} \times \dots}{[M_1'']^{\nu_1''} \times [M_2'']^{\nu_2''} \times \dots} = -\frac{\sum_{\gamma} \nu_\gamma \mu_\gamma^0}{4.575\, T}$$

or

$$\frac{[M_1']^{\nu_1'} \times [M_2']^{\nu_2'} \times \dots}{[M_1'']^{\nu_1''} \times [M_2'']^{\nu_2''} \times \dots} = 10^{-\sum_{\gamma} \nu_\gamma \mu_\gamma^0/4.575T} = K$$

where K is the "equilibrium constant" of the reaction at temperature T.

Relation (3.8) permits calculating, for a given temperature T, the equilibrium constants of all physicochemical transformations if the standard chemical potentials μ_γ^0 (or standard free enthalpies of formation) of all the bodies participating in these transformations are known. The use of this formula makes errors concerning the sign of the constant impossible. Important examples and clarifications of the relationships developed above are discussed in the following paragraphs. In the following discussion we perform calculations for 25°C, but all that applies here can be extended to other temperatures.

When using the formula

$$\sum_{\gamma} \nu_{\gamma} \log [M_{\gamma}] = - \frac{\sum_{\gamma} \nu_{\gamma} \mu_{\gamma}{}^{0}}{1363} \tag{3.10}$$

for the calculation of equilibria, obviously it is necessary to use as far as possible the correct values for the standard chemical potentials $\mu_{\gamma}{}^{0}$ and for the activities and fugacities $[M_{\gamma}]$. The following should not be forgotten.

1. *For solid bodies* $\mu_{\gamma}{}^{0}$ depends on the allotropic state. For example, in the case of $CaCO_3$

$$\mu^{0}_{\text{calcite}} = -\ 269{,}780 \text{ cal}$$

$$\mu^{0}_{\text{aragonite}} = -\ 269{,}530 \text{ cal}$$

When the same compound may exist in two allotropic forms, the more stable of these two forms is that which has the lower value of chemical potential (or free enthalpy of formation). Then, at 25°C, calcite (−269,780 cal) is more stable than aragonite (−269,530 cal) and the affinity of the transformation from aragonite into calcite is +250 cal. At this temperature, while aragonite transforms into calcite, the transformation of calcite into aragonite is impossible.

Obviously the more stable compound is the least soluble in a given solution and the difference between the logarithms of the relative solubility constants of the two forms is equal to the difference between their chemical potentials divided by 1363 (at 25°C). For example, for $CaCO_3$ the solubility product of calcite at 25°C is less than that of aragonite, and the difference between the logarithms of these constants is

$$\frac{269{,}780 - 269{,}530}{1363} = \frac{250}{1363} = 0.18$$

To show this result more clearly, consider the case of the dissolution equilibrium of $CaCO_3 = Ca^{++} + CO_3^{--}$ for which the solubility product is

$$\log [Ca^{++}] \cdot [CO_3^{--}] = - \frac{\mu^{0}_{Ca^{++}} + \mu^{0}_{CO_3^{--}} - \mu^{0}_{CaCO_3}}{1363}$$

Using appropriate values for μ^0's, the solubility product can be calculated for calcite

$$\log[Ca^{++}] \cdot [CO_3^{--}] = - \left(\frac{-\ 132{,}180 - 126{,}220 + 269{,}780}{1363} \right)$$

$$= - \frac{11{,}380}{1363} = -\ 8.35$$

and for aragonite

$$\log[Ca^{++}] \cdot [CO_3^{--}] = -\left(\frac{-132{,}180 - 126{,}220 + 269{,}530}{1363}\right)$$

$$= -\frac{11{,}130}{1363} = -8.17$$

The difference between the logarithms of these two solubility products is 250/1363 = 0.18.

Some hydroxides, e.g., those of Zn, Fe, Al, sometimes "age" with time owing to the progressive transformation to a more stable state. This resulting state will have lower solubility according to the above discussion.

2. *For water* at 25°C $\mu^0_{H_2O}$ has the value $\mu^0_{H_2O} = -56{,}690$ cal only if the water is *pure*, with molarity (number of moles per liter) 1000/18 = 55.5 (18 being the molecular weight of water). However, if the water contains dissolved substances the value of μ_{H_2O} will be less than −56,690 cal. The necessary correction to be applied to give μ_{H_2O} for a certain solution may be simply calculated if the partial pressure of the water vapor in equilibrium with this solution is known.

For the vaporization of liquid water according to $H_2O = \boldsymbol{H_2O}$,

$$\begin{aligned}\mu_{H_2O_l} = \mu_{H_2O_g} &= -54{,}635 + RT \ln p_{H_2O} \\ &= -54{,}635 + 1363 \log p_{H_2O}\end{aligned}$$

For pure water

$$p_{H_2O} = 23{,}756 \text{ mm Hg} = 0.0313 \text{ atm}$$

$$\log p_{H_2O} = -1.505$$

Therefore,

$$\mu^0_{H_2O} = -54{,}635 + 1363 \times (-1.505) = -56{,}690 \text{ cal}$$

Thus, for the chemical potential of water in an aqueous solution where the equilibrium vapor pressure of water is p' (instead of $p = 23.756$ mm Hg for pure water), one has

$$\begin{aligned}\mu_{H_2O_l} &= \mu_{H_2O_g} + 1363 \log p' \\ &= -54{,}635 + 1363 \log p' \\ &= -54{,}635 + 1363 \log p + 1363 \log(p'/p) \\ &= -56{,}690 + 1363 \log(p'/p)\end{aligned}$$

For example, in the case of a solution for which the equilibrium vapor pressure of water is 1% lower than the equilibrium pressure of pure water,

$$p'/p = 0.99$$

and

$$\mu_{H_2O_1} = -\ 56{,}690 + 1363 \log 0.99 = -\ 56{,}690 - 7$$
$$= -\ 56{,}697 \text{ cal}$$

Therefore, in such a solution, the chemical potential of water is −56,697 cal (instead of −56,690 cal for pure water) and the logarithm of the equilibrium constant of all reactions involving H_2O is modified by 7/1363 = 0.005 units per gram-molecule of H_2O taking part in the reaction.

In the case of a solution for which the equilibrium pressure of water vapor is 10% lower than the equilibrium pressure of pure water, then

$$p'/p = 0.90$$

and

$$\mu_{H_2O} = -\ 56{,}690 + 1363 \log 0.90 = -\ 56{,}690 - 63$$
$$= -\ 56{,}753 \text{ cal}$$

So, in such a solution, the chemical potential of water is −56,753 cal (instead of −56,690 cal) and the logarithm of the equilibrium constant of all reactions involving H_2O is modified by 63/1363 = 0.046 units per gram-molecule of H_2O taking part in the reaction.

3. *For gaseous bodies* $[M_\gamma]$ represents *corrected* partial pressures

$$\varphi_\gamma = p_\gamma \cdot f_\gamma$$

4. *For dissolved bodies* $[M_\gamma]$ represents *corrected* concentrations (in gram-ions/liter or in gram-molecules/liter)

$$a_\gamma = c_\gamma \cdot f_\gamma$$

In practice, the values of γ_γ^0 are not always exactly known (the values given in various tables of constants often differ) and very often the correction coefficients f_γ[5] are not exactly known, especially for ions in solution.

[5] A discussion concerning "activity coefficients" f and the concept of "ionic force" which may enable the calculation of these coefficients would be beyond the scope of the present work. Refer to a physical chemistry textbook on this subject.

On account of this, exact calculations of equilibrium constants are often difficult and even impossible; in practice it is sometimes necessary to use *concentrations* c_γ instead of *activities* a_γ. Certainly this is an approximation, but the simplification resulting from it is such that this approximation is often extremely useful and sometimes even indispensable for the study of practical problems. It must be conceded that the results obtained in this manner are only approximate and as far as possible must be verified experimentally.

The second column of Table V indicates the results of some applications of Eq. (3.10) for the calculation of different types of chemical equilibria (dissociation in homogeneous gas phase, dissociation in homogeneous solution, solubility of gas, vapor pressure of condensed bodies, solubility of condensed bodies). For illustrative purposes these results are calculated in detail for the first four equilibrium constants, for which the values of chemical potential given in *CEBELCOR Technical Report* RT. 87 have been used.

1. Reaction $2H_2O = 2H_2 + O_2$ (homogeneous dissociation of gaseous water)

$$\log \frac{(p_{H_2})^2 \cdot p_{O_2}}{(p_{H_2O})^2} = -\frac{2\mu^0_{H_2} + \mu^0_{O_2} - 2\mu^0_{H_2O}}{1363} = -\frac{0 + 0 + 109{,}270}{1363}$$

$$= -80.16$$

2. Reaction $H_2O = H^+ + OH^-$ (homogeneous dissociation of liquid water)[6]

$$\log[H^+] \cdot [OH^-] = -\frac{\mu^0_{H^+} + \mu^0_{OH^-} - \mu^0_{H_2O}}{1363} = -\frac{0 - 37{,}595 + 56{,}690}{1363}$$

$$= -14.00$$

3. Reaction $H_2CO_3 = HCO_3^- + H^+$ (homogeneous dissociation of dissolved H_2CO_3)

$$\log[HCO_3^-] \cdot [H^+] = -\frac{\mu^0_{HCO_3^-} + \mu^0_{H^+} - \mu^0_{H_2CO_3}}{1363}$$

$$= -\frac{-140{,}310 + 0 + 149{,}000}{1363}$$

$$= -6.37$$

[6] The assumption of $\mu^0_{H^+} = 0$ is the basis for the scale of electrochemical potentials which will be discussed in Chapter 4. Wider application of this assumption will become apparent later.

4. Reaction $HCO_3^- = CO_3^{--} + H^+$ (homogeneous dissociation of dissolved HCO_3^-)

$$\log \frac{[CO_3^{--}] \cdot [H^+]}{[HCO_3^-]} = -\frac{\mu^0_{CO_3^{--}} + \mu^0_{H^+} - \mu^0_{HCO_3^-}}{1363}$$

$$= -\frac{-126{,}220 + 0 + 140{,}310}{1363}$$

$$= -10.34$$

It is important to realize that the different equilibrium laws which express the relations obtained in this manner are only particular cases of the same law of physicochemical equilibrium:

$$\sum_\gamma \nu_\gamma \log [M_\gamma] = \log K \qquad \log K = -\frac{\sum_\gamma \nu_\gamma \mu_\gamma{}^0}{1363} \tag{3.11}$$

where, depending on the nature of the transformation being studied, the equilibrium constant K has been denoted in the past by various terms: the constant of Guldberg and Waage, Ostwald's constant, Henry's constant, the vapor pressure, the solubility, and the solubility product. The modern trend is towards a more universal terminology which simply denotes K as the equilibrium constant.

The value of each of these "equilibrium constants," K, is related to the values of "standard chemical potentials," μ^0, of the bodies taking part in the transformation by the same equation

$$\log K = -\frac{\sum_\gamma \nu_\gamma \mu_\gamma{}^0}{1363} \tag{3.9}$$

3.2. INFLUENCE OF PH ON CHEMICAL EQUILIBRIA. GRAPHIC REPRESENTATION

In the case of reactions written in the form

$$aA + cH_2O = bB + mH^+ \tag{3.12}$$

where A and B are, respectively, a relatively acid and alkaline form of the same substance and which share m H^+ ions. The application of the equilibrium relation leads to the relation

$$\log \frac{[B]^b \cdot [H^+]^m}{[A]^a} = \log K$$

or

$$\log \frac{[B]^b}{[A]^a} = \log K - m \log[H^+]$$

or again, since

$$pH = -\log[H^+]$$

$$\log \frac{[B]^b}{[A]^a} = \log K + m\,pH \qquad (3.13)$$

In the particular case where $a = b = 1$, this relation becomes

$$\log \frac{[B]}{[A]} = \log K + m\,pH$$

This equation shows the influence of pH on the relation of the equilibrium activities and fugacities of the reacting species A and B.

To examine the influence of pH on these equilibria three distinct cases are considered successively, according to the state (dissolved, condensed, or gaseous) in which the reacting bodies are found.

3.2.1. Influence of pH on the Equilibrium of Homogeneous Systems (all the reacting bodies being in the dissolved state)

3.2.1.1. Dissociation of Solution of Weak Acids or Weak Bases

In this case, if the coefficients a and b of the reaction equation (3.12) have the same value (e.g., equal to 1) and if $m = 1$, relation (3.13) becomes

$$\log \frac{[B]}{[A]} = \log K + pH \qquad (3.14)$$

where [A] and [B] represent the activities (or approximately the concentrations) of the dissolved bodies A (acid) and B (alkaline).

Recall that the pK of a weak acid whose dissociation constant is K is defined by the relation

$$\mathrm{pK} = -\log K \tag{3.15}$$

Relation (3.14) then may be written

$$\log \frac{[\mathrm{B}]}{[\mathrm{A}]} = \mathrm{pH} - \mathrm{pK} \tag{3.16}$$

Example: For the reaction

$$H_2CO_3 = HCO_3^- + H^+$$

the condition of equilibrium is

$$\frac{[HCO_3^-] \cdot [H^+]}{[H_2CO_3]} = 10^{-6.37}$$

and may be written as

$$\log \frac{[HCO_3^-] \cdot [H^+]}{[H_2CO_3]} = -6.37$$

or

$$\log \frac{[HCO_3^-]}{[H_2CO_3]} = -6.37 - \log[H^+]$$

and finally

$$\boxed{\log \frac{[HCO_3^-]}{[H_2CO_3]} = -6.37 + \mathrm{pH}}$$

Based on the above example, the following equilibrium relations of a few important reactions are calculated.

Reaction	*Equilibrium*
$H_2CO_3 = HCO_3^- + H^+$	$\log \frac{[HCO_3^-]}{[H_2CO_3]} = -6.37 + \mathrm{pH}$
$HCO_3^- = CO_3^{--} + H^+$	$\log \frac{[CO_3^{--}]}{[HCO_3^-]} = -10.34 + \mathrm{pH}$
$NH_4OH + H^+ = NH_4^+ + H_2O$	$\log \frac{[NH_4OH]}{[NH_4^+]} = -9.27 + \mathrm{pH}$

Figures 11a and 11b show graphically the influence of pH on the dissociation of weak acids and weak bases in solution. To calculate the S-shaped curves of Figure 11a, let c represent, on the abscissa of this figure, the proportion of the bodies A and B, expressed as a percentage of the more alkaline form B:

$$c = \frac{100[B]}{[A] + [B]}$$

When pH $= -\log K$, that is, when the pH of the solution is equal to the pK of the substance being considered, then we have

$$\log \frac{[B]}{[A]} = \mathrm{pH} - \mathrm{pK} \qquad (3.16)$$

$$\log \frac{[B]}{[A]} = 0$$

i.e.,

$$[B] = [A]$$

Therefore, when the pH is equal to pK, the concentrations of the alkaline form B and the acid form A are equal and equivalent to 50% of the total concentration of A and B.

When the pH is less than pK by 0.5 unit, according to Eq. (3.16) one has

$$\log \frac{[B]}{[A]} = -0.5$$

$$\frac{[B]}{[A]} = 0.316$$

$$\frac{[A]}{[A] + [B]} = \frac{0.316}{1.316} = 0.24$$

So

$$c = 24\%$$

Thus, when pH = pK − 0.5, 24% of the dissolved bodies are of the more alkaline form B. Similarly, it can be shown that when pH = pK + 0.5, 24% of the dissolved bodies are of the more acid form; the proportion of the alkaline form then is 76%.

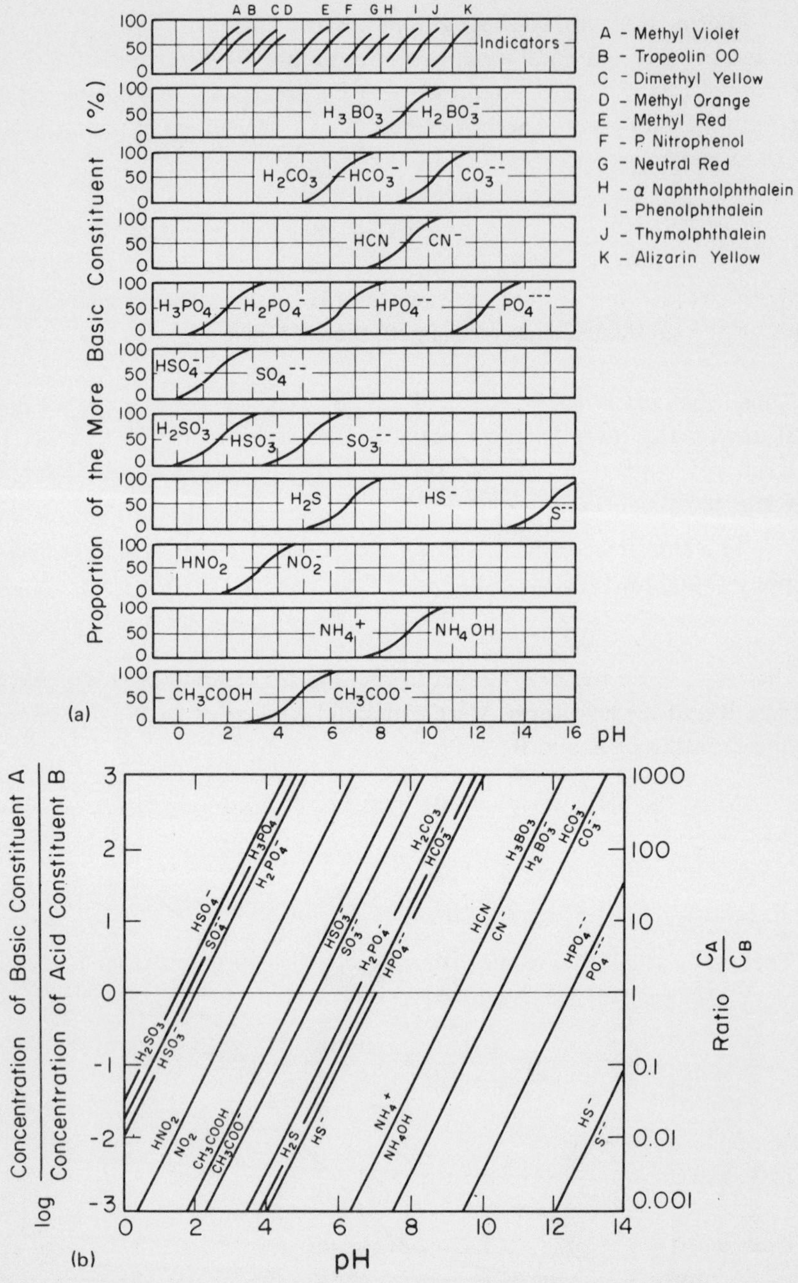

Figure 11. Influence of pH on the equilibrium of homogeneous systems (dissociation of weak acids and weak bases): (a) arithmetic concentration ratio scale; and (b) logarithmic concentration ratio scale.

When pH = pK − 1, then

$$\log \frac{[B]}{[A]} = -1$$

$$\frac{[B]}{[A]} = 0.10$$

$$\frac{[B]}{[A] + [B]} = \frac{0.10}{1.10} = 0.09$$

So

$$c = 9\%$$

Thus, when pH = pK − 1, 9% of the dissolved bodies of the forms A and B are of the more alkaline form B. Similarly, it can be shown that when pH = pK + 1, 9% of the dissolved bodies are of the more acid form; the proportion of the alkaline form then is 91%.

In a similar manner, it is easily shown that for 9 values of pH between pK − 3 and pK + 3, the ordinates c of the S-shaped curve are the following:

pH = pK − 3	c = 0.1%
pK − 2	1%
pK − 1	9%
pK − 0.5	24%
pK	50%
pK + 0.5	76%
pK + 1	91%
pK + 2	99%
pK + 3	99.9%

These values make it possible to draw a prototype of the curves in Figure 11a.

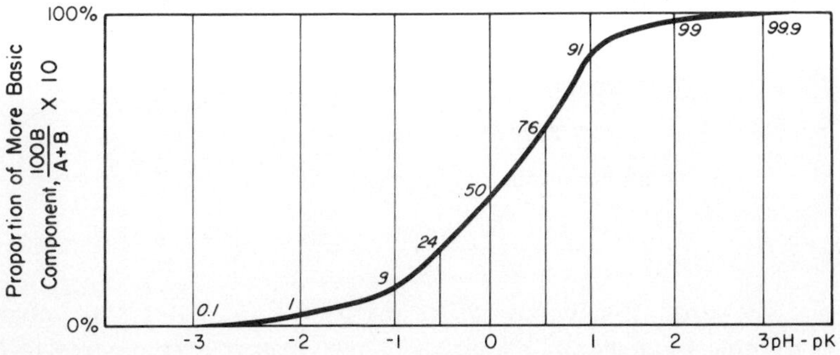

Graph of the S-shaped Curve Relative to the Dissociation of Weak Acids and Weak Bases

3.2.1.2. Applications

3.2.1.2.1. *The Use of pH Indicators*

According to Ostwald, a colored indicator is a weak acid whose more acid form A (undissociated) has a different color from that of the more alkaline form B (dissociated). So for these indicators, like other weak acids and bases, there are S-shaped curves

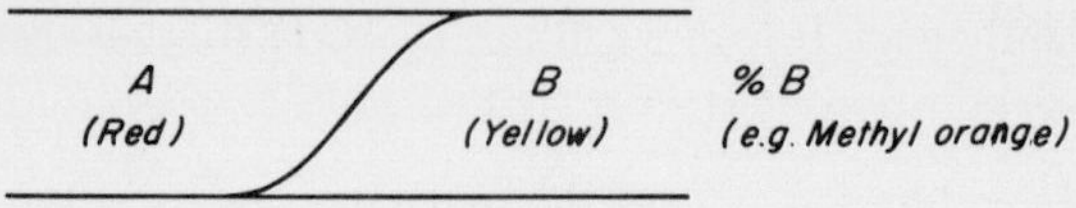

which show the influence of pH on the dissociation of the form A and, as a result, on the coloration of the indicator solutions. In the general case, where 10% of one of the two forms affect the color of the other form, the "curved region" of the indicator extends over two pH units. Note that from the top portion of Figure 11a it is easy to determine which indicators would be convenient for a given acidimetric or alkalimetric titration.

3.2.1.2.2. *Buffering Effect*

Buffers, whose presence in solutions stabilize the pH in the region of a certain value, are made up from two forms (relatively acidic and relatively alkaline) of a weak acid or a weak base. The buffering power of a mixture will be optimum for the pH corresponding to the point of inflection of the S-shaped curve. Listed below are examples of buffering solutions together with their characteristic pH:

acetic acid–acetate	4.8
monophosphate–biphosphate	6.8
ammonium salt–ammonia	9.2
bicarbonate–carbonate	10.4
biphosphate–triphosphate	12.4

3.2.1.2.3. *Hydrolysis*

If a substance dissolving in water reacts with the water and imparts to the solution an acidic or alkaline reaction, then it is considered to be hydrolyzed. Therefore, there is hydrolysis when a substance whose constituents are not absolutely stable at pH 7.0 is dissolved in water.

To illustrate the action of hydrolysis several reactions will be considered. Figure 11a illustrates that boric solutions of pH 7.0 essentially contain boric acid, H_3BO_3, and a small amount of diborate, $H_2BO_3^-$. As a result, when boric acid, H_3BO_3, is dissolved in pure water, a small proportion of H_3BO_3 is hydrolyzed according to the reaction

$$H_3BO_3 \rightarrow H_2BO_3^- + H^+$$

with the formation of a small quantity of H^+ ions. Therefore, the resulting solution is slightly acid. If sodium diborate, NaH_2BO_3, is dissolved in pure water a large proportion of $H_2BO_3^-$ is hydrolyzed according to the reaction

$$H_2BO_3^- + H_2O \rightarrow H_3BO_3 + OH^-$$

with the formation of a relatively large concentration of OH^- ions. Therefore, the resulting solution is significantly alkaline.

For a pH of 7, Figure 11a also shows that solutions of hydrogen cyanide contain essentially HCN (and little CN^-); ammoniacal solutions contain essentially NH_4^+ (and little NH_4OH), and solutions of acetic acid mainly contain $CH_3CO_2^-$ (and little CH_3CO_2H). Solutions of hydrogen cyanide and of ammonium chloride or sulfate then are slightly acidic (pH little below 7) and solutions of acetic acid are appreciably acidic (pH considerably below 7); solutions of sodium or potassiun cyanide and ammonia solutions are appreciably alkaline, while solutions of sodium or potassium acetate are weakly alkaline.

3.2.1.2.4. *pH of Solutions of Weak Acids and Weak Bases*

It is easy to evaluate graphically, and with reasonable precision, the pH of solutions of weak acids or weak bases or their corresponding salts in pure water. One such method of evaluation is outlined below.

3.2.1.2.4.1. *General Remarks*

Consider a weak acid or a weak base likely to exist in solution in two forms A and B (e.g., NH_4^+ and NH_4OH for ammonia) and being able to transform from one form to the other by the hydrolysis reaction

$$B + H^+ = A + H_2O$$

or

$$B = A + OH^-$$

(e.g., $NH_4OH = NH_4^+ + OH^-$). The equilibrium condition is

$$\log \frac{[B]}{[A]} = pH - pK$$

e.g.,

$$\log \frac{[NH_4OH]}{[NH_4^+]} = pH - 9.27$$

If the more alkaline species B (NH_4OH) is dissolved in pure water, this partially hydrolyzes by the reaction

$$B \rightarrow A + OH^-$$

(e.g., $NH_4OH \rightarrow NH_4^+ + OH^-$) and the solution obtained is alkaline. Then the quantity of the acidic form A (NH_4^+) existing in the resulting solution is equal to the quantity of OH^- ions formed by hydrolysis, namely, the quantity of OH^- ions existing in the solution less the quantity of OH^- ions previously present in pure water ($10^{-7.00}$ g-ion/liter at 25°C). The concentrations in the acidic form A (NH_4^+) and OH^- ions in the solution obtained are related as follows:

$$[A] = [OH^-] - 10^{-7.00}$$

e.g.,

$$[NH_4^+] = [OH^-] - 10^{-7.00}$$

Listed in Table VI are values of log $[OH^-]$ and of log $([OH^-] - 10^{-7.00})$ corresponding to different values of pH at 25°C, assuming to the first approximation that the concentration of OH^- ions is equal to their activity. Figure 12a[7] shows the influence of pH on the value of $\log([OH^-] - 10^{-7.00})$ as well as on the value of $\log([H^+] - 10^{-7.00})$. The two lines in Figure 12a are symmetrical about the vertical line from the abscissa at pH 7.0. When an alkaline substance (e.g., NH_4OH, Na_2CO_3, NaCN, CH_3CO_2Na) is hydrolyzed after dissolution in pure water at 25°C, the influence of pH on the concentration of the acid form of this substance (e.g., NH_4^+, HCO_3^-, HCN, CH_3CO_2H) in the resulting solution is represented by the line $\log([OH^-] - 10^{-7.00})$. When an acidic substance (e.g., NH_4Cl, H_2CO_3, HCN, CH_3CO_2H) is hydrolyzed after dissolution in pure water at 25°C,

[7] Figure 12 is similar to a diagram due to H. Flood in *Z. Elektrochem.* **46**, 669, 1940. A discussion on the pH of acid and basic solutions has been given by G. Charlot and R. Gauguin in *Les Méthodes d'Analyse des Réactions en Solution*, Masson, Paris (1951).

TABLE VI. Log Relations for OH^- and Acidic Concentrations

pH	$\log[OH^-]$	$\log([OH^-] - 10^{-7.00})$
7.00	−7.00	−0.00
7.04	−6.96	−8.00
7.14	−6.86	−7.42
7.24	−6.76	−7.13
7.34	−6.66	−6.92
7.44	−6.56	−6.75
7.54	−6.46	−6.60
7.64	−6.36	−6.47
7.74	−6.26	−6.35
7.84	−6.16	−6.23
7.94	−6.06	−6.11
8.04	−5.96	−6.00
8.14	−5.86	−5.89
8.24	−5.76	−5.77
8.34	−5.66	−5.68
8.44	−5.56	−5.57
9.00	−5.00	−5.00
10.00	−4.00	−4.00
11.00	−3.00	−3.00
12.00	−2.00	−2.00
13.00	−1.00	−1.00
14.00	0.00	0.00

the influence of pH on the concentration of the alkaline form of this substance (e.g., NH_4OH, HCO_3^-, CN^-, $CH_3CO_2^-$) in the resulting solution is represented by the line $\log([H^+] - 10^{-7.00})$.

On the other hand, the solutions obtained by dissolving an alkaline hydrolyzable substance (e.g., NH_4OH) contain this substance, not only in the acid form A (e.g., NH_4^+) but also in the alkaline form B (e.g., NH_4OH), and the relation of the concentrations (in the case of ideal solutions) in these two forms varies as a function of pH according to the equilibrium relation

$$\log \frac{[B]}{[A]} = pH - pK$$

For instance,

$$\log \frac{[NH_4OH]}{[NH_4^+]} = pH - 9.27$$

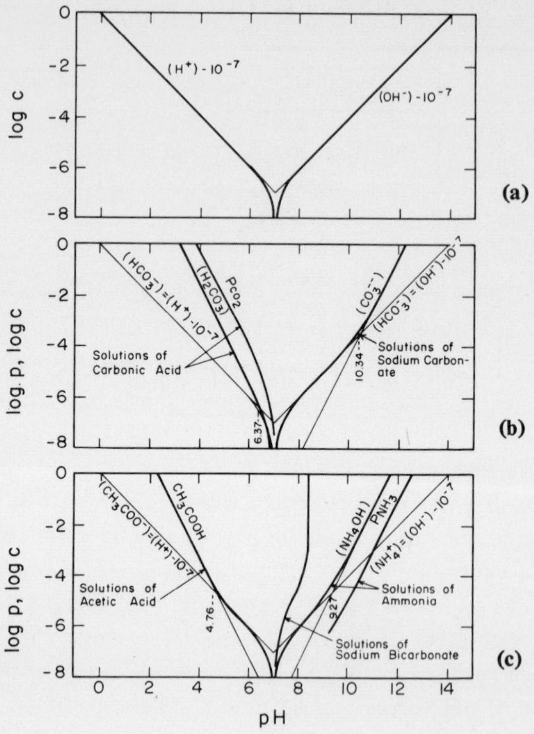

Figure 12. Determination of the pH of solutions of weak acids and weak bases in pure water.

Now the content of the alkaline form B may be expressed as a function of the content of the acid form A and of the relation [B]/[A], i.e.,

$$[B] = [A] \times \frac{[B]}{[A]}$$

Hence

$$\log [B] = \log [A] + \log \frac{[B]}{[A]}$$

or

$$\log[B] = \log[A] + pH - pK$$

e.g.,

$$\log[NH_4OH] = \log[NH_4^+] + pH - 9.27$$

Thus, the contents of the two forms A and B are related to pH (or to the corresponding content of OH^- ions) by the two relations

$$\log[A] = \log([OH^-] - 10^{-7.00}$$
$$\log[B] = \log[A] + pH - pK$$

And, for instance,

$$\log[NH_4^+] = \log([OH^-] - 10^{-7.00})$$
$$\log[NH_4OH] = \log[NH_4^+] + pH - 9.27$$

Therefore, the influence of pH on the logarithm of the content of the alkaline form B will be expressed in Figure 12c by a line whose ordinate value is obtained, for every value of pH, by adding pH − pK (for instance, in the case of ammonia, pH − 9.27) to the ordinate of the line representing the logarithmic values of the content of the acid form A (NH_4^+), namely, to the line which gives the values of $\log([OH^-] - 10^{-7.00})$. The line relative to form B will cut the line relative to form A where pH = pK (pH = 9.27); in the region of pH values greater than 8, where $\log([OH^-] - 10^{-7.00})$ is a straight line of slope +1, the line relating to form B is a straight line of slope +2.

If it is assumed that all of the substance in solution (e.g., NH_4OH) is present in either of the two forms considered (NH_4^+ and NH_4OH), then the quantity of this substance needed to attain a given content of A and B per liter is equal to the sum of the concentrations [A] + [B], e.g., $[NH_4^+] + [NH_4OH]$. The pH of solutions of various concentrations of this substance are given in Figure 12c, for each value of pH, by a line such that

$$\log c = \log([A] + [B])$$

for instance,

$$\log c = \log([NH_4^+] + [NH_4OH])$$

3.2.1.2.4.2. *Graphic Determination of the Influence of pH on the Concentration of Substances Able to Exist in Several Dissolved Forms*

The method outlined below makes it possible to resolve graphically, and in a simple manner, the following general problem which has been outlined in the preceding paragraphs: knowing that the two lines $\log c_1 = f_1(\text{pH})$ and $\log c_2 = f_2(\text{pH})$ graphically express the influence of pH on the logarithms of the concentrations c_1 and c_2 of the two bodies 1 and 2, we can determine the line $\log(c_1 + c_2) = f_{1,2}(\text{pH})$ which graphically expresses the influence of pH on the logarithm of the sum $c_1 + c_2$ of the two concentrations (Figure 13). Five points of the required line will be determined, corresponding to values of pH for which the difference $\log c_1 - \log c_2$ (where $c_1 \geq c_2$) has, respectively, five given values as follows:

(1) $\log c_1 - \log c_2 = 0$

This corresponds to the pH at which there is intersection of the two lines $\log c_1 = f_1(\text{pH})$ and $\log c_2 = f_2(\text{pH})$.

$$\log c_2 = \log c_1$$

$$c_2 = c_1$$

$$\log(c_1 + c_2) = \log 2c_1 = \log c_1 + \log 2$$

$$\log(c_1 + c_2) = \log c_1 + 0.301$$

Hence, for the pH corresponding to the point of intersection of the two lines 1 and 2 the ordinate value of the resulting line will be 0.301 unit above the value common to the two lines.

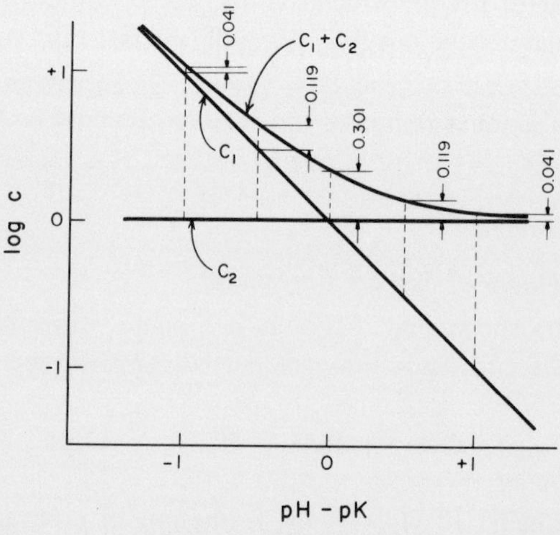

Figure 13. A logarithmic plot relative to the concentration of a substance able to exist in two dissolved forms.

(2) $\log c_1 - \log c_2 = 0.5$

We proceed as follows:

$$\log c_2 = \log c_1 - 0.5$$
$$c_2 = c_1 \times 10^{-0.5} = 0.316\, c_1$$
$$c_1 + c_2 = 1.316\, c_1$$
$$\log(c_1 + c_2) = \log c_1 + \log 1.316$$
$$\log(c_1 + c_2) = \log c_1 + 0.119$$

Thus, for values of pH for which the ordinates of lines 1 and 2 differ by 0.5 unit, the ordinate value of the resulting line will be 0.119 unit above the higher of the two lines.

(3) $\log c_1 - \log c_2 = 1$

As in the previous cases

$$\log c_2 = \log c_1 - 1$$
$$c_2 = c_1 \times 10^{-1} = 0.100\, c_1$$
$$c_1 + c_2 = 1.100\, c_1$$
$$\log(c_1 + c_2) = \log c_1 + \log 1.100$$
$$\log(c_1 + c_2) = \log c_2 + 0.041$$

For values of pH for which the ordinates of lines 1 and 2 differ by 1 unit, the ordinate of the resulting line will be 0.041 unit above the higher of the two lines.

In such a manner points of the required line are easily determined as in Figure 13.

3.2.1.2.4.3. *Examples*

(1) *pH of Solutions of Sodium Carbonate*

Figure 10b shows that if sodium carbonate is dissolved in pure water of pH = 7, the carbonate ions are partially hydrolyzed according to the reaction

$$CO_3^{--} + H_2O \rightarrow HCO_3^- + OH^-$$

the pK value being 10.34, with the formation of equivalent quantities of HCO_3^- and OH^- ions. The line $\log([OH^-] - 10^{-7.00})$ plotted on the right of Figure 12b represents the relation between the pH of the resulting solution and its content of HCO_3^- ions. The relation between pH and the content of CO_3^- ions will be obtained by drawing a straight line of slope +2 through the point on the line relating to HCO_3^- where the pH is 10.34. The pH of Na_2CO_3 solutions of different concentrations, c, will be represented in Figure 12b by a line such that for every value of pH

$$\log c = \log([HCO_3^-] + [CO_3^{--}])$$

This line is plotted in Figure 12b according to the method just described. It leads to the following values of pH of sodium carbonate solutions (for comparison a few values determined experimentally by Auerbach and Pick[8] also are given):

$\log[Na_2CO_3]$	0	−1	−2	−3	−4	−5
$[Na_2CO_3]$ (g/liter)	106	10.6	1.06	0.106	0.011	0.001
pH calculated	12.2	11.7	11.2	10.9	9.9	9.0
pH experimental	—	11.5	11.1	10.4	—	—

The agreement between the calculated and experimental values is satisfactory.

(2) *pH of Solutions of Carbon Dioxide*

Figure 10b shows that if CO_2 is dissolved in pure water of pH = 7 the carbonic acid formed partially dissociates according to the reaction

$$H_2CO_3 \rightarrow HCO_3^- + H^+$$

the pK value being 6.37, with the formation of equivalent quantities of HCO_3^- and H^+ ions.

The line $\log([H^+] - [10^{-7}])$ plotted on the left of Figure 12b represents the relation between the pH of the resulting solution and its content of HCO_3^- ions. The relation between the pH and the content of H_2CO_3 will be obtained by drawing a straight line of slope − 2 through the point on the line relating to HCO_3^- where the pH is 6.37. The pH of solutions of different concentrations of CO_2 content, c, will be represented in Figure 12b by a line such that for every value of pH

$$\log c = \log([H_2CO_3] + [HCO_3^-])$$

This line is drawn in the left portion of Figure 12b.

[8] See *Gmelins Handbuch für Anorganische Chemie*, Vol. **Na**, p. 749 (1928).

Given that the condition of equilibrium of the dissolution reaction of gaseous CO_2 with the formation of undissociated H_2CO_3 is

$$\log p_{CO_2} = \log[H_2CO_3] + 1.48$$

it will suffice to raise by 1.48 units the line relating to the concentration of H_2CO_3 to obtain a line which indicates the pH of solutions saturated with gaseous CO_2 under different partial pressures (where the logarithm is indicated as the ordinate). From this figure we can deduce the values of the pH of solutions of different overall contents of dissolved CO_2 in equilibrium with gaseous CO_2 under different partial pressures:

$\log p_{CO_2}$ (atm)	0	−1	−2	−3	−4	−5
p_{CO_2} (atm)	1	0.1	0.01	0.001	0.0001	0.00001
c	−1.48	−2.48	−3.48	−4.48	−5.48	−6.48
log *c* (mg/liter)	1500	150	15	1.6	0.19	0.03
pH	3.97	4.47	4.97	5.47	5.94	6.37

(3) *pH of Solutions of Acetic Acid*

If acetic acid is dissolved in pure water of pH = 7, this acid partially dissociates according to the reaction

$$CH_3COOH \rightarrow CH_3COO^- + H^+$$

for which the pK value is 4.76.

The line $\log([H^+] - 10^{-7.00})$ plotted in Figure 12a represents the relation between the pH of the resulting solution and the content of acetate ions CH_3COO^-. The relation between the pH and the content of CH_3COOH will be obtained by drawing a straight line of slope −2 through the point on the line relating to CH_3COOH where the pH is 4.76. The pH of solutions of acetic acid of different concentrations *c* will be represented by a line such that for each value of pH,

$$\log c = \log([CH_3COOH] + [CH_3COO^-])$$

This line is drawn in the left portion of Figure 12c. Here are the values of the pH of acetic acid solutions of different concentrations *c* deduced from this figure.

log *c* (mole/liter)	0	−1	−2	−3	−4	−5
c (mg CH_3COOH/liter)	60,000	6,000	600	60	6	0.6
pH	2.4	2.7	3.4	3.9	4.4	5.2

(4) *pH of Ammonia Solutions*

By using the same procedure as above, the following values are obtained for the pH of ammonia solutions of different concentrations c in equilibrium with gaseous NH_3 under different partial pressures (assuming pK = 9.27 for the dissociation of NH_4OH).

$\log p_{NH_3}$ (atm)	0	−1	−2	−3	−4	−5
p_{NH_3} (atm)	1	0.1	0.01	0.001	0.0001	0.00001
$\log c$ (moles/liter)	?	?	−0.25	−1.25	−2.23	−3.18
c (mg/liter)	?	?	9,670	967	99.3	11.2
pH	12.5	12.0	11.5	11.0	10.5	10.0

(5) *pH of Solutions of Sodium Bicarbonate*

To conclude the study of hydrolysis phenomena of weak acids and weak bases we consider the case of solutions of salts of which one of the constituents may simultaneously undergo two hydrolyses: an alkaline hydrolysis and an acid hydrolysis. The case of sodium bicarbonate is taken as an example.

Figure 11a shows that the CO_2 contained in a solution of pH = 7 is principally in the form of bicarbonate HCO_3^- (about 75%), the remainder being mainly carbonic acid H_2CO_3. A solution of sodium bicarbonate in pure water then hydrolyzes by the reaction

$$HCO_3^- + H_2O \rightarrow H_2CO_3 + OH^-$$

and is slightly alkaline due to the formation of OH^- ions.

It may also be seen from Figure 11a that in a slightly alkaline medium CO_2 is present not only in the form of H_2CO_3 and HCO_3^- but also in the form of carbonate CO_3^{--}. Simultaneously there are two hydrolyses

$$HCO_3^- + H_2O \rightarrow H_2CO_3 + OH^-$$

and

$$HCO_3^- \rightarrow CO_3^{--} + H^+$$

Thus, the quantity of OH^- ions formed in the solution is not equal to the quantity of H_2CO_3 (as would be the case if the first hydrolysis alone was considered), rather it is equal to the quantity of H_2CO_3 minus the quantity of CO_3^{--}. Therefore, we obtain the relation

$$[OH^-] - 10^{-7.00} = [H_2CO_3] - [CO_3^{--}]$$

Dividing both sides of the equation by $[H_2CO_3]$, we obtain

$$\frac{[OH^-] - 10^{-7.00}}{[H_2CO_3]} = 1 - \frac{[CO_3^{--}]}{[H_2CO_3]}$$

The conditions of equilibrium of the dissociation reactions of CO_2 in solution are

for $H_2CO_3 = HCO_3^- + H^+$

$$\log \frac{[HCO_3^-]}{[H_2CO_3]} = -6.37 + pH \qquad pK = 6.37$$

for $HCO_3^- = CO_3^{--} + H^+$

$$\log \frac{[CO_3^{--}]}{[HCO_3^-]} = -10.34 + pH \qquad pK = 10.34$$

for $H_2CO_3 = CO_3^{--} + 2H^+$

$$\log \frac{[CO_3^{--}]}{[H_2CO_3]} = -16.72 + 2\,pH \qquad pK = 8.36$$

The preceding relation then may be written

$$\frac{[OH^-] - 10^{-7.00}}{[H_2CO_3]} = 1 - 10^{-16.72+2pH}$$

The influence of pH on the concentrations of H_2CO_3, HCO_3^-, and of CO_3^{--} in solutions obtained by dissolving $NaHCO_3$ in pure water is given by a combination of the three relations

$$\log[H_2CO_3] = \log([OH^-] - 10^{-7.00}) - \log(1 - 10^{2pH-16.72})$$
$$\log[HCO_3^-] = \log[H_2CO_3] + pH - 6.37$$
$$\log[CO_3^{--}] = \log[HCO_3^-] + pH - 10.34$$

For different values of pH values are obtained for the various terms of these relations (see Table VII).

Table VII leads to the values shown in Table VIII for the relation between pH on one hand, and the contents of H_2CO_3, HCO_3, and CO_3 in the total dissolved $NaHCO_3$ ($= H_2CO_3 + HCO_3^- + CO_3^{--}$) on the other hand. It is presumed that the dissolved CO_2 does not exist in other forms such as Na_2CO_{3aq}, $NaHCO_{3aq}$, or $NaCO_3^-$.

TABLE VII. Influence of pH on Concentrations of H_2CO_3, HCO_3^-, and CO_3^{--}

pH	$\log([OH^-]-10^{-7.00})$	$\log(1-10^{2pH-16.72})$	$\log[H_2CO_3]$	$\log[HCO_3^-]$	$\log[CO_3^{--}]$
8.36	−5.66	−∞	+∞	+∞	+∞
8.26	−5.77	−0.40	−5.37	−3.38	−5.40
8.16	−5.87	−0.22	−5.65	−3.86	−6.04
8.06	−5.98	−0.13	−5.85	−4.16	−6.44
7.96	−6.09	−0.07	−6.02	−4.43	−6.81
7.86	−6.21	−0.04	−6.17	−4.68	−7.16
7.76	−6.33	−0.03	−6.30	−4.91	−7.49
7.66	−6.44	−0.02	−6.42	−5.13	−7.81
7.56	−6.58	−0.01	−6.57	−5.38	−8.16
7.46	−6.72	−0.01	−6.71	−5.62	−8.50
7.36	−6.89	−0.01	−6.88	−5.89	−8.87
7.26	−7.10	−0.00	−7.10	−6.21	−9.29
7.16	−7.36	−0.00	−7.36	−6.57	−9.75
7.06	−7.82	−0.00	−7.82	−7.13	−10.41

TABLE VIII. Relation between pH and Concentration in Solutions of $NaHCO_3$

pH	$[H_2CO_3]$	$[HCO_3^-]$	$[CO_3^{--}]$	$[NaHCO_3]$	$\log[NaHCO_3]$
8.36	∞	∞	∞	∞	∞
8.26	0.000 004 2	0.000 416	0.000 003 5	0.000 420	−3.38
8.16	0.000 002 2	0.000 138	0.000 000 9	0.000 140	−3.85
8.06	0.000 001 4	0.000 069	0.000 000 36	0.000 070	−4.15
7.96	0.000 000 95	0.000 037	0.000 000 15	0.000 038	−4.42
7.86	0.000 000 68	0.000 021	0.000 000 07	0.000 022	−4.66
7.76	0.000 000 50	0.000 012 3	0.000 000 03	0.000 013	−4.89
7.66	0.000 000 38	0.000 007 4	0.000 000 02	0.000 007 8	−5.11
7.56	0.000 000 27	0.000 004 2	0.000 000 01	0.000 004 5	−5.35
7.46	0.000 000 19	0.000 002 4	0.000 000 00	0.000 002 6	−5.59
7.36	0.000 000 13	0.000 001 3	0.000 000 00	0.000 001 4	−5.85
7.26	0.000 000 079	0.000 000 62	0.000 000 00	0.000 000 70	−6.15
7.16	0.000 000 044	0.000 000 27	0.000 000 00	0.000 000 31	−6.51
7.06	0.000 000 015	0.000 000 074	0.000 000 00	0.000 000 09	−7.05

The values given in Table VIII make it possible to trace on Figure 12c a line which indicates the values of the pH of sodium bicarbonate solutions of different concentrations.

Among other values the following are derived:

log c (moles $NaHCO_3$/liter)	0	−1	−2	−3	−4	−5
c (mg $NaHCO_3$/liter)	84,000	8400	840	84	8.4	0.84
pH	8.4	8.4	8.4	8.4	8.1	7.7

It may be seen from this figure that the pH of $NaHCO_3$ solutions of increasing concentrations tends toward 8.4, namely, toward the pH for which the contents of both products of hydrolysis, H_2CO_3 and CO_3^{--}, of bicarbonate, HCO_3^-, are equal. This pH is equal to the arithmetic mean of the pK's of the two successive dissociation reactions of H_2CO_3, i.e.,

$$\frac{6.37 + 10.34}{2} = 8.4$$

This is a general rule and applies to all substances undergoing such a double hydrolysis. For example,

The pH of $NaHCO_3$ solutions tends toward $\frac{6.37 + 10.34}{2} = 8.4$

The pH of NaH_2PO_4 solutions tends toward $\frac{2.13 + 7.19}{2} = 4.7$

The pH of Na_2HPO_4 solutions tends toward $\frac{7.19 + 12.03}{2} = 9.6$

3.2.2. Influence of pH on the Equilibrium of Heterogeneous Solid–Solution Systems

3.2.2.1. Solubility of Oxides and Hydroxides

The reaction for the dissolution of an oxide or hydroxide may be written in the form

$$A + cH_2O = \mathbf{B} + mH^+ \qquad (3.17)$$

where **B** is the oxide or hydroxide and A is the dissolved form of the metal or metalloid considered. The equilibrium relation of this reaction is

$$\log \frac{[H^+]^m}{[A]} = \log K \qquad (3.16)$$

or

$$\log [A] = -\log K - m\,\text{pH}$$

where [A] represents the activity (or corrected concentration) of the form A, and where

$$\log K = -\frac{\mu^0_B + m\mu^0_{H^+} - \mu^0_A - c\mu^0_{H_2O}}{1363}$$

An example of this type of equilibrium involves the reaction, $Fe^{+++} + 3H_2O = \mathbf{Fe(OH)_3} + 3H^+$. This is the precipitation of ferric hydroxide from a solution of ferric ions. The equilibrium constant is given by the relation

$$\log K = -\frac{\mu^0_{Fe(OH)_3} + 3\mu^0_{H^+} - \mu^0_{Fe^{+++}} - 3\mu^0_{H_2O}}{1363}$$

$$= -\frac{-166{,}000 + 2530 + 170{,}070}{1363}$$

$$= -4.84$$

which leads to the equilibrium condition

$$\log[Fe^{+++}] = 4.84 - 3\,\text{pH}$$

Similar relationships can be written for several typical reactions:

$Fe^{+++} + 3H_2O = \mathbf{Fe(OH)_3} + 3H^+$	$\log[Fe^{+++}] = 4.84 - 3\,\text{pH}$
$Fe^{++} + 2H_2O = \mathbf{Fe(OH)_2} + 2H^+$	$\log[Fe^{++}] = 13.29 - 2\,\text{pH}$
$2Cu^+ + H_2O = \mathbf{Cu_2O} + 2H^+$	$\log[Cu^+] = -0.84 - \text{pH}$
$2HAsO_{2aq} = \mathbf{As_2O_3} + H_2O$	$\log[HAsO_2] = -0.68$
$\mathbf{Al(OH)_3} = AlO_2^- + H_2O + H^+$	$\log[AlO_2^-] = -12.20 + \text{pH}$
$\mathbf{Zn(OH)_2} = ZnO_2^{--} + 2H^+$	$\log[ZnO_2^{--}] = -28.48 + 2\,\text{pH}$

Such equilibria are represented in Figure 14 by straight lines whose slopes are the inverse of the valencies of the given metallic or metal-containing ions. For instance, the slope is -3 for the precipitation of Fe^{+++} ion, -2 for Fe^{++}, -1 for Cu^+, 0 for $HAsO_2$, $+1$ for AlO_2^-, and $+2$ for ZnO_2^{--}.

Figure 14 shows especially that if an acidic 10^{-2} M zinc solution is made progressively alkaline by the addition of caustic soda, then some

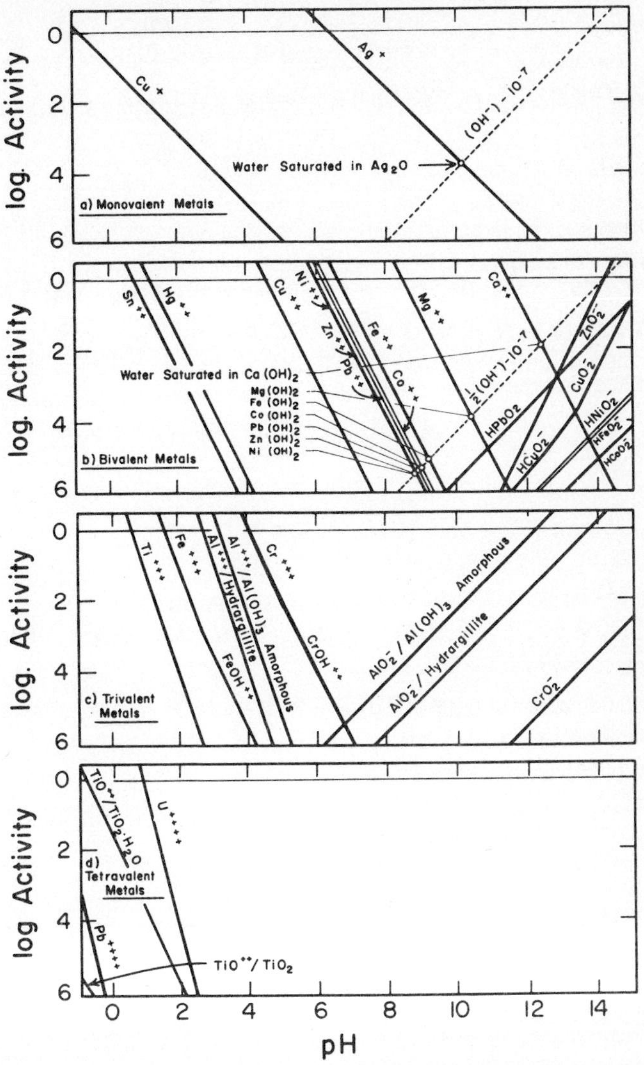

Figure 14. Influence of pH on the equilibrium of heterogeneous solid–solution systems (solubility of oxides and hydroxides).

$Zn(OH)_2$ solid begins to precipitate at a value of pH around 7.2. The precipitation is a maximum for a pH in the region of 10. Finally, the **$Zn(OH)_2$** formed redissolves and its dissolution is complete for a pH value of about 13.2. Note that metal ions tend to change their mode of precipitation as the pH is increased. In the relatively low pH range we are concerned about the Zn^{++}–$\mathbf{Zn(OH)_2}$ equilibrium, while at higher pH the solid product becomes increasingly soluble according to the $\mathbf{Zn(OH)_2}$–ZnO_2^{--} equilibrium. Starting from the lower pH, precipitation is essentially complete at pH 7.2; the $\mathbf{Zn(OH)_2}$ is essentially dissolved at pH 13.2.

3.2.2.2. pH of Oxide and Hydroxide Solutions

Consider, for example, that neutral aqueous solutions of arsenic salts, silver salts, or calcium salts contain these substances in the form of, respectively, undissociated $HAsO_2$, Ag^+ ions of +1 valence, and Ca^{++} ions of +2 valence. In pure neutral water the dissolution of arsenious oxide, silver oxide, or calcium hydroxide thus takes place essentially according to the reactions

$$\mathbf{As_2O_3} + H_2O \rightarrow 2HAsO_2$$

$$\mathbf{Ag_2O} + H_2O \rightarrow 2Ag^+ + 2OH^-$$

$$\mathbf{Ca(OH)_2} \rightarrow Ca^{++} + 2OH^-$$

Hence, in the case of *undissociated molecules* ($HAsO_2$), the dissolution of the oxide or hydroxide does not modify the value of pH; the pH of the resulting solution remains equal to 7.

For *univalent ions* (Ag^+) dissolution leads to the simultaneous formation of metal and hydroxide ions; one OH^- ion for every metal ion. The pH of solutions of different concentrations, c, resulting from the dissolution of such an oxide or hydroxide in pure water is, in the case of "ideal" solutions, governed by the equation

$$\log c = \log([OH^-] - 10^{-7})$$

The values of pH associated with this equilibrium will be represented by a straight line drawn on the right of Figure 13a and also reproduced in Figure 14a. The pH and the concentration of the solution saturated with oxide or hydroxide will be given in this figure by the coordinates of the point of intersection of the line $[OH^-] - 10^{-7}$ and the line representing the conditions of saturation of the oxide or hydroxide. From Figure 14a, the pH and amount of Ag^+ dissolved are found to be 10.2 and $10^{-3.8}$ g-atom/liter (or 18 mg Ag_2O/liter), respectively, for water saturated with Ag_2O.

In the case of divalent ions (Ca^{++}) dissolution occurs with the formation of metal and hydroxide ions to the extent of 2 OH^- ions for every metal ion. The pH of solutions of different concentrations, c, resulting from the dissolution of such an oxide or hydroxide in pure water is, in the case of ideal solutions, governed by the equation

$$\log c = \log \tfrac{1}{2}([\mathrm{OH^-}] - 10^{-7}) = \log([\mathrm{OH^-}] - 10^{-7}) - 0.30$$

The values of pH for a divalent ion will be given by a line 0.3 unit below the line relating to $\log([\mathrm{OH^-}] - 10^{-7})$ which has already been discussed above. In Figure 14b, this is drawn as a dotted line. The pH and concentration of solutions completely saturated by the oxide or hydroxide of the given divalent metal or metalloid will be represented on this figure by the coordinates of the point of intersection of the line representing the conditions of saturation by this oxide or hydroxide with the line $\tfrac{1}{2}([\mathrm{OH^-}]-10^{-7})$ for divalent ions, or the line $[\mathrm{OH^-}] - 10^{-7}$ for univalent ions. Using such an analysis the following characteristics are obtained when pure water is saturated by the hydroxides of calcium, magnesium, ferrous ion, cobalt, lead, zinc, and nickel:

$Ca(OH)_2$	pH = 12.4	$[Ca^{++}] = 10^{-1.9}$	or	933.0 mg $Ca(OH)_2$/liter
$Mg(OH)_2$	pH = 10.4	$[Mg^{++}] = 10^{-3.9}$	or	7.3 mg $Mg(OH)_2$/liter
$Fe(OH)_2$	pH = 9.2	$[Fe^{++}] = 10^{-5.1}$	or	0.7 mg $Fe(OH)_2$/liter
$Co(OH)_2$	pH = 9.0	$[CO^{++}] = 10^{-5.3}$	or	0.5 mg $Co(OH)_2$/liter
$Pb(OH)_2$	pH = 8.9	$[Pb^{++}] = 10^{-5.4}$	or	1.0 mg $Pb(OH)_2$/liter
$Zn(OH)_2$	pH = 8.8	$[Zn^{++}] = 10^{-5.5}$	or	0.3 mg $Zn(OH)_2$/liter
$Ni(OH)_2$	pH = 8.8	$[Ni^{++}] = 10^{-5.5}$	or	0.3 mg $Ni(OH)_2$/liter

These calculated values of solubility are in reasonably good general agreement with the experimental values, with the exception of the hydroxides of zinc and nickel for which the experimental results are higher: 2 instead of 0.3 mg/liter for $Zn(OH)_2$ and 1.3 instead of 0.3 mg/liter for $Ni(OH)_2$. These discrepancies are probably due to the existence of the little known monovalent ions $ZnOH^+$ and $NiOH^+$, or to the existence of allotropic varieties of $Zn(OH)_2$ and $Ni(OH)_2$ different from those considered at Figure 14.

3.2.2.3. pH of Solutions of Metallic Salts

The oxides of alkali and alkaline earth metals are very soluble in solutions of pH = 7, and such solutions contain these metals as simple ions (Na^+, K^+, Ca^{++}, Mg^{++}, ...). The salts of strong acids of these metals then are not hydrolyzed in aqueous solutions and their solutions in pure water are virtually neutral. Conversely, salts of metals, the oxide or hydroxide of which is scarcely soluble in solutions of pH = 7, tend to hydrolyze in solution with the formation of oxide or hydroxide (or basic salt). For example consider the following reactions:

$$\mathbf{CuSO_4} + 2H_2O \rightarrow \quad Cu^{++} + SO_4^{--} + 2H_2O \rightarrow \mathbf{Cu(OH)_2} + SO_4^{--} + 2H^+$$

$$\mathbf{FeCl_3} + 3H_2O \rightarrow \quad Fe^{+++} + 3Cl^- + 3H_2O \rightarrow \mathbf{Fe(OH)_3} + 3Cl^- + 3H^+$$

$$\mathbf{Al_2(SO_4)_3} + 6H_2O \rightarrow 2Al^{+++} + 3SO_4^{--} + 6H_2O \rightarrow \mathbf{2Al(OH)_3} + 3SO_4^{--} + 6H^+$$

These solutions, which are in fact suspensions of oxide or hydroxide (or basic salt) are acidic from the formation of H^+ ions; their pH is that of a solution saturated with oxide or hydroxide (or basic salt). In the absence of basic salts these values of pH depend on the concentration of metallic salt as indicated by the lines drawn on the left of Figure 14.

3.2.2.4. Solubility of Salts of Weak Acids

The dissolution of a weak acid salt in aqueous solution may be written in various ways depending on the different dissolved forms of the weak acid. For example, the solution of $\mathbf{CaCO_3}$ may give rise to the formation of CO_3^{--} and HCO_3^- ions, of undissociated H_2CO_3 and of gaseous CO_2 by the reactions

$$\mathbf{CaCO_3} = Ca^{++} + CO_3^{--} \quad (1)$$

$$\mathbf{CaCO_3} + H^+ = Ca^{++} + HCO_3^- \quad (2)$$

$$\mathbf{CaCO_3} + 2H^+ = Ca^{++} + H_2CO_3 \quad (3)$$

$$\mathbf{CaCO_3} + 2H^+ = Ca^{++} + \mathit{CO_2} + H_2O \quad (4)$$

The equilibrium conditions of these reactions are, respectively,

$$\log[Ca^{++}] = -8.35 - \log[CO_3^{--}] \quad (1')$$

$$\log[Ca^{++}] = 1.99 - \log[HCO_3^-] - pH \quad (2')$$

$$\log[Ca^{++}] = 8.36 - \log[H_2CO_3] - 2\,pH \quad (3')$$

$$\log[Ca^{++}] = 9.79 - \log p_{CO_2} - 2\,pH \quad (4')$$

In the particular case of solutions obtained by dissolving $\mathbf{CaCO_3}$ in a solution initially free from calcium and CO_2, the content of Ca^{++} is equal to the sum of the contents of CO_3^{--}, HCO_3^-, and H_2CO_3 and the first three of the four relations then become

$$2 \log[Ca^{++}] = -8.35$$

$$2 \log[Ca^{++}] = 1.99 - pH$$

$$2 \log[Ca^{++}] = 8.36 - 2\,pH$$

or again

$$\log[Ca^{++}] = -4.17 \quad (1'')$$

$$\log[Ca^{++}] = 0.99 - 0.50\,pH \quad (2'')$$

$$\log[Ca^{++}] = 4.18 - pH \quad (3'')$$

These three relations make it possible to draw in Figure 15 vertical lines representing for the particular cases considered here the influence of pH on the solubility of $\mathbf{CaCO_3}$, with the formation of respectively CO_3^{--}, HCO_3^-, and H_2CO_3. The lines all cut the axis at pH values corresponding to the pK of the two stages of dissociation of carbonic acid (6.37 and 10.34). The solubility of $\mathbf{CaCO_3}$, in all these three dissolved forms of CO_2, is illustrated in Figure 15 by a line drawn between the three vertical sections which may

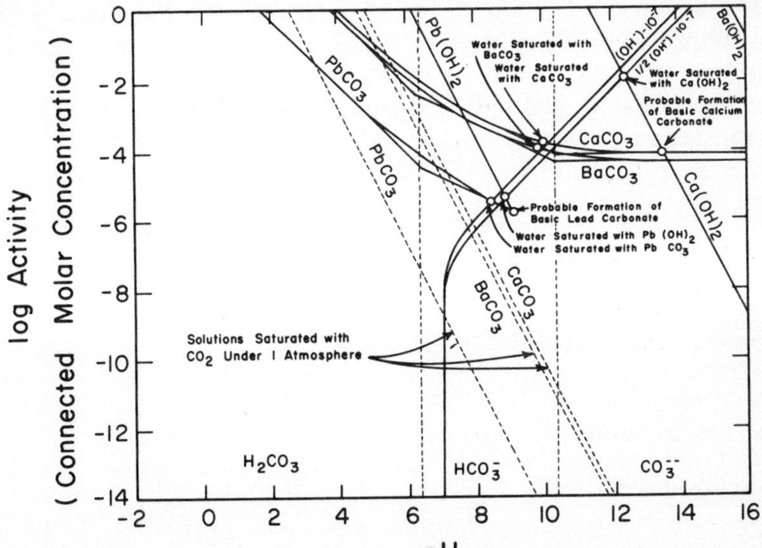

Figure 15. Influence of pH on the solubility of carbonates and hydroxides (calcium, barium, and lead).

easily be plotted by the method indicated in Section 3.2.1.2.4.2 of this chapter.

The characteristics of the solution obtained by dissolving **$CaCO_3$** in pure water according to the reactions (1), (2), and (3) will be indicated by the point of that line for which the quantity of OH^- ions formed during dissolution will be equal to the HCO_3^- concentration plus twice the concentration of H_2CO_3. As for the values of pH considered here, practically no H_2CO_3 is formed, then the pH of this solution will be the intersection of the right-hand section relating to Eq. (2″) and the line indicating the content of $[OH^-] - 10^{-7}$. The resulting pH is 10.0. Therefore the characteristics of a saturated solution of **$CaCO_3$** in pure water are, at 25°C

$$\mathrm{pH} = 10.0$$

$$\log[\mathrm{Ca}^{++}] = -3.8$$

$$[\mathrm{Ca}^{++}] = 0.00016 \text{ g-atom/liter } (16 \text{ ppm } \mathrm{CaCO_3})$$

On another part of Figure 15, a straight line indicates the circumstances of equilibrium of **$Ca(OH)_2$**; at pH = 13.5 this line cuts the line relating to the solubility of **$CaCO_3$**. This means that, in the presence of solutions containing equal quantities in g-ion/liter of calcium, Ca^{++}, and of carbonate, CO_3^{--}, the stable body would be pure **$CaCO_3$** for pH values less than 13.5.[9] At pH 13.5 and above, the stable material is a mixture of **$CaCO_3$** and **$Ca(OH)_2$**.

Applying the reasoning just outlined for calcium carbonate to barium carbonate and lead carbonate, Figure 15 presents, in the same way, the solubility of **$BaCO_3$** and **$Ba(OH)_2$**, of **$PbCO_3$** and **$Pb(OH)_2$**, as well as the probable circumstances of formation of basic lead carbonate. Characteristics derived from Figure 15 are listed below:

	Calcium	*Barium*	*Lead*
• Solution saturated with carbonate in pure water			
pH	10.0	9.9	8.5
log concentration	−3.8	−3.9	−5.5
• Solution saturated with hydroxide in pure water			
pH	12.4	—	8.9
log concentration	−1.9	—	−5.4
• Probable formation of basic carbonate			
pH	13.5	—	9.1
log concentration	−4.2	—	−5.8

[9] The affinity of formation of basic carbonate from **$CaCO_3$** and **$Ca(OH)_2$**, which is not known, has been neglected here. This affinity would have the effect of lowering the pH of formation of basic carbonate.

Also in Figure 15, the dotted straight lines illustrate the influence of pH on the solubility of calcium, barium, and lead carbonates in solutions saturated with CO_2 gas under a pressure of 1 atmosphere.

Similarly, it would be possible to examine the influence of pH on the solubility of other salts of weak acids, such as sulfides, sulfites, phosphates, etc.

3.2.2.5. Application: Saturation Equilibria of Calcium Carbonate and Treatment of Aggressive Water[10]

From classic work which they carried out in 1912 Tillmans and Heublein[11] considered that when water contains $CaCO_3$ in solution, it is possible to distinguish in it effectively four kinds of CO_2:[12]

1. CO_2 present in the $CaCO_3$.
2. CO_2 necessary for the conversion of $CaCO_3$ into $Ca(HCO_3)_2$; this CO_2 is termed *semicombined* CO_2 and is equal to half of the CO_2 existing in the form of bicarbonate ions, HCO_3^-.
3. Additional CO_2 necessary to maintain the $Ca(HCO_3)_2$ in solution ("zugehörige Kohlensäure" or "equilibrating CO_2"); this CO_2 is in the form of undissociated carbonic acid, H_2CO_3 (free CO_2).
4. Free CO_2 in excess of the "equilibrating CO_2"; this excess free CO_2 is called "aggressive CO_2."

Tillmans has shown that water generally is corrosive or noncorrosive with respect to iron depending on whether or not it contains aggressive CO_2. The absence of corrosion is due to the fact that water forms on iron a protective layer consisting of a mixture of calcium carbonate and iron oxide. The "Tillmans curve," well-known by water specialists and reproduced in Figure 16, plots as a function of the "semicombined" CO_2 the content of free "equilibrating" CO_2.

Water in which the content of free CO_2 is greater than the content of equilibrating CO_2 indicated by this curve, is then considered as "aggressive."

[10] Some detailed discussions of this subject may be found in M. Pourbaix, Étude graphique du traitement des eaux par la chaux. Application au conditionnement des eaux et à la fabrication de magnésie hydratée. Thesis, Brusselles, 1945 (extract) *Bull. Soc. Chim. Belge* **54**, 10–42 (1945); Étude graphique du traitement des eaux par la chaux et par la soude, Mémoires "Corrosion et Protection des Matériaux," CEBELCOR Publication F. 24 (1952).

[11] J. Tillmans and D. Heublein, *Gesundheitsingenieur* **35**, 669 (1912).

[12] See U. R. Evans, *Metallic Corrosion, Passivity, and Protection*, Arnold, London (1946), p. 287, 2nd ed.

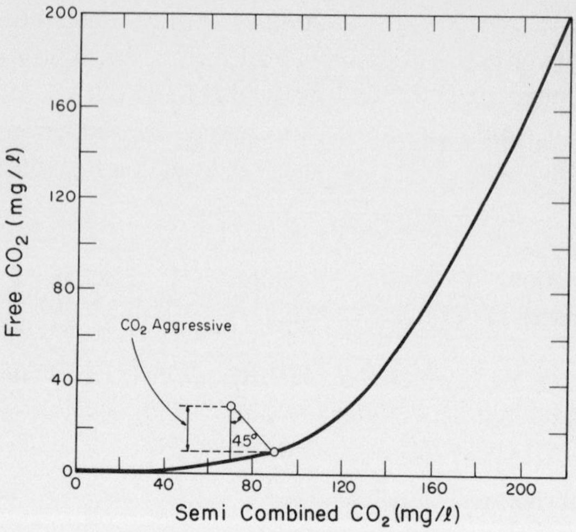

Figure 16. Tillmans' diagram.

According to Tillmans, it is possible to determine the content of aggressive CO_2 by drawing on the diagram, from a point representative of the water, a straight line descending with a slope of -1 (that is at 45°) which corresponds to the contents of free CO_2 and semicombined CO_2 established when the water reacts with $\mathbf{CaCO_3}$. The content of aggressive CO_2 is the difference between the initial content of CO_2 and the content of free CO_2 corresponding to the point where this straight line intersects the Tillmans' curve.

In practice, the indications given by this Tillmans' curve are sometimes correct and sometimes erroneous; the real content of aggressive CO_2 may be different from the content determined by the curve. Furthermore, corrosion by iron may also occur when the water does not contain aggressive CO_2.

We now proceed to examine the significance of this diagram and indicate a few precautions to be taken in its use.

The Tillman's curve gives the conditions under which water may form on iron a protective film composed of a mixture of $\mathbf{CaCO_3}$ and iron oxide. Therefore, it must really express the circumstances in which water is saturated with $\mathbf{CaCO_3}$: that is, the conditions of equilibrium between $\mathbf{CaCO_3}$ and an aqueous solution. The ordinate axis of the diagram represents a content of free CO_2 (i.e., H_2CO_3) and the abscissa represents a content of semicombined CO_2 (i.e., half of the content of HCO_3^-). Therefore, the Tillman's curve may be compared with the relation which expresses the condition of equilibrium between solid $\mathbf{CaCO_3}$ and dissolved H_2CO_3, namely, with the condition of equilibrium of the reaction

$$\mathbf{CaCO_3} + H_2CO_3 = 2HCO_3^- + Ca^{++}$$

this condition is of the form

$$\frac{[H_2CO_3]}{[HCO_3^-]^2 \cdot [Ca^{++}]} = K$$

and the value of the constant K is given by the general equilibrium relation

$$\log K = -\frac{\mu^0_{CaCO_3} + \mu^0_{H_2CO_3} - 2\mu^0_{HCO_3} - \mu^0_{Ca^{++}}}{1363}$$

Substituting the values of free enthalpy μ^0 given elsewhere[13] and assuming $\mathbf{CaCO_3}$ to be in the form of calcite, then

$$\log K = -\frac{-267{,}780 - 149{,}000 + 280{,}620 + 132{,}180}{1363}$$
$$= 4.38$$

Hence the relation

$$\log \frac{[H_2CO_3]}{[HCO_3^-]^2 \cdot [Ca^{++}]} = 4.38$$

which relates not only the concentrations[14] of H_2CO_3 and HCO_3^- which are implicitly involved in the Tillmans' curve but also the content of Ca^{++} which is not involved.

Now, as U. R. Evans[15] has pointed out, Tillmans' curve applies to water containing only CaO and CO_2. In practice this applies only to solutions of calcium bicarbonate, $Ca(HCO_3)_2$, resulting, for example, from the flow of water containing free CO_2 through soil containing $CaCO_3$. In such water the bicarbonate and the calcium are present in the ratio of two HCO_3^- ions for every Ca^{++} ion.

Thus

$$[Ca^{++}] = \tfrac{1}{2}\,[HCO_3^-]$$

[13] Standard free enthalpies of formation, at 25°C, *Rapports Techniques CEBELCOR* **72** RT. 87 (1960) (in French).

[14] Remembering that, in the present study, it is assumed that all the solutions are "ideal," that is the activities of the dissolved bodies are equal to their concentrations.

[15] U. R. Evans, *Metallic Corrosion, Passivity, and Protection* (2nd ed.), Arnold, London (1946), p. 287.

and the formula above then becomes

$$\log \frac{2[H_2CO_3]}{[HCO_3^-]^3} = 4.38$$

or

$$\log \frac{[H_2CO_3]}{[HCO_3^-]^3} = 4.38 - \log 2 = 4.38 - 0.30 = 4.08$$

In the Tillmans' diagram the abscissa is not the content of HCO_3^-, but half of this value. The relation above may then be written

$$\log \frac{[H_2CO_3]}{\left[\frac{HCO_3^-}{2}\right]^3} = 4.08 + \log 8 = 4.08 + 0.90 = 4.98$$

These contents of H_2CO_3 and HCO_3^- are expressed by Tillmans not in g-ions/liter or g-molecules/liter but in milligrams of CO_2 per liter. Since 1 g-molecule of CO_2 corresponds to 44,000 mg, the relation becomes

$$\log \frac{[\text{free } CO_2]}{[\text{semicombined } CO_2]^3} = 4.98 - 2 \log 44{,}000 = 4.98 - 9.28 = -4.30$$

or finally

$$\frac{[\text{free } CO_2]}{[\text{semicombined } CO_2]^3} = 10^{-4.30} = 0.000050$$

Hence, for concentration expressed in ppm,

$$\boxed{[\text{free } CO_2] = 0.000050 \ [\text{semicombined } CO_2]^3} \qquad (3.18)$$

For various values of the content of semicombined CO_2, the table below shows the contents of free equilibrating CO_2 given by Tillmans and Heublein (see footnote 15, U. R. Evans, p. 288) and those corresponding to Eq. (3.18).

The values given by Eq. (3.18) are noticeably higher than those given by Tillmans' curve. The agreement between the theoretical relation and the experimental curve is more exact if it is assumed that the value for the equilibrium constant of the reaction

$$\mathbf{CaCO_3 + H_2CO_3 = 2HCO_3^- + Ca^{++}}$$

is log $K = 4.13$ (instead of 4.38). Then the formula (expressed in ppm) becomes

$$[\text{free } CO_2] = 0.0000282\ [\text{semicombined } CO_2]^3 \qquad (3.19)$$

Table IX is a comparison between Tillmans' values, and the values given by Eqs. (3.18) and (3.19).

It follows that the experimental Tillmans' curve corresponds well to the equilibrium condition of the reaction,

$$\mathbf{CaCO_3} + H_2CO_3 = 2HCO_3^- + Ca^{++}$$

in the particular case where the content of HCO_3^- is double that of Ca^{++}. But the equilibrium constant considered for this reaction differs appreciably from that which results when the values of standard free enthalpy at 25°C, tabulated in report RT. 87, are used. This lack of agreement may be due to two factors; either the experiments of Tillman and Heublein were carried out at a temperature below 25°C, or the solutions used in these experiments probably were not ideal.

On the other hand, from work carried out by Mayne and Pryor, it has been found that the deposits which form on iron in water free from "aggressive CO_2" really are only protective if they contain ferric oxide $\mathbf{Fe_2O_3}$ in addition to $\mathbf{CaCO_3}$.

TABLE IX. A Comparison of Values of Equilibrating CO_2 between the Tillmans Curve and Eqs. (3.18) and (3.19)

Semicombined CO_2 (ppm)	Equilibrating CO_2 (ppm)		
	Tillmans	Equation (3.18)	Equation (3.19)
20	0.5	0.4	0.2
40	1.7	3.2	1.8
60	4.8	10.8	6.1
80	11.5	25.6	14.4
100	25.0	50.0	28.2
120	47.0	85.4	48.7
140	76.4	136.0	77.2
160	112.5	200.0	115.0
180	154.5	290.0	164.0
200	199.5	400.0	225.0

M. C. Bloom and L. Goldenberg[16] have shown that the protective iron oxide γFe_2O_3 does not exist without hydrogen in its lattice. When its lattice is filled with hydrogen its formula is $\mathbf{HFe_5O_8}$. This compound has the same structure and lattice parameters as the γFe_2O_3 found by C. L. Foley, J. Kruger, and C. J. Bechtold[17] to be the passive film on iron. Support for the presence of hydrogen in the γFe_2O_3 lattice has been given by H. T. Yolken, J. Kruger, and J. P. Calvert.[18]

Besides this, CEBELCOR has shown that this $\mathbf{Fe_2O_3}$ is protective only if the water does not contain significant quantities of certain active ions, particularly chlorides.

Thus, there are four possible causes of error when using Tillmans' curve.

1. The content of bicarbonate, HCO_3^-, must be double the content of calcium, Ca^{++} (these two contents being expressed in g-ions per liter); in other words, the temporary hardness, TAC, must be equal to the calcic hardness (the hardness being expressed in "degrees"). Failing this, the *position of the curve is not correct.*

2. The conditions of temperature and salinity of water must be such that the equilibrium constant of the reaction

$$\mathbf{CaCO_3} + H_2CO_3 = 2HCO_3^- + Ca^{++}$$

is

$$\frac{[H_2CO_3]}{[HCO_3^-]^2 \cdot [Ca^{++}]} = 10^{4.13}$$

When condition 1 is satisfied, this gives rise to the following formula (expressed in ppm) for Tillmans' curve:

$$[\text{free } CO_2] = 0.0000282\ [\text{semicombined } CO_2]^3$$

If this condition is not satisfied the *position of the curve is incorrect.*

3. The water must contain oxygen or another oxidant in sufficient quantity so that its action on iron leads to the formation of oxide, $\mathbf{Fe_2O_3}$, in direct contact with the metal. Otherwise the carbonate deposit which forms

[16] M. C. Bloom and L. Goldenberg, Fe_2O_3 and the passivity of iron, *Corr. Sci.* **5**, 623–630 (1965).

[17] C. L. Foley, J. Kruger, and C. J. Bechtoldt, Electron diffraction studies on active, passive, and transpassive oxide films formed on iron, *JECS* **114**, 994–1001 (1967).

[18] H. Thomas Yolken, Jerome Kruger, and Joan P. Calvert, Hydrogen in passive films on Fe, *Corr. Sci.* **8**, 103–108 (1968).

on the metal is not protective. Alghough not containing aggressive CO_2, water may then cause an *appreciable general corrosion* of the metal.

4. The water must not contain significant quantities of chloride or other active ions. Otherwise the oxide, $\mathbf{Fe_2O_3}$, of the deposit formed on the metal contains pores and is unprotective; then the water causes a *localized corrosion* of the metal.

The Tillmans theory may be generalized for water for which the temporary hardness TAC is not equal to the calcic hardness as follows. (Often the terms below, expressed in "french degrees" are employed to define the chemical composition of a water).

1. The *calcic hardness* T_{Ca}, measures the calcium content; 1 french degree corresponds to 10 ppm $CaCO_3$, that is 0.1×10^{-3} g-ion Ca^{++}/liter (4.0 mg Ca^{++}/liter).

2. The *magnesia hardness* T_{Mg}, measures the magnesium content; 1 french degree corresponds to 0.1×10^{-3} g-ion Mg^{++}/liter (2.4 ppm Mg^{++}).

3. The *alkalimetric titration* ("*titre alcalimétrique*," TA) measures the alkalinity by titration to the final end-point of phenolphthalein (pH = 8.3). Figure 11a shows that for water not containing weak bases other than carbonates (and especially not phosphates), this titration corresponds to the two reactions

$$OH^- + H^+ \rightarrow H_2O$$

$$CO_3^{--} + H^+ \rightarrow HCO_3^-$$

Hence, the TA measures the content of $OH^- + CO_3^{--}$; 1 french degree corresponds to 0.2×10^{-3} g-ion $OH^- + CO_3^{--}$/liter or 0.2 mg-ion $OH^- + CO_3^{--}$/liter.

4. The *complete alkalimetric titration* ("*titre alcalimétrique complet*," TAC) measures the alkalinity by titration up to the onset of the curve of methylorange (pH = 4.5). Figure 11a shows that, for the water just considered in paragraph 3 above, this corresponds to the following three reactions:

$$OH^- + H^+ \rightarrow H_2O$$

$$CO_3^{--} + H^+ \rightarrow HCO_3^-$$

$$HCO_3^- + H^+ \rightarrow H_2CO_3$$

Then the TAC measures the content of $OH^- + 2CO_3^{--} + HCO_3^-$; 1 french degree corresponds to 0.2×10^{-3} g-ion of $OH^- + 2CO_3^{--} + HCO_3^-$/liter. The values of these titrations in french degrees are thus related as follows to the values of concentration in g-ion/liter:

$$1^\circ\ T_{Ca} = 10{,}000\ [Ca^{++}]$$
$$1^\circ\ T_{Mg} = 10{,}000\ [Mg^{++}]$$
$$1^\circ\ TA = 5{,}000\ ([OH^-] + [CO_3^{--}])$$
$$1^\circ\ TAC = 5{,}000\ ([OH^-] + 2[CO_3^{--}] + [HCO_3^-])$$

On the other hand, treatments employed industrially to ensure nonaggressive water formation are based generally on the fixation of free carbonic acid by means of calcium carbonate or calcium hydroxide up to the limit of saturation of the water with $CaCO_3$. These treatments correspond to the following reactions:

1. *In the case of treatment by* $\mathbf{CaCO_3}$

$$H_2CO_3 + \mathbf{CaCO_3} \rightarrow 2HCO_3^- + Ca^{++}$$

The increase in the content of HCO_3^- is double the decrease of the content of H_2CO_3; the increase of the semicombined CO_2 content (which is half of the content of CO_2 in the form HCO_3^-) then is equal to the decrease of the content of free CO_2; water equilibrates according to the line of slope -1 in Tillmans' diagram. If the volume remains practically constant during the treatment the contents of HCO_3^- and Ca^{++} in water increase in the proportion of two ions HCO_3^- for one ion Ca^{+++}: i.e., T_{Ca} and TAC increase by the same value and the *difference* T_{Ca} — TAC *remains constant.*

2. *In the case of treatment by* $\mathbf{Ca(OH)_2}$

$$2H_2CO_3^- + \mathbf{Ca(OH)_2} \rightarrow 2HCO_3^- + Ca^{++} + 2H_2O$$

The increase of the content of HCO_3^- balances the decrease of the content of H_2CO_3; the increase of the content of semicombined CO_2 is equal to half the decrease of the content of free CO_2; water equilibrates according to the line of slope -2 in the diagram. But, if the volume of water remains practically constant (that is, if the $Ca(OH)_2$ is employed as a concentrated limemilk rather than as limewater), then T_{Ca} and TAC increase (as in the case of $\mathbf{CaCO_3}$ treatment) by the same value and the *difference* T_{Ca} — TAC *remains constant.*

If the addition of $Ca(OH)_2$ is continued until the precipitation of $\mathbf{Mg(OH)_2}$ occurs one has not

$$T_{Ca} - TAC \qquad = \text{constant}$$

but rather

$$T_{Ca} + T_{Mg} - \text{TAC} = \text{constant}$$

As shown in the above analysis, treatments of aggressive water by the addition of $CaCO_3$ or $Ca(OH)_2$ do not modify the difference T_{Ca} – TAC between the calcic hardness and the temporary hardness of the water. The graphical method of Tillmans, which is only valid for water whose calcic hardness is equal to the temporary hardness (that is for which $T_{Ca} - \text{TAC} = 0$), may then be generalized if a family of lines, each one of which is relative to a particular value of the difference T_{Ca} – TAC, is used instead of merely a single line as in this particular case. To determine the aggressive CO_2 content of the water a line is drawn through the point representing the water with a slope of -1 (in the case of treatment by calcium carbonate) or with a slope of -2 (in the case of treatment by calcium hydroxide) to intersect with the equilibrium curve relating to the value T_{Ca} – TAC for the given water. The content of aggressive CO_2 of the water will be equal to the difference between the contents of free CO_2 corresponding, respectively, to the initial and final state of the water.

The Table X gives an example of the application of this method, corresponding to the particular case represented in Figure 17. Also, the value of the "Langelier saturation index" is indicated which is the difference between the initial and final pH in the case of treatment by $CaCO_3$.

TABLE X. An Example for Generalizing Tillmans' Method and the Langelier Index

	Untreated water	Water treated with $CaCO_3$	Water treated with $Ca(OH)_2$
TAC	16.0°	20.0°	18.2°
T_{Ca}	10.0°	14.0°	12.2°
Free CO_2	30 mg/liter	12 mg/liter	9 mg/liter
pH	7.00	7.48	7.56
T_{Ca}–TAC	−6.0°	−6.0°	−6.0°
Aggressive CO_2[a]	18 mg/liter	18 mg/liter	—
Langelier index[a]	−0.48	−0.48	—

[a] The aggressive CO_2 and the Langelier index are calculated by comparing the untreated water with the water treated with $CaCO_3$.

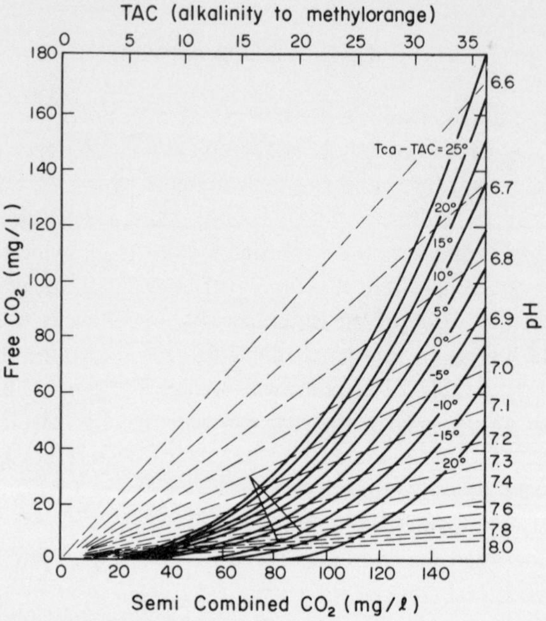

Figure 17. Diagram for the study of aggressive water (a generalization of Tillmans' diagram).

The increases of hardness caused by treatment with $\mathbf{Ca(OH)_2}$ are less than those caused by treatment with $\mathbf{CaCO_3}$; in the case of treatment by $\mathbf{CaCO_3}$ the quantity of reagent required, calculated assuming 10 mg $CaCO_3$/degree-liter, is $(20.0 - 16.0) \times 10 = 40$ mg $\mathbf{CaCO_3}$/liter; in the case of treatment by $Ca(OH)_2$ the quantity of reagent calculated, assuming 5.6 mg $Ca(OH)_2$/degree-liter, is $(18.2 - 16.0) \times 5.6 = 12.3$ mg $\mathbf{Ca(OH)_2}$/liter.

Figure 18[19] indicates the conditions of saturation of aqueous solutions with $\mathbf{Ca(OH)_2}$, $\mathbf{Mg(OH)_2}$, and $\mathbf{CaCO_3}$. Studies relating to the use of this diagram for various problems concerning the treatment of water by lime

[19] This figure, established in 1942, was based on the following values of equilibrium conditions:

saturation with $\mathbf{Ca(OH)_2}$	$\log[Ca^{++}] = 23.60 - 2\,pH$
saturation with $\mathbf{Mg(OH)_2}$	$\log[Mg^{++}] = 17.02 - 2\,pH$
saturation with $\mathbf{CaCO_3}$	$\log[Ca^{++}] = -7.964 - \log[CO_3^{--}]$
	$\log[Ca^{++}] = 2.231 - \log[HCO_3^-] - pH$
	$\log[Ca^{++}] = 8.557 - \log[H_2CO_3] - 2\,pH$
	$\log[Ca^{++}] = 10.04 - \log p_{CO_2} - 2pH$

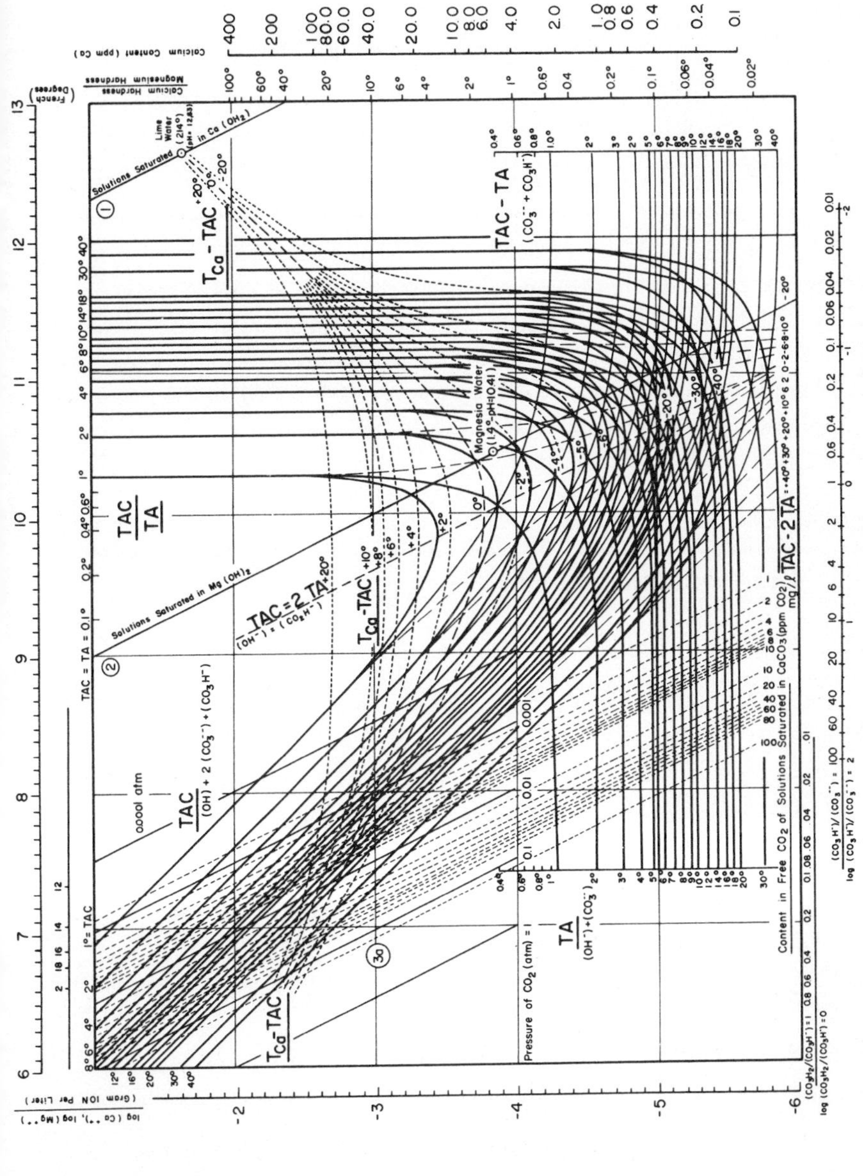

Figure 18. Characteristics of saturation by $Ca(OH)_2$, $Mg(OH)_2$, and $CaCO_3$.

TABLE XI. Solubility of $CaCO_3$ in Water Saturated with Gaseous CO_2 Under Various Partial Pressures

$\log p_{CO_2}$ (atm)	—	−5	−4	−3	−2	−1	0
p_{CO_2} (atm)	—	0.00001	0.0001	0.001	0.01	0.1	1
pH	10.00	9.47	8.85	8.16	7.52	6.85	6.17
T_{Ca}	1.7	1.8	2.6	9.5	18.0	26.0	95.0
mg $CaCO_3$/liter	17.0	18.0	26.0	95.0	180.0	260.0	950.0

and sodium carbonate have been published elsewhere.[20] This diagram gives rise to the values of the solubility of $\mathbf{CaCO_3}$ in pure water and in water saturated with gaseous CO_2 under various partial pressures (see Table XI). For example, it may be seen that the presence of 1% CO_2 in the air under atmospheric pressure, increases from 17 to 180 mg/liter the solubility of $\mathbf{CaCO_3}$ in aerated distilled water.

3.2.3. Influence of pH on the Equilibria of Heterogeneous Gas–Solution Systems

3.2.3.1. Solubility of Acidic and Alkaline Gases

The dissolution in water of gases which may form a weak acid or weak base leads to reactions where the equilibrium condition mainly depends on the partial pressure of the gas and the pH of the solution. Discussed below are several examples of reactions and the equivalent equilibrium formula involving CO_2, SO_2, H_2S, and NH_3:

$$CO_2 + H_2O = H_2CO_3 \qquad \log [H_2CO_3] = -1.43 + \log p_{CO_2}$$

$$CO_2 + H_2O = HCO_3^- + H^+ \qquad \log [HCO_3^-] = -7.80 + \log p_{CO_2} + pH$$

$$CO_2 + H_2O = CO_3^{--} + 2H^+ \qquad \log [CO_3^{--}] = -18.14 + \log p_{CO_2} + 2\,pH$$

$$SO_2 + H_2O = H_2SO_3 \qquad \log [H_2SO_3] = 0.08 + \log p_{SO_2}$$

$$SO_2 + H_2O = HSO_3^- + H^+ \qquad \log [HSO_3^-] = -1.82 + \log p_{SO_2} + pH$$

$$SO_2 + H_2O = SO_3^{--} + 2H^+ \qquad \log [SO_3^{--}] = -9.08 + \log p_{SO_2} + 2\,pH$$

[20] CEBELCOR Publication F. 24, p. 63 (1952).

$\boldsymbol{H_2S}$	$= H_2S$	$\log [H_2S]$	$= -0.99 + \log p_{H_2S}$
$\boldsymbol{H_2S}$	$= HS^- + H^+$	$\log [HS^-]$	$= -8.00 + \log p_{H_2S} + pH$
$\boldsymbol{H_2S}$	$= S^{--} + 2H^+$	$\log [S^{--}]$	$= -22.00 + \log p_{H_2S} + 2\,pH$
$\boldsymbol{NH_3} + H_2O$	$= NH_4OH$	$\log [NH_4OH]$	$= 1.75 + \log p_{NH_3}$
$\boldsymbol{NH_3} + H^+$	$= NH_4^+$	$\log [NH_4^+]$	$= 11.02 + \log p_{NH_3} - pH$

These equilibrium formulas make it possible to establish Figure 19. Here, the influence of pH on the solubility of $\boldsymbol{CO_2}$, $\boldsymbol{SO_2}$, $\boldsymbol{H_2S}$, and $\boldsymbol{NH_3}$ in all the dissolved forms considered above is represented for different values of partial pressure of the gas. For each of these gases the solubility is represented by an identical family of lines. These families of lines all intersect at values of pH corresponding to the pK values of the dissociation reactions of the dissolved gas; i.e., 6.37 and 10.34 for $\boldsymbol{CO_2}$, 1.90 and 7.26 for $\boldsymbol{SO_2}$, 7.01 and 14.01 for $\boldsymbol{H_2S}$, and 9.27 for $\boldsymbol{NH_3}$.

3.2.3.2. Applications

3.2.3.2.1. *Action of Strong Acids on Solutions of Carbonates, Sulfites, and Sulfides*

For a definite content of dissolved gas as shown in Figure 19, the partial pressure of gas increases when the pH diminishes. In the case of solutions containing 0.1 g-molecule/liter of dissolved gas, the pH for which the pressure of gas becomes 1 atm is indicated on the abscissa by the point of intersection of the horizontal line from the ordinate value of -1 with the zero isobar curve; i.e., pH $= 6.5$ is found for $\boldsymbol{CO_2}$ and pH $= 6.0$ for $\boldsymbol{S_2H}$. It is not possible to attain a pressure of 1 atm for $\boldsymbol{SO_2}$.

The carbonates and bicarbonates and the sulfides and bisulfides are relatively easily decomposed by strong acids; the decomposition of sulfites and bisulfites is more difficult.

3.2.3.2.2. *Action of Strong Bases on Ammoniacal Solutions*

The bottom diagram of Figure 19 shows that it is not possible to reach a partial pressure of 1 atmosphere of ammonia at 25°C by the elevation of pH using NaOH or KOH. The evolution of ammonia gas is only obtained under vacuum or by heating highly alkaline solutions.

3.2.3.2.3. *Absorption of Acidic or Alkaline Gases*

If it is desired to remove $\boldsymbol{CO_2}$, $\boldsymbol{H_2S}$, and $\boldsymbol{NH_3}$ by washing at 25°C using a solution containing 0.01 g-molecule/liter of carbonate, sulfite, sulfide or

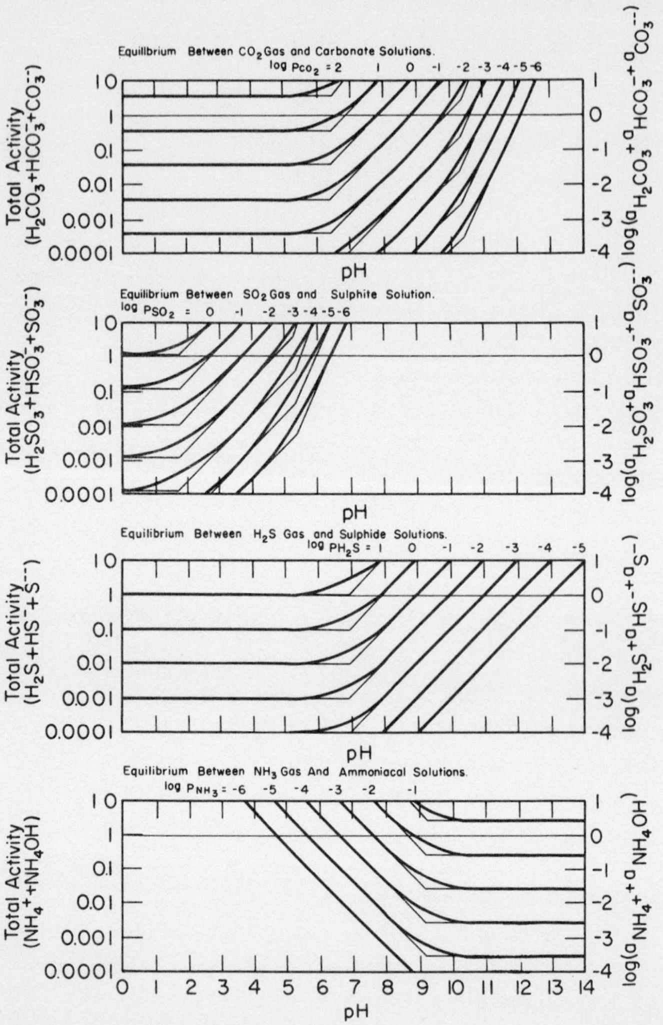

Figure 19. Influence of pH on the equilibrium of heterogeneous gas–solution systems (solubility of acidic and alkaline gases).

ammonium salt, the values of pH needed to ensure that the partial pressure of CO_2, SO_2, H_2S, or NH_3 in the treated gas becomes less than one millionth of an atmosphere (0.0001% under atmospheric pressure or, for example, 1.4 mg H_2S/m^3) will be indicated on the abscissa by the intersection of the horizontal line from the ordinate value of -2 with the -6 isobar curve.

Such values are for

$\boldsymbol{CO_2}$	pH = 11.1
$\boldsymbol{SO_2}$	pH = 5.8
$\boldsymbol{H_2S}$	pH = 12.0
$\boldsymbol{NH_3}$	pH = 6.0

Therefore, the purification of $\boldsymbol{H_2S}$ calls for a highly alkaline solution, e.g., a solution of NaOH or a weakly bicarbonated solution of Na_2CO_3. A solution of Na_2CO_3 is sufficiently active for the purification of $\boldsymbol{CO_2}$ provided that it is not allowed to become too enriched with bicarbonate. For purification of $\boldsymbol{SO_2}$ a solution of Na_2CO_3 remains active even when it is fairly strongly bicarbonated. Washing with water absorbs a considerable quantity of $\boldsymbol{NH_3}$ but this does not generally achieve complete purification of the gas; a weak acid is required for this purification.

Chapter 4

ELECTROCHEMICAL EQUILIBRIA

4.1. ELECTROCHEMICAL OXIDATIONS AND REDUCTIONS

An electrochemical reaction has been defined as a reaction in which both chemical bodies M_γ (neutral molecules and/or ions) and free electric charges, or electrons e^-, participate

$$\sum_\gamma \nu_\gamma M_\gamma + ne^- = 0 \tag{4.1}$$

Typical of electrochemical reactions are the important examples which follow below (see Figure 20).

4.1.1. Decomposition of Water

The decomposition of water into hydrogen and oxygen, $2H_2O \rightarrow 2\boldsymbol{H_2} + \boldsymbol{O_2}$, which occurs *chemically* by heating water vapor to a high temperature, may be carried out *electrochemically* by *electrolysis* of certain aqueous solutions with specific metal electrodes: e.g., electrolysis of sulfuric acid solutions with platinum electrodes (Figure 20a). Here the following reactions operate:

Negative pole	$4H^+ + 4e^-$	$\rightarrow 2\boldsymbol{H_2}$
Positive pole	$2H_2O$	$\rightarrow \boldsymbol{O_2} + 4H^+ + 4e^-$
	$2H_2O$	$\rightarrow 2\boldsymbol{H_2} + \boldsymbol{O_2}$

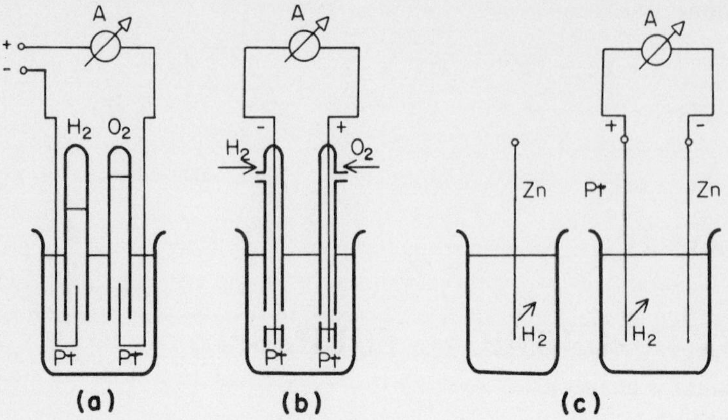

Figure 20. Chemical reactions produced electrochemically: (a) decomposition of water (electrolysis); (b) synthesis of water (fuel cell); and (c) corrosion of zinc with evolution of hydrogen.

4.1.2. Synthesis of Water

The synthesis of water, $2\boldsymbol{H}_2 + \boldsymbol{O}_2 \rightarrow 2H_2O$, which may be achieved *chemically* by the combustion of hydrogen, may be accomplished *electrochemically* in an hydrogen–oxygen fuel cell (Figure 20b) where the following reactions occur:

Negative pole	$2\boldsymbol{H}_2$	$\rightarrow 4H^+ + 4e^-$
Positive pole	$\boldsymbol{O}_2 + 4H^+ + 4e^-$	$\rightarrow 2H_2O$
	$2\boldsymbol{H}_2 + \boldsymbol{O}_2$	$\rightarrow 2H_2O$

4.1.3. Corrosion of Iron or Zinc by an Acid with Evolution of Hydrogen

The corrosion of iron or zinc with the evolution of hydrogen which may be expressed as

$$\mathbf{Fe} + 2H^+ \rightarrow Fe^{++} + \boldsymbol{H}_2$$

or as

$$\mathbf{Zn} + 2H^+ \rightarrow Zn^{++} + \boldsymbol{H}_2$$

results, in fact, from the completion of two simultaneous electrochemical reactions

$$\mathbf{Fe} \rightarrow Fe^{++} + 2e^- \qquad \text{or} \qquad \mathbf{Zn} \rightarrow Zn^{++} + 2e^-$$

$$2H^+ + 2e^- \rightarrow \boldsymbol{H_2} \qquad\qquad 2H^+ + 2e^- \rightarrow \boldsymbol{H_2}$$

$$\mathbf{Fe} + 2H^+ \rightarrow Fe^{++} + \boldsymbol{H_2} \qquad\qquad \mathbf{Zn} + 2H^+ \rightarrow Zn^{++} + \boldsymbol{H_2}$$

The existence of these electrochemical reactions is apparent if a platinum wire is connected through a galvanometer to the corroding metal (Figure 20c). Then an electric current is observed to flow between the iron or zinc (the negative electrode on which the corrosion reaction proceeds) and the platinum (the positive electrode on which the evolution of hydrogen occurs).[1]

4.1.4. Oxidation of Ferrous Salts by Permanganate

The oxidation of ferrous salts by permanganate which proceeds in acid solution according to the equation

$$MnO_4^- + 5Fe^{++} + 8H^+ \rightarrow Mn^{++} + 5Fe^{+++} + 4H_2O$$

or in neutral solution by the equation

$$MnO_4^- + 3Fe^{++} + 4H^+ \rightarrow \mathbf{MnO_2} + 3Fe^{+++} + 2H_2O$$

also results from the completion of two simultaneous electrochemical reactions

$$5Fe^{++} \rightarrow 5Fe^{+++} + 5e^-$$

$$MnO_4^- + 8H^+ + 5e^- \rightarrow Mn^{++} + 4H_2O$$

$$MnO_4^- + 5Fe^{++} + 8H^+ \rightarrow Mn^{++} + 5Fe^{+++} + 4H_2O$$

or

$$3Fe^{++} \rightarrow 3Fe^{+++} + 3e^-$$

$$MnO_4^- + 4H^+ + 3e^- \rightarrow \mathbf{MnO_2} + 2H_2O$$

$$MnO_4^- + 3Fe^{++} + 4H^+ \rightarrow \mathbf{MnO_2} + 3Fe^{+++} + 2H_2O$$

[1] In general hydrogen ion reduction will occur on both the corroding metal and the platinum. However, the exclusive reaction on platinum is the hydrogen reduction.

These electrochemical reactions are *oxidation–reduction reactions* which proceed in the direction of an *oxidation* if they correspond to a *liberation of negative electric charge* and in the direction of a *reduction* if they correspond to a *consumption of negative electric charge.*

Examples:

Oxidations

$\mathbf{Fe} \rightarrow Fe^{++} + 2e^-$	Corrosion of iron with the formation of ferrous ions
$Fe^{++} \rightarrow Fe^{+++} + e^-$	Oxidation of ferrous ions to ferric ions
$\boldsymbol{H_2} \rightarrow 2H^+ + 2e^-$	Oxidation of gaseous hydrogen into hydrogen ions
$2H_2O \rightarrow \boldsymbol{O_2} + 4H^+ + 4e^-$	Oxidation of water with the formation of hydrogen ions and gaseous oxygen

Reductions

$Fe^{++} + 2e^- \rightarrow \mathbf{Fe}$	Electrodeposition of iron
$Fe^{+++} + e^- \rightarrow Fe^{++}$	Reduction of ferric ions to ferrous ions
$2H^+ + 2e^- \rightarrow \boldsymbol{H_2}$	Reduction of hydrogen ions to hydrogen gas
$\boldsymbol{O_2} + 4H^+ + 4e^- \rightarrow 2H_2O$	Reduction of oxygen gas and hydrogen ions with formation of water

For the study of electrochemical reactions we begin with the particular case of simple galvanic cells, in which the two electrochemical reactions proceed simultaneously on two distinct but coupled electrodes. Then the general case of any electrochemical reaction is examined.

4.2. GALVANIC CELLS

4.2.1. General Remarks

Electrochemical reactions will be studied which proceed on electrodes of a galvanic cell, such as a fuel cell, battery, or electrolysis cell, as schematically illustrated on the left of Figure 21. This particular example is the Daniell cell, constructed of a zinc electrode, 1, and a copper electrode, 2, which are immersed, respectively, in solutions of zinc sulfate and copper sulfate

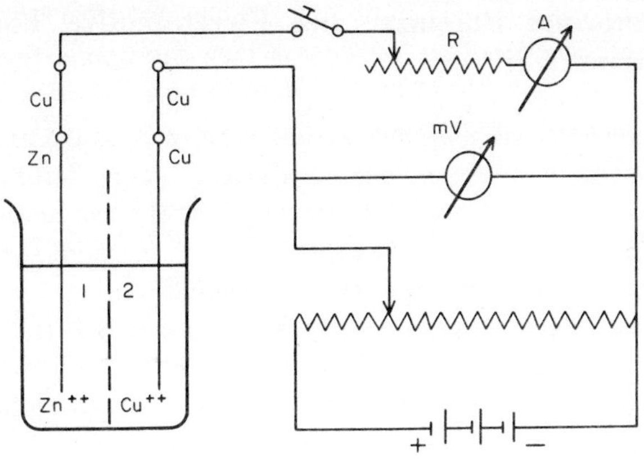

Figure 21. Daniell cell.

separated by a porous membrane. The terminals of the two electrodes are made of the same metal (copper, for example). The two electrodes of such a galvanic cell are connected by an external circuit which incorporates, besides a resistance and an ammeter, a potentiometer. This makes it possible to vary the potential across the ends of the cell; and it is then possible to attain two associated electrochemical reactions on the two electrodes for which the equations involve equal numbers of electrons produced or consumed.

If the species involved in the reaction on the surface of electrode 1 (zinc) are designated by M_1 and the species taking part in the reaction on the surface of electrode 2 (copper) by M_2 then these reactions are

$$\begin{array}{llcll} \text{on zinc:} & \sum \nu_1 M_1 + n_1 e^- = 0 & \text{or} & \mathbf{Zn} = Zn^{++} + 2e^- \\ \text{on copper:} & \sum \nu_2 M_2 - n_1 e^- = 0 & & Cu^{++} + 2e^- = \mathbf{Cu} \\ \hline & \sum \nu_1 M_1 + \sum \nu_2 M_2 = 0 & \text{or} & \mathbf{Zn} + Cu^{++} = Zn^{++} + \mathbf{Cu} \end{array}$$

Thus, the overall chemical reaction is accomplished electrochemically; this reaction may be carried out in either direction according to the value of the potential of the terminals. Also, it is well known that the quantity of electricity passed during such reactions is 1 faraday (96,484 coulombs or amp-sec)/g-eq; i.e., 2 faradays/g-atom of zinc dissolved or g-atom of copper deposited.

4.2.2. Equilibrium Potential and Electromotive Force of a Galvanic Cell[2]

The equilibrium potential of a galvanic cell, such as the one described above (whose absolute value is equal to the electromotive force of the cell), is the difference $\varphi_1 - \varphi_2$ between the potentials of the two terminals of the electrodes 1 and 2 on open circuit (open circuit potential of the terminals) *when the whole system is in electrochemical equilibrium.*

This condition implies than no net reaction proceeds at either electrode and that there is no difference if potential due to diffusion between the two solutions in which the electrodes are immersed. The following equation relates this reversible potential to the affinity, $A_{1.2}$, of overall chemical reaction

$$\sum \nu_1 M_1 + \sum \nu_2 M_2 = 0 \qquad (4.2)^3$$

$$\varphi_1 - \varphi_2 = -\frac{A_{1.2}}{n_1 F} \qquad (4.3)^4$$

Now

$$A_{1.2} = -\left(\sum \nu_1 \mu_1 + \sum \nu_2 \mu_2\right) \qquad (4.4)$$

[2] According to a report published in 1954 by Commission No. 2 of CITCE. "Nomenclature and Electrochemical Definitions" (see *Proceedings of the 6th Meeting of CITCE, Poitiers 1954*, Sections 4.8 and 4.9, p. 33, Butterworths, London (1955)) the *reversible potential* (or *equilibrium potential*) of a galvanic cell is equal, but of opposite sign, to the chemical affinity of the cell reaction divided by the reaction charge. The *electromotive force* of the cell is equal, both in magnitude and sign, to the quotient of the chemical affinity divided by the reaction charge.

[3] With regard to the sign given to the stoichiometric coefficients, ν, it is usual, according to the school of De Donder, to consider as positive the coefficients relating to constituents which appear during the reaction (those which are written on the right-hand side of the reaction equation) and to consider as negative the coefficients relating to constituents which are consumed during the reaction. For example, in the case of the reaction **Zn** = $Zn^{++} + 2e^-$ proceeding in the direction **Zn** $\rightarrow Zn^{++} + 2e^-$, it is assumed $\nu_{Zn^{++}} = +1$, n = +2, $\nu_{Zn} = -1$; where the reaction proceeds in the direction $Zn^{++} + 2e^- \rightarrow$ **Zn** it is assumed that $\nu_{Zn^{++}} = -1$, $n = -2$, and $\nu_{Zn} = +1$.

It is noted that this sign convention is by no means necessary in the present case where frequently reactions are considered independently of the direction in which they proceed, and where a reaction written as **Zn** = $Zn^{++} + 2e^-$ may be considered as proceeding from right to left as **Zn** $\leftarrow Zn^{++} + 2e^-$, or from left to right as **Zn** $\rightarrow Zn^{++} + 2e^-$. Thus, concerning the choice of sign of the stoichiometric coefficients, there is complete freedom and either the right- or left-hand side of the equation may be arbitrarily considered as positive.

[4] R. Defay, *Lectures in Physical Chemistry*, Université Libre de Bruxelles, Formula 20–21, p. 218 (in French).

Then the equilibrium potential of the cell is related to the chemical potentials, μ, of the constituents of the overall chemical reaction as follows:

$$\varphi_1 - \varphi_2 = \frac{\sum \nu_1\mu_1 + \sum \nu_2\mu_2}{n_1 F} \tag{4.5}$$

If the electrical potentials are expressed in volts, and the chemical potentials, μ, in calories per mole, relation (4.5) may be written in the form

$$\varphi_1 - \varphi_2 = \frac{\sum \nu_1\mu_1 + \sum \nu_2\mu_2}{23{,}060\ n_1} \tag{4.6}$$

where

$$\begin{aligned} 1\ \text{faraday} &= 96{,}484\ \text{coulombs} \\ 1\ \text{volt} \times 1\ \text{faraday} &= 96{,}484\ \text{volts} \times \text{coulombs (or joules)} \\ &= 96{,}484 \times 0.239\ \text{calories} \\ &= 23{,}060\ \text{calories} \end{aligned}$$

Concerning the *sign* of the reversible potential, it is noted that this relation gives the difference of potential, $\varphi_1 - \varphi_2$ (that is, the potential of terminal 1 with respect to terminal 2), *where, of course, the stoichiometric coefficient n_1 of the electron involved has the same sign as in the equation of the reaction which is proceeding on electrode 1.* If it is desired to express the potential difference, $\varphi_2 - \varphi_1$ (that is, the potential of terminal 2 with respect to terminal 1), the stoichiometric coefficient n_2 of the electron must be given the sign that it possesses in the equation of the reaction which is proceeding on electrode 2; as $n_2 = -n_1$, obviously

$$(\varphi_2 - \varphi_1) = -(\varphi_1 - \varphi_2)$$

Example. Consider the calculation of the equilibrium potential of the Daniell cell at 25°C in the particular case where the electrolytic solutions of, respectively, Zn^{++} and Cu^{++} ions both have unit activity. Let us assume that the reactions occurring at the electrodes of this galvanic cell may be written in either of the two following forms:

$\mathbf{Zn} = Zn^{++} + 2e^-$	or	$Zn^{++} + 2e^- = \mathbf{Zn}$
$Cu^{++} + 2e^- = \mathbf{Cu}$		$\mathbf{Cu} = Cu^{++} + 2e^-$
$\mathbf{Zn} + Cu^{++} = Zn^{++} + \mathbf{Cu}$		$Zn^{++} + \mathbf{Cu} = \mathbf{Zn} + Cu^{++}$

For the sake of definition, if the stoichiometric coefficients, ν and n, of the species written on the left-hand side of the equation are considered positive and those written on the right-hand side are considered negative, the reversible potential, $\varphi_{Zn} - \varphi_{Cu}$, is given by the relations

$$\varphi_{Zn} - \varphi_{Cu} = \frac{\mu_{Zn} - \mu_{Zn^{++}} + \mu_{Cu^{++}} - \mu_{Cu}}{23{,}060 \times (-2)}$$

or

$$\varphi_{Zn} - \varphi_{Cu} = \frac{\mu_{Zn^{++}} - \mu_{Zn} + \mu_{Cu} - \mu_{Cu^{++}}}{23{,}060 \times (+2)}$$

which, in practice, are identical.

Assuming, for the chemical potentials, the following standard μ^0 values, indicated in tables of RT. 87 of CEBELCOR (and abstracted from *Oxidation Potentials*, by W. Latimer)

$$\begin{aligned} \mu^0_{Zn} &= 0 \text{ cal} \\ \mu^0_{Zn^{++}} &= -35{,}184 \text{ cal} \\ \mu^0_{Cu} &= 0 \text{ cal} \\ \mu^0_{Cu^{++}} &= 15{,}530 \text{ cal} \end{aligned}$$

the above relations become

$$\varphi_{Zn} - \varphi_{Cu} = \frac{-35{,}184 - 0 + 0 - 15{,}530}{46{,}120} = \frac{-50{,}714}{46{,}120} = -1.099 \text{ V}$$

Thus, at equilibrium the potential between the terminals of the Daniell cell is 1.099 V; zinc is the negative electrode of this cell.

In conclusion, considering the general case of an overall chemical reaction, $\sum \nu M = 0$, carried out electrochemically by two electrodes, 1 and 2, of a galvanic cell, the potential of electrode 1 estimated with respect to electrode 2 (or more exactly, the *electrode potential* of electrode 1 measured with respect to electrode 2), for the condition of electrochemical equilibrium of the cell, is given both in magnitude and sign by the relation

$$\boxed{\varphi_1 - \varphi_{2_{equ}} = \frac{\sum \nu\mu}{23{,}060\, n_1}} \tag{4.7}$$

The stoichiometric coefficient, n_1, of the electron maintains the same sign as it possessed in the equation of the electrochemical reaction which is produced on electrode 1.

The use of such a formula avoids all errors of sign concerning electrode potentials.

4.2.3. Fuel Cells, Batteries, Electrolysis Cells[5]

If a galvanic cell is allowed to discharge current spontaneously through an external resistance, this cell (fuel cell or battery) discharges by effecting an *oxidation at the negative electrode* and a *reduction at the positive electrode*; free electrons flow from the negative pole to the positive pole.

Examples:

	Daniell cell zinc–copper
Negative electrode (oxidations)	$\mathbf{Zn} \rightarrow Zn^{++} + 2e^{-}$ ↓
Positive electrode (reductions)	$\mathbf{Cu} \leftarrow Cu^{++} + 2e^{-}$
Total reaction	$\mathbf{Zn} + Cu^{++} \rightarrow Zn^{++} + \mathbf{Cu}$

	Fuel cell hydrogen–oxygen
Negative electrode (oxidations)	$2\boldsymbol{H_2} \rightarrow 4H^{+} + 4e^{-}$ ↓
Positive electrode (reductions)	$2H_2O \leftarrow \boldsymbol{O_2} + 4H^{+} + 4e^{-}$
Total reaction	$2H_2 + \boldsymbol{O_2} \rightarrow 2H_2O$

	Lead battery
Negative electrode (oxidations)	$\mathbf{Pb} \rightarrow Pb^{++} + 2e^{-}$ ↓
Positive electrode (reductions)	$Pb^{++} + 2H_2O \leftarrow \mathbf{PbO_2} + 4H^{+} + 2e^{-}$
Total reaction	$\mathbf{Pb} + \mathbf{PbO_2} + 4H^{+} \rightarrow 2Pb^{++} + 2H_2O$

[5] We consider that a *fuel cell* is a galvanic cell generating electric current in which the overall chemical reaction proceeds in the direction permitted by its affinity; an *electrolysis cell* is a galvanic cell consuming current in which the overall chemical reaction proceeds in the inverse direction to that permitted by its affinity; a *battery* is a galvanic cell which may function, either as a fuel cell or as an electrolysis cell.

Conversely, if a sufficiently large potential is applied to the terminals of the galvanic cell so that the electric current flows in the opposite direction, there is *electrolysis* of the solution and possibly a charging of the battery; then there is *reduction at the negative electrode* and *oxidation at the positive electrode*; free electrons flow from the positive pole to the negative pole.

Examples:

Daniell cell zinc–copper

Negative pole $\quad \mathbf{Zn} \leftarrow Zn^{++} + 2e^-$

$\uparrow$

Positive pole $\quad \mathbf{Cu} \rightarrow Cu^{++} + 2e^-$

$$\mathbf{Cu} + Zn^{++} \rightarrow Cu^{++} + \mathbf{Zn}$$

Electrolysis of water

Negative pole $\quad 2\boldsymbol{H}_2 \leftarrow 4H^+ \qquad + 4e^-$

$\uparrow$

Positive pole $\quad 2H_2O \rightarrow \boldsymbol{O}_2 + 4H^+ + 4e^-$

$$2H_2O \rightarrow 2\boldsymbol{H}_2 + \boldsymbol{O}_2$$

Lead battery

Negative pole $\quad \mathbf{Pb} \leftarrow Pb^{++} \qquad + 2e^-$

$\uparrow$

Positive pole $\quad Pb^{++} + 2H_2O \rightarrow \mathbf{PbO_2} + 4H^+ + 2e^-$

$$2Pb^{++} + 2H_2O \rightarrow \mathbf{Pb} + \mathbf{PbO_2} + 4H^+$$

Figure 22 illustrates especially the direction of current flow during the charge and discharge of a lead battery. If the *anode* of a galvanic system is defined as the electrode where there is oxidation (or where several simultaneous electrochemical reactions are produced with a predominance of oxidations relative to reductions) and if the *cathode* is defined as being the electrode where there is *reduction* (or a predominance of reductions relative to oxidations), the *negative electrode* will be the *anode* during the discharge of a fuel cell and the *cathode* in an electrolysis cell; the *positive electrode* will be the *cathode* during the discharge of a fuel cell and the *anode* in an electrolysis cell.

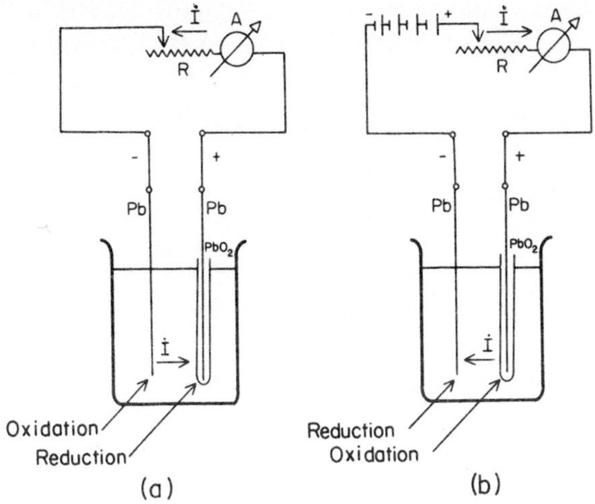

Figure 22. Lead battery: (a) discharge; and (b) charge.

4.3. ANY ELECTROCHEMICAL REACTION

Up to this point we have considered the particular case of electrochemical reactions associated with two electrodes of a galvanic cell. Now the general case of any electrochemical reaction, not necessarily associated with the function of an electrolysis cell, is considered.

4.3.1. Equilibrium Potential of an Electrochemical Reaction

Let us examine, not the combination of two associated electrochemical reactions, such as the reactions

$$\mathbf{Zn} = Zn^{++} + 2e^-$$

$$\mathbf{Cu} = Cu^{++} + 2e^-$$

which may be produced in a Daniell cell, but rather a given electrochemical reaction proceeding at the interface between a metal and an aqueous solution, independent of the fact that this reaction is associated with one or more other electrochemical reactions (e.g., the reaction $\mathbf{Zn} = Zn^{++} + 2e^-$).

A special kind of galvanic cell is considered for this study (Figure 23) which consists, on one side, of the electrode on which the reaction to be studied is produced, and on the other side, of a *given reversible reference*

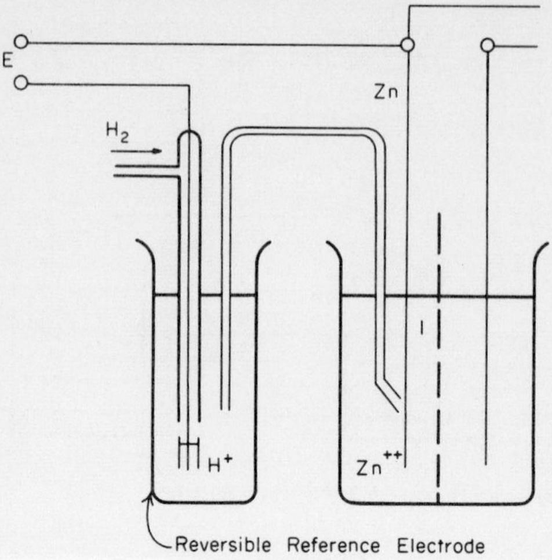

Figure 23. Measurement of electrode potential (Haber and Luggin's method).

electrode on which the state of equilibrium of a given reversible electrochemical reaction is obtained. For example, such a reference electrode may be a *standard hydrogen electrode*, made of a platinized platinum wire, saturated with hydrogen gas under atmospheric pressure and immersed in a solution of hydrochloric acid of pH = 0. At this surface the equilibrium state of the reversible reaction $\boldsymbol{H_2} = 2H^+ + 2e^-$ operates.

Of course, it is essential that this reference electrode is connected to the reaction solution by a conducting bridge with minimal mass transport thereby avoiding the existence of a potential drop caused by diffusion between the two solutions (i.e., agar–agar gel saturated with KCl). *During the passage of electric current this salt bridge must be placed in the solution in close proximity to the metal surface on which the reaction under study is proceeding* (Haber and Luggin or Piontelli probe).

The electrodes of this particular galvanic cell may be the site of the two following electrochemical reactions:

Electrode under study:

$$\sum \nu_1 M_1 + n_1 e^- = 0 \qquad\qquad \mathbf{Zn} = Zn^{++} + 2e^-$$

Reference electrode:

$$\sum \nu_{\text{ref}} M_{\text{ref}} - n_1 e^- = 0 \qquad 2H^+ + 2e^- = \boldsymbol{H_2}$$

Total reaction:

$$\sum \nu_1 M_1 + \sum \nu_{\text{ref}} M_{\text{ref}} = 0 \qquad \mathbf{Zn} + 2H^+ = Zn^{++} + \boldsymbol{H_2}$$

and the equilibrium potential of the cell has a value

$$\varphi_1 - \varphi_{\text{ref}} = \frac{\sum \nu_1 \mu_1 + \sum \nu_{\text{ref}} \mu_{\text{ref}}}{23{,}060\, n_1} \tag{4.8}$$

e.g.,

$$\varphi_1 - \varphi_{\text{ref}} = \frac{\mu_{Zn^{++}} - \mu_{Zn} + \mu_{H_2} - 2\mu_{H^+}}{23{,}060 \times 2}$$

If, as is usual, the chemical potentials of constituents of the standard hydrogen electrode (hydrogen gas under a pressure of 1 atm and H^+ ions in solution of pH = 0) are considered as zero, then

$$\mu_{H_2} = \mu_{H_2}^0 = 0$$

$$\mu_{H^+} = \mu_{H^+}^0 = 0$$

and

$$\sum \nu_{\text{ref}} \mu_{\text{ref}} = 0$$

Then equation (4.8) becomes

$$\varphi_1 - \varphi_{\text{ref}} = \frac{\sum \nu_1 \mu_1}{23{,}060\, n_1} \tag{4.9}$$

e.g.,

$$\varphi_1 - \varphi_{\text{ref}} = \frac{\mu_{Zn^{++}} - \mu_{Zn}}{46{,}120}$$

Now this relation gives the value of the equilibrium potential of the cell constituted by the coupling of the electrode under study and the reference electrode; this is the value of potential in the case where the whole system is in electrochemical equilibrium. By definition, the reference electrode is in a state of equilibrium; *therefore Eq. (4.9) expresses the condition of equilibrium of the electrode under study; that is, the condition of equilibrium of the electrochemical reaction under study.*

As a result, any electrochemical reaction r, $\sum \nu_r M_r + n_r e^- = 0$, is in a state of equilibrium if the difference between the potential, φ, of the electrode on which this reaction is produced and the potential, φ_{ref}, of a given reversible reference electrode[6] has a definite value which is given both

[6] Of course, this potential difference is measured by the method of Haber and Luggin or of Piontelli cited above (Figure 23).

in magnitude and sign by the relation

$$\varphi - \varphi_{\text{ref}} = \frac{\sum \nu_r \mu_r - \sum \nu_{\text{ref}} \mu_{\text{ref}}}{23{,}060\, n_r} \tag{4.10}$$

The potential difference measured between a certain electrode and a reversible reference electrode is called the *electrode potential, E*, of this electrode. In the particular case where this corresponds to the state of equilibrium of the electrochemical reaction proceeding on this electrode it is called the *equilibrium potential of the reaction.*

If the equilibrium potential of a reaction, r, written in the form, $\sum \nu_r M_r + n_r e^- = 0$, is represented by the symbol E_{0r}, the Eq. (4.10) may be written

$$E_{0r} = \frac{\sum \nu_r \mu_r - \sum \nu_{\text{ref}} \mu_{\text{ref}}}{23{,}060\, n_r} \tag{4.11}$$

If the standard hydrogen electrode is used as reference electrode, where the chemical potentials of the constituents H^+ and H_2 are considered as zero, the equilibrium condition (4.11) has the simpler form

$$E_{0r} = \frac{\sum \nu_r \mu_r}{23{,}060\, n_r} \tag{4.12}$$

Hence, it follows that *for all given electrochemical reactions carried out under definite physicochemical conditions* [that is, for given values of chemical potential of the reactive constituents, under definite conditions of temperature, activity (or concentration) of the dissolved reacting species, and fugacity (or partial pressure) of the gaseous reacting species], *there exists a definite value of the electrode potential for which the state of equilibrium of the reaction is obtained.* Consider a metal–solution system at the interface of which an electrochemical reaction may be produced (for instance, zinc in a solution of zinc sulfate, where the reaction $\mathbf{Zn} = Zn^{++} + 2e^-$ may occur;[7] or platinum in a solution of ferrous and ferric ions where the reaction $Fe^{++} = Fe^{+++} + e^-$ may take place; or any metal in an aqueous solution saturated with hydrogen gas where the reaction $H_2 = 2H^+ + 2e^-$ may occur), there then exists a value of electrode potential of the metal for which the state of equilibrium of the reaction is attained. When the potential of the metal has this value the reaction may occur neither in the direction of an oxidation nor in the direction of a reduction. For example, in the

[7] In principle this reaction may proceed in the direction $\mathbf{Zn} \rightarrow Zn^{++} + 2e^-$ as well as in the direction $\mathbf{Zn} \leftarrow Zn^{++} + 2e^-$.

foregoing reactions it is possible to have neither corrosion nor electrodeposition of zinc, neither oxidation of ferrous ions nor reduction of ferric ions, and neither evolution nor oxidation of hydrogen. For all other values of electrode potential the equilibrium state is not attained and, from the viewpoint of energetics, the reaction can proceed as an oxidation or as a reduction.

4.3.2. General Formula of Electrochemical Equilibria

The relation

$$E_{0r} = \frac{\sum \nu_r \mu_r}{23{,}060\, n_r} \tag{4.12}$$

or more simply

$$E_0 = \frac{\sum \nu\mu}{23{,}060\, n} \qquad (4.13)^{8}$$

which gives the value of the equilibrium potential of a given electrochemical reaction, $\sum \nu M + ne^- = 0$, in short expresses the condition of thermodynamic equilibrium of this reaction. This relation may also be written as

$$\boxed{\sum \nu\mu - 23{,}060 n E_0 = 0} \tag{4.14}$$

which then makes it possible to compare the conditions of equilibrium of chemical and electrochemical reactions as follows:

	Chemical reactions	*Electrochemical reactions*
Reaction equation	$\sum \nu M = 0$	$\sum \nu M + ne^- = 0$
Condition of equilibrium	$\sum \nu\mu = 0$	$\sum \nu\mu - 23{,}060 n E_0 = 0$

The equation of an electrochemical reaction only differs from that of a chemical reaction by the presence of a term which expresses the existence of electron transfer (oxidation or reduction). The condition of equilibrium of an electrochemical reaction only differs from that of a chemical reaction by the presence of a term which expresses the energetic influence of these electrons in the form of an electrode potential, E.

[8] This relation is Eq. (4.12) where, to simplify the discussion, all the subscripts r have been omitted. Note also that this is a general relation and can be applied for the case of several reactions.

As was done in Chapter 3 for the study of chemical equilibria, the chemical potential, μ, of each of the constituents of the reaction is represented, for a given temperature T, by the relation

$$\mu = \mu^0 + RT \ln [M] \tag{3.6}$$

where μ^0 is the "standard chemical potential" of the constituent under consideration; $[M]$ is the activity (or corrected concentration) of this constituent if it is in the dissolved state and is the fugacity (or corrected partial pressure) if it is in the gaseous state. Now relation (4.14) may be written as follows:

$$\boxed{\sum \nu\mu^0 + RT \sum \nu \ln [M] - 23{,}060 nE_0 = 0} \tag{4.15}$$

or, replacing R by its value of 1.985 cal/deg and converting to decimal logarithms,

$$\boxed{\sum \nu\mu^0 + 4{,}575T \sum \nu \log [M] - 23{,}060 nE_0 = 0} \tag{4.16}$$

At 25°C (298.1°K) this relation becomes

$$\sum \nu\mu^0 + 1363 \sum \nu \log [M] - 23{,}060 nE_0 = 0 \tag{4.17}$$

In the case of *chemical reactions*, in which no electron transfer occurs and in which case $n = 0$, this relation may be expressed in the following form, which has already been established:

$$\sum \nu \log [M] = \log K \tag{3.8}$$

$$\log K = -\frac{\sum \nu\mu^0}{1363} \tag{3.9}$$

In the case of *electrochemical reactions* for which n is not equal to zero the relation (4.17) may be expressed as follows:

$$E_0 = E_0^0 + \frac{0.0591}{n} \sum \nu \log [M] \tag{4.18}$$

$$E_0^0 = \frac{\sum \nu\mu^0}{23{,}060\, n} \tag{4.19}$$

where E_0 is the *equilibrium potential* of the electrochemical reaction and $E_0{}^0$ is the *standard equilibrium potential* of the reaction. The latter is the equilibrium potential in the particular case where all the constituents of the reaction are in a standard state (activity of 1 g-molecule/liter or 1 g-ion/liter for dissolved bodies; fugacity of 1 atm for gaseous bodies).[9]

4.3.3. Influence of Electrode Potential on Electrochemical Equilibria

Below are examples of the calculation of equilibrium potential together with a discussion of the influence of electrode potential on equilibria in homogeneous and heterogeneous systems.

4.3.3.1. Influence of the Potential on the Equilibrium of Homogeneous Systems: "Oxidation–Reduction Potentials"

Consider the reaction, $Fe^{++} = Fe^{+++} + e^-$, for which the standard chemical potentials (or standard free enthalpies of formation) have the following values[10]:

$$\mu^0_{Fe^{++}} = -20{,}300 \text{ cal}$$

$$\mu^0_{Fe^{+++}} = -\ 2{,}530 \text{ cal}$$

Equations (4.19) and (4.18) above give, respectively,

$$E_0{}^0 = \frac{\mu^0_{Fe^{+++}} - \mu^0_{Fe^{++}}}{23{,}060} = \frac{-\ 2{,}530 + 20{,}300}{23{,}060} = +\ 0.771 \text{ V}$$

$$E_0 = +\ 0.771 + 0.0591 \log \frac{[Fe^{+++}]}{[Fe^{++}]} \text{ V}$$

This is the classic formula of the "*oxidation–reduction potential* of a solution." Note that this equilibrium potential increases when the activity of the oxidized form, Fe^{+++}, increases and it decreases when the activity of the reduced form, Fe^{++}, increases. If the logarithmic term is written with a positive sign, the activity of the oxidized form is the numerator and the

[9] It is recalled that each of these equilibrium potentials is relative to a given reaction: to reactions *a*, *b*, . . . , *r*, . . . correspond the equilibrium potentials $E_{0a}, E_{0b}, \ldots, E_{0r}$, and the standard equilibrium potentials $E^0_{0a}, E^0_{0b}, \ldots, E^0_{0r}$.

[10] Standard free enthalpies of formation, at 25°C, *Rapports Techniques CEBELCOR* **72**, RT. 87 (1960) (in French).

activity of the reduced form is the denominator:

$$E_0 = +0.771 + 0.0591 \log \frac{[Fe^{+++}]}{[Fe^{++}]} \quad \begin{matrix} \swarrow \text{oxidized form} \\ \nwarrow \text{reduced form} \end{matrix}$$

As a second example of equilibria of homogeneous systems consider the case of the reaction

$$MnO_4^- + 8H^+ + 5e^- = Mn^{++} + 4H_2O$$

for which

$$\begin{aligned} \mu^0_{MnO_4^-} &= -107{,}400 \text{ cal} \\ \mu^0_{H^+} &= 0 \text{ cal} \\ \mu^0_{Mn^{++}} &= -\ 54{,}400 \text{ cal} \\ \mu^0_{H_2O} &= -\ 56{,}690 \text{ cal} \end{aligned}$$

Equations (4.19) and (4.18) give, respectively,

$$\begin{aligned} E_0{}^0 &= \frac{\mu^0_{MnO_4^-} + 8\mu^0_{H^+} - \mu^0_{Mn^{++}} - 4\mu^0_{H_2O}}{23{,}060 \times 5} \\ &= \frac{-107{,}400 + 54{,}000 + 226{,}700}{115{,}300} \\ &= +1.507 \text{ V} \end{aligned}$$

$$E_0 = +1.507 - 0.0946 \text{ pH} + 0.0118 \log \frac{[MnO_4^-]}{[Mn^{++}]} \text{ V}$$

The activity of the oxidized form, MnO_4^-, is the numerator of the logarithmic term; that of the reduced form, Mn^{++}, is the denominator.

4.3.3.2. Influence of the Potential on the Equilibrium of Heterogeneous Solid–Solution Systems: "Dissolution Potentials" of Solid Bodies

As an example of this type of system, consider the oxidation of solid iron into ferrous ions

$$\mathbf{Fe} = Fe^{++} + 2e^-$$

for which

$$\begin{aligned} \mu^0_{Fe} &= 0 \text{ cal} \\ \mu^0_{Fe^{++}} &= -20{,}300 \text{ cal} \end{aligned}$$

Equations (4.19) and (4.18) give, respectively,

$$E_0^0 = \frac{\mu^0_{Fe^{++}} - \mu^0_{Fe}}{23{,}060 \times 2} = \frac{-20{,}300}{46{,}120} = -0.440 \text{ V}$$

$$E_0 = -0.440 + 0.0295 \log [Fe^{++}] \text{ V}$$

Note that the activity of ferrous ions, the oxidized form, occurs with the positive sign in this formula.

Further, consider the reaction

$$\mathbf{Zn} = Zn^{++} + 2e^-$$

for which

$$\mu^0_{Zn} = 0 \text{ cal}$$
$$\mu^0_{Zn^{++}} = -35{,}184$$

Equations (4.19) and (4.18) then give

$$E_0^0 = \frac{\mu^0_{Zn^{++}} - \mu^0_{Zn}}{23{,}060 \times 2} = \frac{-35{,}184}{46{,}120} = -0.763 \text{ V}$$

$$E_0 = -0.763 + 0.0295 \log [Zn^{++}] \text{ V}$$

Notice that the activity of zinc ions, the oxidized form, occurs with a positive sign in this formula.

Now, consider the reaction

$$\mathbf{Fe_3O_4} + 8H^+ + 2e^- = 3Fe^{++} + 4H_2O$$

for which

$$\mu^0_{Fe_3O_4} = -242{,}400 \text{ cal}$$
$$\mu^0_{H^+} = 0 \text{ cal}$$
$$\mu^0_{Fe^{++}} = -20{,}300 \text{ cal}$$
$$\mu^0_{H_2O} = -56{,}690 \text{ cal}$$

Equations (4.19) and (4.18) give, respectively,

$$E_0^0 = \frac{\mu^0_{Fe_3O_4} + 8\mu^0_{H^+} - 3\mu^0_{Fe^{++}} - 4\mu^0_{H_2O}}{23{,}060 \times 2}$$

$$= \frac{-242{,}400 + 60{,}900 + 226{,}760}{46{,}120} = +0.981 \text{ V}$$

$$E_0 = +0.981 - 0.2364 \text{ pH} - 0.0885 \log [Fe^{++}] \text{ V}$$

Note that the activity of ferrous ions, the reduced form, occurs with the negative sign in this formula. The classic formulas of the "*dissolution potentials of metals*" are included in each of these equilibrium relations.

4.3.3.3. Influence of the Potential on the Equilibrium of Heterogeneous Gas–Solution Systems: "Dissolution Potentials" of Gaseous Bodies

By way of example in this instance consider the reaction

$$\boldsymbol{H_2} = 2H^+ + 2e^-$$

for which

$$\mu^0_{H_2} = 0 \text{ cal}$$
$$\mu^0_{H^+} = 0 \text{ cal}$$

Equations (4.19) and (4.18) then give

$$E_0{}^0 = \frac{2\mu^0_{H^+} - \mu^0_{H_2}}{23{,}060 \times 2} = 0.000 \text{ V}$$

$$E_0 = 0.000 - 0.0591 \text{ pH} - 0.0295 \log p_{H_2} \text{ V}$$

The standard equilibrium potential of this reaction is equal to zero. This follows since it was found convenient to measure all values of electrode potential with respect to this potential. The pressure of gaseous hydrogen occurs with a negative sign in the logarithmic term since $\boldsymbol{H_2}$ is the reduced form of the system H_2–H^+.

Further, consider the reaction

$$2H_2O = \boldsymbol{O_2} + 4H^+ + 4e^-$$

for which

$$\mu^0_{H_2O} = -56{,}690 \text{ cal}$$
$$\mu^0_{O_2} = 0 \text{ cal}$$
$$\mu^0_{H^+} = 0 \text{ cal}$$

Equations (4.19) and (4.18) give, respectively,

$$E_0{}^0 = \frac{\mu^0_{O_2} + 4\mu^0_{H^+} - 2\mu^0_{H_2O}}{23{,}060 \times 4} = \frac{+\,113{,}380}{92{,}240} = +\,1.228 \text{ V}$$

$$E_0 = +\,1.228 - 0.0591 \text{ pH} + 0.0148 \log p_{O_2} \text{ V}$$

The pressure of gaseous oxygen occurs with a positive sign in the logarithmic term since O_2 is the oxidized form of the system O_2–H_2O.

Also, consider the reaction

$$Cl_2 + 2e^- = 2Cl^-$$

for which

$$\mu^0_{Cl_2} = 0 \text{ cal}$$

$$\mu^0_{Cl^-} = -31{,}350 \text{ cal}$$

The Eqs. (4.19) and (4.18) give, respectively,

$$E_0{}^0 = \frac{\mu^0_{Cl_2} - 2\mu^0_{Cl^-}}{23{,}060 \times 2} = \frac{+\,62{,}700}{46{,}120} = +\,1.359 \text{ V}$$

$$E_0 = +\,1.359 + 0.0295 \log \frac{p_{Cl_2}}{[Cl^-]^2} \text{ V}$$

The partial pressure of chlorine, the oxidized form, appears as the numerator of the logarithmic term; the activity of chloride, the reduced form, occurs as the denominator.

The equilibrium equations above for chlorine and oxygen are typical of classical formulas of the "potentials of a gas electrode."

The preceding discussion in Sections 4.3.3.1–4.3.3.3 demonstrate that the "*oxidation–reduction potentials*," the "*dissolution potentials of metals*" and the "*potentials of gas electrodes*" are only three particular cases of "*equilibrium potentials of electrochemical reactions.*"

If the reacting species are *all in the dissolved state* and therefore constitute a homogeneous system, the equilibrium potential generally is called an *oxidation–reduction potential*;[11] a typical example is the ferrous–ferric equilibrium, $Fe^{++} = Fe^{+++} + e^-$.

If the reactive bodies constitute a heterogeneous system containing a *solid body* and a *dissolved species*, the equilibrium potential generally is known as a *dissolution potential* (e.g., the reaction $\mathbf{Fe} = Fe^{++} + 2e^-$).

If the reactive bodies constitute a heterogeneous system containing a

[11] We will see later (Chapter 5—Electrochemical Kinetics) that, as suggested by P. Van Rysselberghe, it is reasonable to modify this usual idea of oxidation–reduction potential and to define the oxidation–reduction potential as being the equilibrium potential of a reversible electrochemical reaction, which reaction may be homogeneous or heterogeneous.

gaseous body and a *dissolved species*, the equilibrium potential generally is known as *a potential of a gas electrode* (e.g., the reaction $Cl_2 + 2e^- = 2Cl^-$).

All of the conditions and qualifications discussed earlier concerning the "equilibrium potential" apply to each of these three particular cases.

4.3.4. Graphic Representation of Electrochemical Equilibria

In Chapter 3 the influence of pH on chemical equilibria was presented graphically for three particular cases: homogeneous systems (Figures 11a and 11b: the influence of pH on the dissociation of weak acids and bases); *heterogeneous solid–solution systems* (Figure 14: influence of pH on the solubility of oxides and hydroxides); and *heterogeneous gas–solution systems* (Figure 19: influence of pH on the solubility of acidic and alkaline gases).

In an analogous manner Figures 24–26 represent the influence of *electrode potential on electrochemical equilibria* in three similar cases: homogeneous systems (Figures 24a and 24b: the influence of the potential on the oxidation states of the dissolved body); *heterogeneous solid–solution systems* (Figure 25: the influence of the potential on the solubility of solid metals and metalloids, as well as of mercury); and *heterogeneous gas–solution systems* (Figure 26: the influence of potential on the solubility of oxidizing and reducing gases).

It may be seen that Figures 24–26 are similar to Figures 11a, 11b, 14, and 15 respectively, which represent the influence of pH on chemical equilibria. Accordingly, an important conclusion follows: *the electrode*

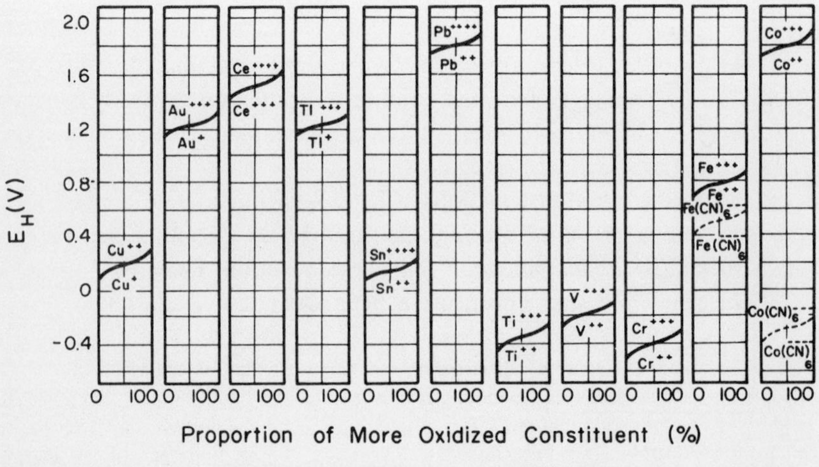

Figure 24. Influence of electrode potential on the equilibrium of homogeneous systems.

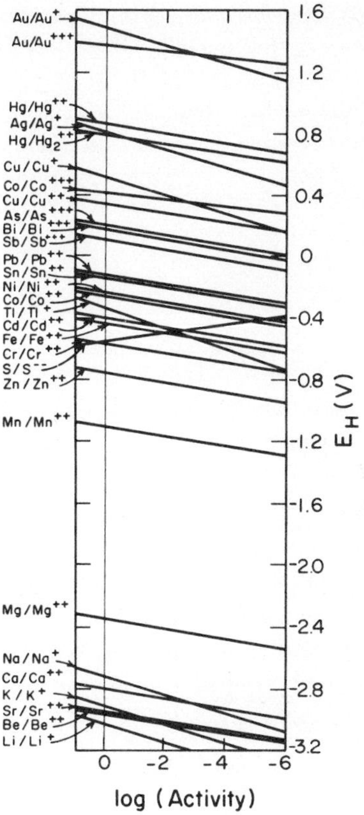

Figure 25. Influence of electrode potential on the equilibrium of heterogeneous solid–solution systems (dissolution potentials of metals and solid metalloids).

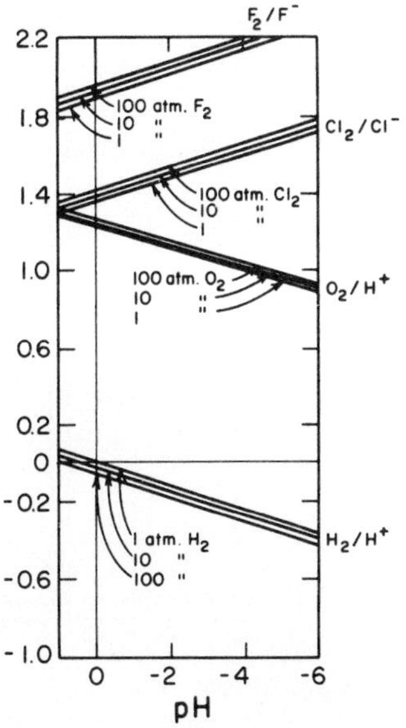

Figure 26. Influence of electrode potential on the equilibrium of heterogeneous gas–solution systems (electrode potentials of gases).

potential governs oxidation–reduction equilibria as the pH governs the acid–base equilibria. The influence of potential on the oxidation state of dissolved bodies is illustrated in Figure 24b by S-shaped curves analogous to those of Figure 11a, which outline the influence of pH on the dissociation of weak acids and bases. *Then, to dissolve cupric hydroxide* $\mathbf{Cu(OH)_2}$, *in a solution containing* 10^{-2} *g-atom of copper/liter, it is necessary, according to Figure 14,*

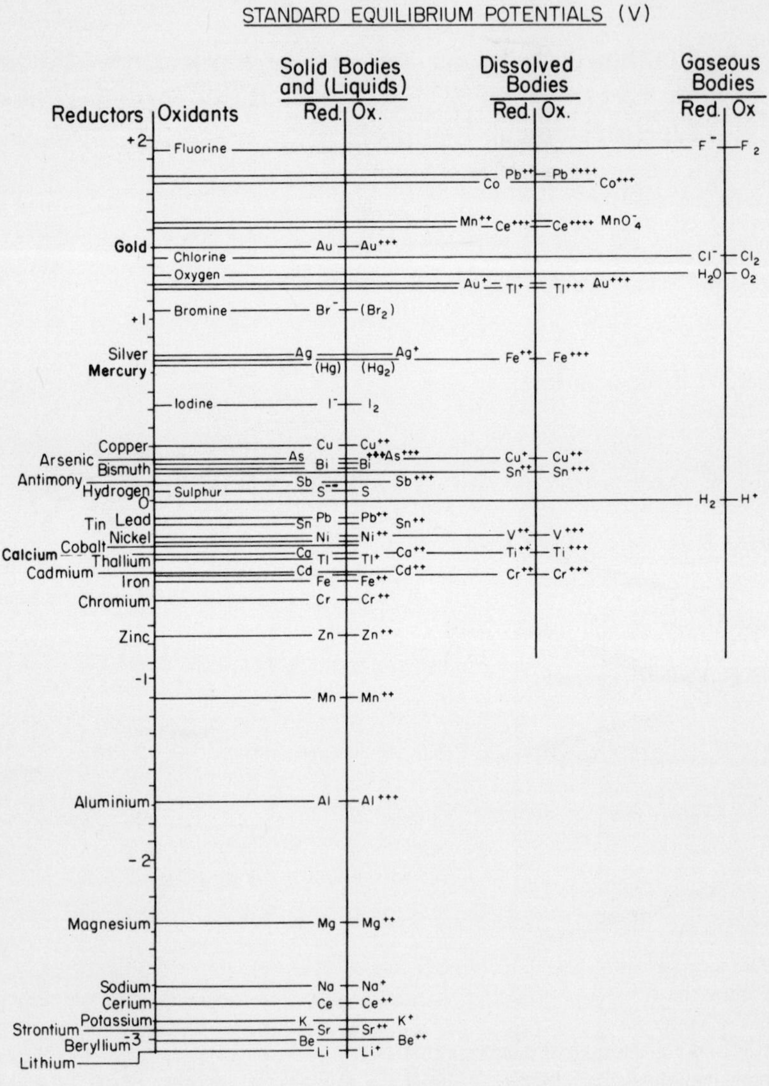

Figure 27. Standard equilibrium potentials.

to lower the pH of the solution to a value below 5.6; and to dissolve metallic copper, **Cu**, *in such solution it is necessary, according to Figure 25, to raise the electrode potential above +0.30 V.* This potential may be exceeded either by application of an external potential or by adding an oxidizing redox couple such as Fe^{+++}–Fe^{++}.

Figure 27 indicates the standard equilibrium potential of oxidation–reduction systems in which metals and metalloids take part.

4.3.5. Combined Influence of pH and Electrode Potential on Electrochemical Equilibria. Basis for Diagrams of Electrochemical Equilibria

Above, we have seen that if chemical and electrochemical reactions are considered, they can be written in the following forms[12]:

Chemical reactions

$$\sum \nu M = 0 \qquad \text{or} \qquad a\mathrm{A} + c\mathrm{H_2O} = b\mathrm{B} + m\mathrm{H^+}$$

where A is the acid form.

Electrochemical reactions

$$\sum \nu M + ne^- = 0 \qquad \text{or} \qquad a\mathrm{A} + c\mathrm{H_2O} + ne^- = b\mathrm{B} + m\mathrm{H^+}$$

where A is the oxidized form.[13]

The equilibrium conditions of these reactions at 25°C may be written respectively as follows:

Chemical reactions

$$\sum \nu \log [M] = \log K \tag{3.8}$$

where

$$\log K = \frac{\sum \nu\mu^0}{1363} \tag{3.9}$$

[12] The transformation reactions of body *A* to body *B* are written in this manner to emphasize the H^+ ions and the electric charges which may be involved in these reactions (see Chapter 2).

[13] It is only for the sake of definition that H_2O is represented in those parts of the equations containing the electrons e^- (and the H^+ ions in the other parts). The inverse also may occur.

or

$$\log \frac{[B]^b}{[A]^a} = \log K + m\,\text{pH}$$

where

$$\log K = \frac{b\mu_B{}^0 - a\mu_A{}^0 - c\mu^0_{H_2O}}{1363}$$

Electrochemical reactions

$$E_0 = E_0{}^0 + \frac{0.0591}{n} \sum \nu \log M \tag{4.18}$$

where

$$E_0{}^0 = \frac{\sum \nu\mu^0}{23{,}060\, n} \tag{4.19}$$

or

$$E_0 = E_0{}^0 + \frac{0.0591}{n}\,\text{pH} + \frac{0.0591}{n} \log \frac{[A]^a}{[B]^b}$$

where

$$E_0{}^0 = \frac{a\mu_A{}^0 + c\mu^0_{H_2O} - b\mu_B{}^0}{23{,}060\, n}$$

For the particular case of electrochemical reactions not involving H^+ ions, the coefficient, *m*, in the reaction equation is equal to zero,[14] and the condition of equilibrium simplifies to

$$E_0 = E_0{}^0 + \frac{0.0591}{n} \log \frac{[A]^a}{[B]^b}$$

Returning to the classification of reactions which was given in Table IV (Chapter 2, p. 27), it is convenient to consider successively, among the

[14] Such are the reactions considered in Figures 24 to 27 with the exception of the reactions $\boldsymbol{H_2} = 2H^+ + 2e^-$ and $2H_2O = \boldsymbol{O_2} + 4H^+ + 4e^-$.

reactions outlined above, the chemical reactions in which H^+ ions take part, the electrochemical reactions in which H^+ ions do not take part, and the electrochemical reactions in which H^+ ions participate.

4.3.5.1. Chemical Reactions in Which H^+ Ions Participate

Homogeneous:

$$H_2CO_3 = HCO_3^- + H^+_{aq} \qquad \log\frac{[HCO_3^-]}{[H_2CO_3]} = -6.37 + pH$$

Solid–Solution:

$$\mathbf{Fe(OH)_2} + 2H^+ = Fe^{++} + 2H_2O \qquad \log[Fe^{++}] = 13.29 - 2\,pH$$

Gas–Solution:

$$\mathit{CO_2} + H_2O = HCO_3^- + H^+ \qquad \log\frac{p_{CO_2}}{[HCO_3^-]} = 7.80 - pH$$

These equilibria depend on the pH and not on the electrode potential. In a diagram where the pH is plotted horizontally and the potential vertically, they are represented by a family of *vertical* lines, each one of which corresponds to a definite value of the logarithm of the function of partial pressures (or fugacities) of gaseous bodies and/or of concentrations (or activities) of dissolved bodies.

4.3.5.2. Electrochemical Reactions in Which H^+ Ions Do Not Participate

Homogeneous:

$$Fe^{++} = Fe^{+++} + e^- \qquad E_0 = +0.771 + 0.0591\log\frac{[Fe^{+++}]}{[Fe^{++}]}$$

Solid–Solution:

$$\mathbf{Fe} = Fe^{++} + 2e^- \qquad E_0 = -0.440 + 0.0295\log[Fe^{++}]$$

Gas–Solution:

$$2Cl^- = \mathit{Cl_2} + 2e^- \qquad E_0 = +1.359 + 0.0295\log\frac{p_{Cl_2}}{[Cl^-]^2}$$

These equilibria do not depend on pH but depend on the electrode potential; in a similar potential–pH diagram, as mentioned above, they are represented by a family of *horizontal* lines each one of which corresponds to a definite value of the logarithm of the function of pressures (or fugacities) and/or of concentrations (or activities).

4.3.5.3. Electrochemical Reactions in Which H^+ Ions Participate

Homogeneous:

$$Mn^{++} + 4H_2O = MnO_4^- + 8H^+ + 5e^-$$

$$E_0 = +1.507 - 0.0946\ \text{pH} + 0.0118 \log \frac{[MnO_4^-]}{[Mn^{++}]}$$

Solid–Solutions:

$$Fe^{++} + 3H_2O = \mathbf{Fe(OH)_3} + 3H^+ + e^-$$

$$E_0 = +1.507 - 0.1773\ \text{pH} - 0.0591 \log [Fe^{++}]$$

Gas–Solutions:

$$H_2 = 2H^+ + 2e^-$$

$$E_0 = 0.000 - 0.0591\ \text{pH} - 0.0295 \log p_{H_2}$$

These equilibria depend both on the pH and the electrode potential; in the same potential–pH diagram they are represented by a family of *oblique parallel* lines each one of which corresponds to a definite value of the logarithm of the function of pressures (or fugacities) and/or of concentrations (or activities).

When no gaseous or dissolved bodies other than the H^+ ion take part in the reaction, these families of lines reduce to a single line of slope -0.0591

$$\mathbf{Fe} + 2H_2O = \mathbf{Fe(OH)_2} + 2H^+ \qquad E_0 = -0.047 - 0.0591\ \text{pH}$$

If, in the same potential–pH diagram, the equilibrium conditions of all the chemical and electrochemical reactions likely to be produced in a given system are represented according to the above cases, then a profusion of lines and families of lines is obtained which represents all of these equilibrium conditions. This *diagram of electrochemical equilibria* makes it possible to represent the circumstances in which each of these given bodies is thermodynamically stable or unstable regardless of the complexity of the given system.

Below we establish and analyze, as examples, a few of these diagrams, considering first of all the principal constituent of aqueous solutions—namely, water.

4.4. DIAGRAM OF ELECTROCHEMICAL EQUILIBRIA OF WATER

4.4.1. Thermodynamic Stability of Water: Acidic and Alkaline Media; Oxidizing and Reducing Media

As is well known, water dissociates into H^+ ions and OH^- ions according to the reaction $H_2O = H^+ + OH^-$, where the condition of equilibrium is

$$\log([H^+] \cdot [OH^-]) = \log K$$

and where at 25°C

$$\log K = \frac{\mu^0_{H^+} + \mu^0_{OH^-} - \mu^0_{H_2O}}{1363}$$

which gives

$$\log K = - \frac{-37{,}595 + 56{,}690}{1363} = -14.00$$

Therefore

$$[H^+] \cdot [OH^-] = 10^{-14.00}$$

or

$$\log [H^+] + \log [OH^-] = -14.00$$

Since

$$pH = -\log [H^+]$$

for

$pH = 7.00$	$[H^+] = [OH^-]$	neutral solutions
$pH < 7.00$	$[H^+] > [OH^-]$	acid solutions
$pH > 7.00$	$[H^+] < [OH^-]$	alkaline solutions

Thus, the vertical line at pH = 7.00 drawn in Figure 28 separates a domain where there is predominance of H^+ ions with respect to OH^- ions (acidic solutions) from a domain where there is predominance of OH^- ions with respect to H^+ ions (alkaline solutions).

On the other hand water, H_2O, and its constituents, H^+ and OH^-, may be reduced with the evolution of hydrogen or oxidized with the evolution of oxygen according to the electrochemical reactions

$$2H^+ + 2e^- \rightarrow \boldsymbol{H_2} \qquad \text{reduction} \qquad \text{(a)}$$

$$2H_2O \rightarrow \boldsymbol{O_2} + 4H^+ + 4e^- \qquad \text{oxidation} \qquad \text{(b)}$$

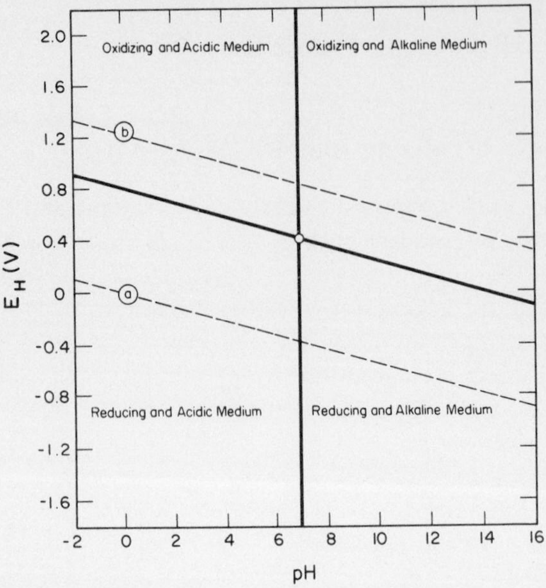

Figure 28. Acid, alkaline, oxidizing, and reducing media.

whose standard equilibrium potentials, according to the formula

$$E_0{}^0 = \frac{\sum \nu\mu^0}{23{,}060\, n} \tag{4.19}$$

are, respectively,

$$E_{0a}^0 = \frac{2\mu_{H^+}^0 - \mu_{H_2}^0}{23{,}060 \times 2} = \frac{0}{46{,}120} = 0.000 \text{ V}$$

$$E_{0b}^0 = \frac{\mu_{O_2}^0 + 4\mu_{H^+}^0 - 2\mu_{H_2O}^0}{23{,}060 \times 4} = \frac{+\ 113{,}380}{92{,}240} = +\ 1.228 \text{ V}$$

This leads to the following conditions of equilibrium:

for the reaction $\boldsymbol{H_2} = 2H^+ + 2e^-$,

$$E_{0a} = \quad 0.000 - 0.0591 \text{ pH} - 0.0295 \log p_{H_2} \tag{a}$$

for the reaction $2H_2O = \boldsymbol{O_2} + 4H^+ + 4e^-$,

$$E_{0b} = +1.228 - 0.0591 \text{ pH} + 0.0147 \log p_{O_2} \tag{b}$$

If $p_{H_2} = 1$ atm,

$$E_{0a} = \quad 0.000 - 0.0591 \text{ pH}$$

If $p_{O_2} = 1$ atm,

$$E_{0b} = +1.228 - 0.0591 \text{ pH}$$

These two conditions of equilibrium are represented in Figure 29, respectively, by the parallel lines (a) and (b) of slope −0.0591. Between these two lines the equilibrium pressures of hydrogen and oxygen are both less than 1 atm; the region between these two lines then is the *area of thermodynamic stability of water* under a pressure of 1 atm.

Below the line *a*, corresponding to $p_{H_2} = 1$ atm, water under a pressure of hydrogen of 1 atm will tend to decompose by reduction according to the reaction $2H^+ + 2e^- \rightarrow \boldsymbol{H_2}$.

Above the line *b*, corresponding to $p_{O_2} = 1$ atm, water under a pressure of oxygen of 1 atm tends to decompose by oxidation according to the reaction $2H_2O \rightarrow \boldsymbol{O_2} + 4H^+ + 4e^-$.

If, in the same way as

$$\text{pH} = -\log[H^+]$$

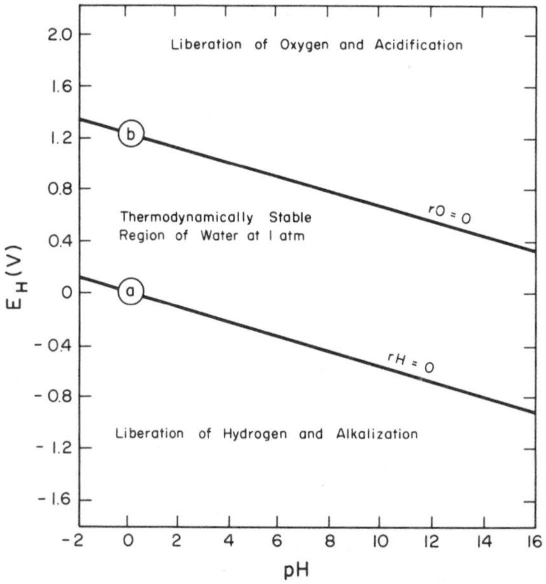

Figure 29. Region of thermodynamic stability of water under a pressure of 1 atmosphere at 25°C

one puts

$$\text{rH} = -\log p_{H_2}$$

and

$$\text{rO} = -\log p_{O_2}$$

then both of the above equilibrium relations may be written, respectively, as

$$E_{0a} = 0.000 - 0.0591\ \text{pH} + 0.0295\ \text{rH}$$

and

$$E_{0b} = +1.228 - 0.0591\ \text{pH} - 0.0147\ \text{rO}$$

Lines corresponding to different values of rH and rO are drawn in Figure 30 in the region of stability of water.

Now, in the same way as water may dissociate into H^+ and OH^- ions according to the reaction $H_2O = H^+ + OH^-$ and is considered to be neutral from the acid–alkali standpoint if $[H^+] = [OH^-]$, water may decompose into hydrogen and oxygen gas according to the reaction $2H_2O = 2\boldsymbol{H_2} + \boldsymbol{O_2}$ and may be considered to be neutral from the oxidation–reduction viewpoint if $p_{H_2} = 2p_{O_2}$, that is if

$$\log p_{H_2} = \log p_{O_2} + \log 2$$
$$= \log p_{O_2} + 0.30$$

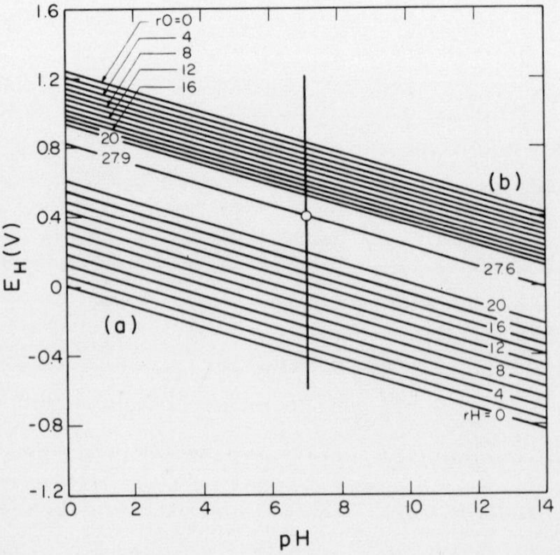

Figure 30. Electrochemical equilibria of water, rH and rO. Absolute neutrality at 25°C.

or if

$$rH = rO - 0.30$$

If this relation is combined with the previous relations

$$E_{0a} = 0.000 - 0.0591\ pH + 0.0295\ rH$$

and

$$E_{0b} = +1.228 - 0.0591\ pH - 0.0147\ rO$$

it is found that, as a condition of neutrality of water from the oxidation–reduction point of view at 25°C,

$$rH = 27.56$$

$$rO = 27.86$$

Combining these values with the relation pH = 7.00 which expresses the neutrality of an aqueous solution from the acid–alkali standpoint, the following characteristics are obtained for the *condition of absolute neutrality* of a dilute aqueous solution at 25°C

$$pH = 7.00$$

$$rH = 27.56$$

$$rO = 27.86$$

$$E = +0.40\ V$$

However, it is emphasized that although the idea of neutrality from the acid–alkali viewpoint has considerable practical interest, the interest in the notion of neutrality from the oxidation–reduction viewpoint is mainly academic.

The characteristics of neutrality defined above make it possible to divide a potential–pH diagram into the following four regions which are indicated in Figure 28:

top left	acidic and oxidizing media
bottom left	acidic and reducing media
bottom right	alkaline and reducing media
top right	alkaline and oxidizing media

4.4.2. Decomposition of Water. Formation of Hydrogen, Oxygen, Ozone, and Hydrogen Peroxide

In the portion of Figure 29 situated below line *a*, relative to the equilibrium of the reaction $\boldsymbol{H_2} = 2H^+ + 2e^-$ for a pressure of hydrogen of 1 atm, it is thermodynamically possible to reduce water with the evolution of gaseous hydrogen under atmospheric pressure. Practically this is achieved by the action of an electrolytic cathode or by means of sufficiently reducing substances, such as metals of low electrode potentials (iron, zinc, magnesium, etc.).

In the portion of Figure 29 situated above line *b*, relative to the equilibrium of the reaction $2H_2O = \boldsymbol{O_2} + 4H^+ + 4e^-$ for a pressure of oxygen of 1 atm, it is thermodynamically possible to oxidize water with evolution of gaseous oxygen under atmospheric pressure. This is achieved practically by using an electrolytic anode, or by the action of sufficiently oxidizing substances (fluorine, permanganate, etc.).

It is well known that the oxidation of water may give rise not only to the formation of diatomic oxygen $\boldsymbol{O_2}$ but also to the formation of triatomic oxygen $\boldsymbol{O_3}$ (ozone) and of hydrogen peroxide H_2O_2 according to the reaction equations and conditions of equilibrium indicated below:

for $\boldsymbol{O_3}$

$$3H_2O = \boldsymbol{O_3}+6H^++6e^- \qquad E_0 = +1.501 - 0.0591\ \text{pH}+0.0100 \log p_{O_3}$$

for H_2O_2

$$2H_2O = H_2O_2+2H^++2e^- \qquad E_0 = +1.776 - 0.0591\ \text{pH}+0.0295 \log[H_2O_2]$$

Extremely vigorous oxidizing actions associated with values of electrode potential higher than the values indicated by these relations may simultaneously produce formation of hydrogen peroxide and evolution of ozone as well as oxygen. For example, according to the potential values of the equilibrium $\boldsymbol{F_2}$–F^- illustrated at the top of Figure 26, it is possible to simultaneously bring about, by bubbling fluorine into water, the reduction

$$\boldsymbol{F_2} + 2e^- \rightarrow 2F^-$$

and three oxidations

$$2H_2O \rightarrow H_2O_2 + 2H^+ + 2e^-$$

$$2H_2O \rightarrow \boldsymbol{O_2} + 4H^+ + 4e^-$$

$$3H_2O \rightarrow \boldsymbol{O_3} + 6H^+ + 6e^-$$

which correspond to the three simultaneous reactions

$F_2 + 2H_2O \rightarrow 2F^- + 2H_2O_2 + 2H^+$ (formation of hydrofluoric acid and hydrogen peroxide)

$2F_2 + 2H_2O \rightarrow 4F^- + O_2 + 4H^+$ (formation of hydrofluoric acid and oxygen)

$3F_2 + 3H_2O \rightarrow 6F^- + O_3 + 6H^+$ (formation of hydrofluoric acid and ozone)

4.5. DIAGRAMS OF ELECTROCHEMICAL EQUILIBRIA OF HYDROGEN PEROXIDE. OXIDATION, REDUCTION AND DECOMPOSITION OF HYDROGEN PEROXIDE. REDUCTION OF OXYGEN

The electrochemical behavior of solutions of hydrogen peroxide, however, is not as simple as the single reaction

$$2H_2O = H_2O_2 + 2H^+ + 2e^-$$

suggests. Here the condition of equilibrium is

$$E_0 = +1.776 - 0.0591\ \text{pH} + 0.0295 \log[H_2O_2]$$

In fact it is well known to analysts that hydrogen peroxide may act, not only as an oxidant but also as a reducing agent; for example, when it reacts with permanganate to form oxygen according to the reaction

$$2MnO_4^- + 5H_2O_2 + 6H^+ \rightarrow 2Mn^{++} + 5O_2 + 8H_2O$$

Hydrogen peroxide may dissociate with the formation of HO_2^- ion according to the reaction, $H_2O_2 = HO_2^- + H^+$, with the condition of equilibrium being

$$\log \frac{[HO_2^-]}{[H_2O_2]} = -11.63 + \text{pH}$$

It may be formed not only by the action of a vigorous oxidant but also during the corrosion of certain metals (zinc, aluminum) in aqueous solutions.

The reductions and oxidations of hydrogen peroxide and its ion, HO_2^-, correspond to the equilibria below[15]:

Reductions with formation of water

$$H_2O_2 + 2H^+ + 2e^- = 2H_2O$$

$$E_0 = +1.776 - 0.0591\,\text{pH} + 0.0295 \log [H_2O_2] \quad (2)$$

$$HO_2^- + 3H^+ + 2e^- = 2H_2O$$

$$E_0 = +2.119 - 0.0886\,\text{pH} + 0.0295 \log [HO_2^-] \quad (3)$$

Oxidations with formation of oxygen

$$H_2O_2 = \boldsymbol{O_2} + 2H^+ + 2e^-$$

$$E_0 = +0.682 - 0.0591\,\text{pH} + 0.0295 \log \frac{p_{O_2}}{[H_2O_2]} \quad (4)$$

$$HO_2^- = \boldsymbol{O_2} + H^+ + 2e^-$$

$$E_0 = +0.338 - 0.0295\,\text{pH} + 0.0295 \log \frac{p_{O_2}}{[HO_2^-]} \quad (5)$$

These equilibria make it possible to establish Figure 31, which represents the conditions of electrochemical equilibria of solutions of hydrogen peroxide. The two families of lines (2,3), and (4,5) plotted on this figure represent, respectively, the equilibria of oxidation by hydrogen peroxide (or of reduction of hydrogen peroxide in water) and the equilibria of reduction by hydrogen peroxide (or of oxidation of hydrogen peroxide in oxygen) for different values of the activity of the solution of $H_2O_2 + HO_2^-$ (2,3) and for different values of the term $p_{O_2}/([H_2O_2] + [HO_2^-])$ (4,5).

Below the family of lines (2,3), hydrogen peroxide in solution may theoretically be reduced to water; above the family of lines (4,5) theoretically it may be oxidized to oxygen; therefore solutions of hydrogen peroxide are always thermodynamically unstable at 25°C, whatever the values of pH and potential. In the region between the two families of lines which have just been considered hydrogen peroxide is *doubly unstable*; it tends to be reduced to H_2O according to the reaction $H_2O_2 + 2H^+ + 2e^- \rightarrow 2H_2O$ (2) and to be oxidized to $\boldsymbol{O_2}$ according to the reaction $H_2O_2 \rightarrow \boldsymbol{O_2} + 2H^+ + 2e^-$ (4), that is, to decompose into H_2O and $\boldsymbol{O_2}$ according to the overall chemical reaction

$$2H_2O_2 \rightarrow 2H_2O + \boldsymbol{O_2}$$

[15] The numbers given to these reactions are those used in the *Atlas*, Ref. 34.

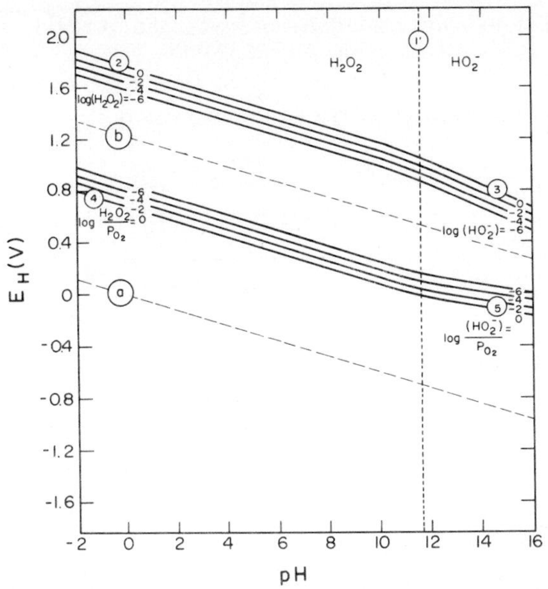

Figure 31. Electrochemical equilibria of solutions of hydrogen peroxide at 25°C.

As a result, if a solution of hydrogen peroxide contacts a metallic surface whose electrode potential is in this region of *double instability*, hydrogen peroxide may spontaneously decompose into water and oxygen by the reaction $2H_2O_2 \rightarrow 2H_2O + O_2$. Conversely, if hydrogen peroxide does spontaneously decompose into water and oxygen on a metallic surface, this surface must possess a potential in the region of double instability. Here then is an *example of electrochemical catalysis of a chemical reaction* which is discussed at greater length in Section 5.6.

However, in the presence of metallic surfaces where the value of potential does not lie within this region of double instability the decomposition of hydrogen peroxide into water and oxygen is impossible.

Hydrogen peroxide then is often practically stable due to the irreversibility of this reaction. Two of these facts have been verified in the particular instance of iron immersed in hydrogen peroxide.[16]

Figure 31 also shows that the electrochemical reduction of oxygen in aqueous solution may occur not only with the formation of water according

[16] Application of potential/pH diagrams relating to iron and to hydrogen peroxide. Demonstration experiments, CEBELCOR's *Rapport Technique* RT. 12 (in French). A detailed discussion concerning the conditions of stability and metastability of hydrogen peroxide has been given in the *Atlas of Electrochemical Equilibria* (Ref. 34).

to the four-electron reaction, $O_2 + 4H^+ + 4e^- \rightarrow 2H_2O$, but also with formation of hydrogen peroxide by the two-electron reaction, $O_2 + 2H^+ + 2e^- \rightarrow H_2O_2$. The reduction of oxygen then may be realized in two stages, corresponding, respectively, to the two reactions

$$
\begin{array}{c}
O_2 + 2H^+ + 2e^- \rightarrow H_2O_2 \\
H_2O_2 + 2H^+ + 2e^- \rightarrow 2H_2O \\
\hline
O_2 + 4H^+ + 4e^- \rightarrow 2H_2O
\end{array}
$$

Occasionally such reactions are established experimentally. For instance, during the polarographic reduction of oxygen on a mercury cathode two polarographic waves are observed corresponding respectively to these two reactions. Sometimes even the second of these two waves is itself split into two single electron steps corresponding to the transitory formation of an OH radical according to the reactions

$$
\begin{array}{c}
H_2O_2 + H^+ + e^- \rightarrow H_2O + OH \\
OH + H^+ + e^- \rightarrow H_2O \\
\hline
H_2O_2 + 2H^+ + 2e^- \rightarrow 2H_2O
\end{array}
$$

Furthermore, when certain metals (e.g., zinc, aluminum) corrode, sometimes it is possible to detect the formation of hydrogen peroxide in the solution. This hydrogen peroxide results from the electrochemical reduction of dissolved oxygen by the metal corroding at a potential less than those indicated by the lines (4,5) of Figure 31.

To end this discussion concerning the electrochemical behavior of water, hydrogen peroxide and oxygen, it is worth noting that most of these reactions we have just considered are more or less strongly *irreversible.* They do not necessarily occur even though they are thermodynamically possible.

Generally, it is not sufficient that the potential of a metal corresponds to a point within the region of stability of hydrogen in order that water be reduced with the formation of hydrogen: *overpotential* is necessary for the reaction $2H^+ + 2e^- \rightarrow H_2$ to proceed at a significant rate. The value of this overpotential depends mainly on the nature of the metal. It is very low for platinum and palladium, very high for lead and mercury. The term "hydrogen overpotential" refers to that departure from equilibrium, expressed for instance in millivolts, required for hydrogen to form at a given rate on a given substrate material. The subject of overpotential will be examined later during the study of electrochemical kinetics (Section 5.3).

4.6. DIAGRAMS OF ELECTROCHEMICAL EQUILIBRIA OF METALS AND METALLOIDS

As we have already pointed out in Section 4.3.4. (Diagrams of Electrochemical Equilibria), if the conditions of equilibrium of all the chemical and electrochemical reactions likely to occur in a given system are represented on the same potential–pH diagram then a profusion of lines and families of lines is obtained which represent all equilibrium conditions. A method for the establishment of such a diagram is described below. The system chosen is that of copper and its oxides and hydroxides in contact with aqueous solutions at 25°C and free from substances that form soluble complexes or insoluble salts with copper. The general case of copper in the presence of such complexes and salts will be considered under Section 7.2 of the present "lectures."

4.6.1. Diagram of Electrochemical Equilibria of Copper

4.6.1.1. Establishment of the Diagram

To establish a diagram of electrochemical equilibrium it is convenient to proceed as follows below.[17]

4.6.1.1.1. *Standard Free Enthalpies of Formation*

A list of all the species to be considered is tabulated and the values of their standard free enthalpies of formation (or standard chemical potential, μ^0) are abstracted from the literature. A table of these values (in calories) is drawn up which groups separately the solvent and dissolved species, the condensed bodies (solids and possibly liquids other than water), and the gaseous bodies. The precision of each of the values of free enthalpy of

[17] M. Pourbaix, Thermodynamics of dilute aqueous solutions: graphical representation of the role of pH and potential, Thesis, Delft (1945) (Preface by F. E. C. Scheffer. Foreword by U. R. Evans) CEBELCOR Publication F. 227 (reprinted 1963) (in French). A detailed statement on the establishment of this diagram has been published in *Rapports Techniques CEBELCOR* **85**, RT. 100, 101 (1962).

TABLE XII. Free Enthalpies of Formation for Copper in Water

Solvent and dissolved bodies		Condensed bodies		Gaseous bodies	
H_2O	−56,690 cal			*O_2*	*0* cal
H^+	0 cal			*H_2*	*0* cal
OH^-	−37,595 cal				
Cu^+	+12,*000* cal	**Cu**	0 cal	*CuH*	+*64,000* cal
Cu^{++}	+15,53*0* cal	**Cu_2O**	−34,980 cal		
		CuOH	−43,*500*		
CuO_2H^-	−61,42*0* cal	**CuO**	−30,400 cal		
CuO_2^{--}	−43,5*00* cal	**$Cu(OH)_2$**	−85,300 cal		
Cu^{+++}	+72,*500* cal	**Cu_2O_3** hydrated	−41,5*00* cal		
CuO_2^-	−26,*800* cal				

formation is indicated by italic type for the figures which reasonably may not be considered to be exact, the last of the preceding figures always being subject to some caution. Such a tabulation is shown in Table XII for copper in water.

4.6.1.1.2. *Reactions*

The equations of the different reactions in which the above bodies may participate are written in the manner indicated in Chapter 2, p. 26, i.e., along with the bodies A and B the H_2O, H^+ ion, and the transfer charge, e^-, are included where appropriate. The reaction equations then have the general form

$$aA + cH_2O + ne^- = bB + mH^+$$

These various reactions are next grouped as follows:

(*a*) *Homogeneous reactions* (all dissolved bodies)
- chemical reactions (without oxidation, for which n is equal to zero)
- electrochemical reactions (with oxidation, for which n is different from zero)

(*b*) *Heterogeneous reactions in which two condensed bodies participate* (generally solid bodies)
- chemical reactions (without oxidation)
- electrochemical reactions (with oxidation)

(*c*) *Heterogeneous reactions in which only one condensed body takes part* (generally a solid body)

- chemical reactions (without oxidation)
- electrochemical reactions (with oxidation)

Should the occasion arise, in addition to the reactions in which dissolved bodies participate, one or two gaseous bodies would be considered.

4.6.1.1.3. *Conditions of Equilibrium and Graph of the Equilibrium Diagram*

Having tabulated the possible equilibrium reactions above, it is now possible to plot them graphically in a potential–pH diagram. The values listed in Section 4.6.1.1.1 are applied to the respective reactions written in Section 4.6.1.1.2 and the conditions of equilibrium are calculated according to the following formulas which have been already established[18]:

For chemical reactions

$$\sum \nu \log [M] = \log K \tag{3.8}$$

where

$$\log K = -\frac{\sum \nu\mu^0}{1363} \tag{3.9}$$

For electrochemical reactions

$$E_0 = E_0^{\,0} \times \frac{0.0591}{n} \sum \nu \log [M] \tag{4.18}$$

where

$$E_0^{\,0} = \frac{\sum \nu\mu^0}{23{,}060\, n} \tag{4.19}$$

From these calculations one obtains the following for the three groups of reactions considered in Section 4.6.1.1.2 above.

[18] Remember that in these formulas, valid for a temperature of 25°C, the standard free enthalpies, μ^0, are expressed in calories per molar group, the activities (or concentrations) and fugacities (or partial pressures) $[M]$ are expressed, respectively, in gram-ion/liter (or gram-molecule/liter) and in atmospheres, and, finally, the electrode potentials, E_0 and $E_0^{\,0}$, are expressed in volts, with respect to the standard hydrogen electrode.

4.6.1.1.3.1. *Homogeneous Reactions. Regions of Relative Predominance of Dissolved Bodies*

The conditions of equilibrium of these reactions have the following form, where [A] and [B] represent the activities (or corrected concentrations) of the two dissolved bodies A and B:

for chemical reactions

$$\log \frac{[A]^a}{[B]^b} = \log K + m\,\mathrm{pH}$$

for electrochemical reactions

$$E_0 = E_0{}^0 - \frac{0.0591\,m}{n}\,\mathrm{pH} + \frac{0.0591}{n} \log \frac{[A]^a}{[B]^b}$$

Each one of these equilibrium conditions may be represented on a potential–pH diagram by a family of parallel straight lines (vertical, horizontal, or oblique) corresponding to different values of the term $\log([A]^a/[B]^b)$.

When the stoichiometric coefficients $a = b = 1$, then

for chemical reactions

$$\log \frac{[A]}{[B]} = \log K + m\,\mathrm{pH}$$

for electrochemical reactions

$$E_0 = E_0{}^0 - \frac{0.0591}{n}\,m\,\mathrm{pH} + \frac{0.0591}{n} \log \frac{[A]}{[B]}$$

Here, when the ratio $[A]/[B] = 1$, then the log term is zero. The lines so defined separate *regions of relative predominance* of the two dissolved bodies under consideration.

Tabulated below are the equilibrium conditions of *homogeneous reactions* in aqueous solution relative to copper, which indicate the *conditions of relative stability of the dissolved forms.* The "oxidation number," Z (in this case equal to the valency of copper), is given for each of the chemical reactions, as well as the changes of valency involved in each of the electrochemical reactions. We consider successively the chemical reactions (not involving electron transfer and at a constant valency) and the electrochemical reactions (involving both electron transfer, e^-, and changes of valency).

Chemical reactions

$Z = +2$

$Cu^{++} + 2H_2O = HCuO_2^{-} + 3H^{+}$ $\quad \log \frac{[HCuO_2^{-}]}{[Cu^{++}]} = -26.72 + 3\,pH$ (1)

$Cu^{++} + 2H_2O = CuO_2^{--} + 4H^{+}$ $\quad \log \frac{[CuO_2^{--}]}{[Cu^{++}]} = -39.88 + 4\,pH$ (2)

$HCuO_2^{-} = CuO_2^{--} + H^{+}$ $\quad \log \frac{[CuO_2^{--}]}{[HCuO_2^{-}]} = -13.15 + pH$ (3)

$Z = +3$

$Cu^{+++} + 2H_2O = CuO_2^{-} + 4H^{+}$ $\quad \log \frac{[CuO_2^{-}]}{[Cu^{+++}]} = -10.28 + 4\,pH$ (4)

Electrochemical reactions

$Z = +1$ or $+2$

$Cu^{+} = Cu^{++} + e^{-}$

$E_0 = +0.153 + 0.0591 \log \frac{[Cu^{++}]}{[Cu^{+}]}$ (5)

$Cu^{+} + 2H_2O = HCuO_2^{-} + 3H^{+} + e^{-}$

$E_0 = +1.733 - 0.1773\,pH + 0.0591 \log \frac{[HCuO_2^{-}]}{[Cu^{+}]}$ (6)

$Cu^{+} + 2H_2O = CuO_2^{--} + 4H^{+} + e^{-}$

$E_0 = +2.510 - 0.2364\,pH + 0.0591 \log \frac{[CuO_2^{--}]}{[Cu^{+}]}$ (7)

$Z = +2$ or $+3$

$Cu^{++} = Cu^{+++} + e^{-}$

$E_0 = +2.475 + 0.0591 \log \frac{[Cu^{+++}]}{[Cu^{++}]}$ (8)

$Cu^{++} + 2H_2O = CuO_2^{-} + 4H^{+} + e^{-}$

$E_0 = +3.078 - 0.2364\,pH + 0.0591 \log \frac{[CuO_2^{-}]}{[Cu^{++}]}$ (9)

$HCuO_2^{-} = CuO_2^{-} + H^{+} + e^{-}$

$E_0 = +1.498 - 0.0591\,pH + 0.0591 \log \frac{[CuO_2^{-}]}{[HCuO_2^{-}]}$ (10)

$CuO_2^{--} = CuO_2^{-} + e^{-}$

$E_0 = +0.721 + 0.0591 \log \frac{[CuO_2^{-}]}{[CuO_2^{--}]}$ (11)

The following relations, which express the circumstances for which the concentrations (or activities) of the two dissolved forms are equal, make it possible, as illustrated in Figure 32, to represent the *regions of relative predominance* of the ions Cu^+, Cu^{++}, CuO_2H^-, CuO_2^{--}, and CuO_2^-. (The region of relative predominance of trivalent Cu^{+++} ions is beyond the scope of this figure; it is included with Figures 34 and 37.)

Cu^{++}–$HCuO_2^-$	$pH = 8.91$	(1′)
$HCuO_2^-$–CuO_2^{--}	$pH = 13.15$	(3′)
Cu^{+++}–CuO_2^-	$pH = 2.57$	(4′)
Cu^+–Cu^{++}	$E_0 = +0.153$	(5′)
Cu^+–$HCuO_2^-$	$E_0 = +1.733 - 0.1773pH$	(6′)
Cu^+–CuO_2^{--}	$E_0 = +2.510 - 0.2364pH$	(7′)
Cu^{++}–Cu^{+++}	$E_0 = +2.475$	(8′)
Cu^{++}–CuO_2^-	$E_0 = +3.078 - 0.2364pH$	(9′)
$HCuO_2^-$–CuO_2^-	$E_0 = +1.498 - 0.0591pH$	(10′)
CuO_2^{--}–CuO_2^-	$E_0 = +0.721$	(11′)

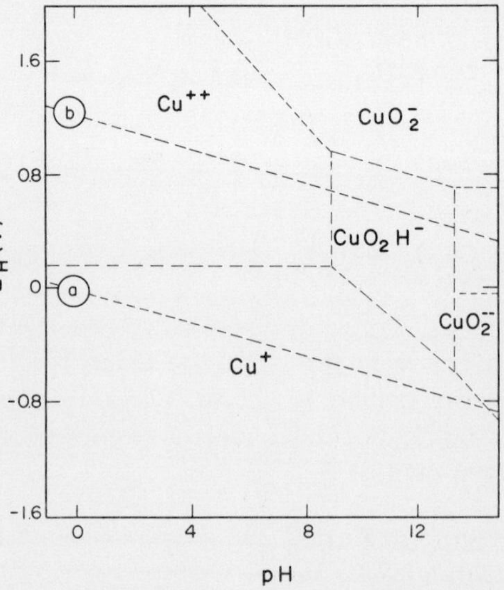

Figure 32. Cu–H_2O equilibrium. Regions of relative predominance of dissolved bodies (Cu^+, Cu^{++}, $HCuO_2^-$, CuO_2^{--}, and CuO_2^-) at 25°C.

4.6.1.1.3.2. *Heterogeneous Reactions in Which Two Solid Bodies Participate. Regions of Stability of Solid Bodies*

In the general case of *electrochemical* reactions, the conditions of equilibrium of these reactions have the form

$$E_0 = E_0{}^0 - 0.0591\text{pH}$$

Each one of these conditions of equilibrium corresponds on a potential–pH diagram to a straight line whose slope is 59 mV/pH unit. These lines separate the *regions of relative stability of condensed bodies.*

Listed below are conditions of equilibrium of such reactions relative to copper.

$$2\mathbf{Cu} + H_2O = Cu_2O + 2H^+ + 2e^- \quad E_0 = +0.471-0.0591\text{pH} \tag{12}$$

$$\mathbf{Cu} + H_2O = CuO + 2H^+ + 2e^- \quad E_0 = +0.570-0.0591\text{pH} \tag{13a}$$

$$\mathbf{Cu} + 2H_2O = Cu(OH)_2 + 2H^+ + 2e^- \quad E_0 = +0.609-0.0591\text{pH} \tag{13b}$$

$$\mathbf{Cu_2O} + H_2O = \mathbf{2CuO} + 2H^+ + 2e^- \quad E_0 = +0.669-0.0591\text{pH} \tag{14a}$$

$$\mathbf{Cu_2O} + 3H_2O = \mathbf{2Cu(OH)_2} + 2H^+ + 2e^- \quad E_0 = +0.747-0.0591\text{pH} \tag{14b}$$

$$2\mathbf{CuO} + H_2O = \mathbf{Cu_2O_3} + 2H^+ + 2e^- \quad E_0 = +1.648-0.0591\text{pH} \tag{15a}$$

$$2\mathbf{Cu(OH)_2} = \mathbf{Cu_2O_3} + H_2O + 2H^+ + 2e^- \quad E_0 = +1.578-0.0591\text{pH} \tag{15b}$$

As shown in Figure 33 these relations make it possible to represent the regions of relative stability of copper, **Cu**, and its oxides $\mathbf{Cu_2O}$, **CuO**, and $\mathbf{Cu_2O_3}$. For simplification the conditions of stability of cupric hydroxide $\mathbf{Cu(OH)_2}$ have been omitted.

4.6.1.1.3.3. *Heterogeneous Reactions in Which One Solid Body Participates. Solubility of Solid Bodies*

If the reactions are written in such a way that the stoichiometric coefficient *a* of dissolved species A is equal to 1 the conditions of equilibrium

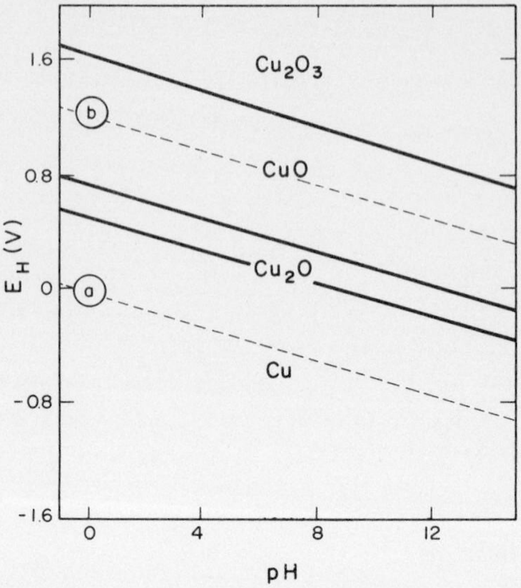

Figure 33. **Cu–H_2O** equilibrium. Regions of relative stability of solid bodies (**Cu**, **Cu_2O**, **CuO**, and **Cu_2O_3**, not considering **$Cu(OH)_2$**) (at 25°C).

of these reactions have the following forms, where the activity [A] of dissolved species A does not require explanation.

Chemical reactions

$$\log[A] = \log K + m\text{pH}$$

If the dissolved body A is alkaline (negatively charged ion, e.g., CuO_2H^-) the coefficient m is positive; if this dissolved body is acidic (positively charged ion, e.g., Cu^{++}) the coefficient m is negative; if this dissolved body is an unionized molecule (e.g., AsO_2H), the coefficient m is equal to zero.

Electrochemical reactions

If the reactions are written in such a manner that the stoichiometric coefficient n of the electron is positive and if the stoichiometric coefficient a or b of the dissolved oxidized A or reduced B is equal to unity in absolute value, then the conditions of equilibrium of these reactions are one or other of the two following types depending on whether the dissolved body is the oxidized form A (involving the activity with the plus sign) or the reduced form B (involving the activity with the minus sign):

$$E_0 = E_0{}^0 - \frac{0.0591m}{n}\,\mathrm{pH} + \frac{0.0591}{n}\log\,[\mathrm{A}]$$

$$E_0 = E_0{}^0 - \frac{0.0591m}{n}\,\mathrm{pH} - \frac{0.0591}{n}\log\,[\mathrm{B}]$$

All of these equilibrium conditions may be represented on a potential–pH diagram by a family of parallel straight lines each corresponding to a given value of the term log[A] or log[B]. Then, if the solution is ideal, each of these straight lines corresponds to a definite value of the *solubility* of the considered body in the dissolved form A or B.

Listed below are conditions of equilibrium relative to the solubility of copper, its oxides $\mathbf{Cu_2O}$, $\mathbf{CuO}$, and $\mathbf{Cu_2O_3}$, and cuprous and cupric hydroxides $\mathbf{CuOH}$ and $\mathbf{Cu(OH)_2}$ in the various dissolved forms under consideration.

Chemical reactions

$$2Cu^{+} + H_2O = \mathbf{Cu_2O} + 2H^{+} \qquad \log[Cu^{+}] = -\ 1.68 - \mathrm{pH} \qquad (16a)$$

$$Cu^{+} + H_2O = \mathbf{Cu(OH)} + H^{+} \qquad \log[Cu^{+}] = -\ 0.84 - \mathrm{pH} \qquad (16b)$$

$$Cu^{++} + H_2O = \mathbf{CuO} + 2H^{+} \qquad \log[Cu^{++}] = 7.89 - 2\mathrm{pH} \qquad (17a)$$

$$Cu^{++} + 2H_2O = \mathbf{Cu(OH)_2} + 2H^{+} \qquad \log[Cu^{++}] = 9.21 - 2\mathrm{pH} \qquad (17b)$$

$$\mathbf{CuO} + H_2O = HCuO_2 + H^{+} \qquad \log[HCuO_2^{-}] = -18.83 + \mathrm{pH} \qquad (18a)$$

$$\mathbf{Cu(OH)_2} = HCuO_2^{-} + H^{+} \qquad \log[HCuO_2^{-}] = -17.52 + \mathrm{pH} \qquad (18b)$$

$$\mathbf{CuO} + H_2O = CuO_2^{--} + 2H^{+} \qquad \log[CuO_2^{--}] = -31.98 + 2\mathrm{pH} \qquad (19a)$$

$$\mathbf{Cu(OH)_2} = CuO_2 + 2H^{+} \qquad \log[CuO_2^{--}] = -30.67 + 2\mathrm{pH} \qquad (19b)$$

$$2Cu^{+++} + 3H_2O = \mathbf{Cu_2O_3} + 6H^{+} \qquad \log[Cu^{+++}] = -\ 6.09 - 3\mathrm{pH} \qquad (20)$$

$$\mathbf{Cu_2O_3} + H_2O = 2CuO_2^{-} + 2H^{+} \qquad \log[CuO_2^{-}] = -16.31 + \mathrm{pH} \qquad (21)$$

Electrochemical reactions

$$\mathbf{Cu} = Cu^{+} + e^{-}$$
$$E_0 = +0.520 + 0.0591 \log[Cu^{+}] \quad (22)$$

$$\mathbf{Cu} = Cu^{++} + 2e^{-}$$
$$E_0 = +0.337 + 0.0295 \log[Cu^{++}] \quad (23)$$

$$\mathbf{Cu} + 2H_2O = HCuO_2^{-} + 3H^{+} + 2e^{-}$$
$$E_0 = +1.127 - 0.0886pH + 0.0295 \log[HCuO_2^{-}] \quad (24)$$

$$\mathbf{Cu} + 2H_2O = CuO_2^{--} + 4H^{+} + 2e^{-}$$
$$E_0 = +1.515 - 0.1182pH + 0.0295 \log[CuO_2^{--}] \quad (25)$$

$$\mathbf{Cu_2O} + 2H^{+} = 2Cu^{++} + H_2O + 2e^{-}$$
$$E_0 = +0.203 + 0.0591pH + 0.0591 \log[Cu^{++}] \quad (26a)$$

$$\mathbf{Cu(OH)} + H^{+} = Cu^{++} + H_2O + e^{-}$$
$$E_0 = +0.105 + 0.0591pH + 0.0591 \log[Cu^{++}] \quad (26b)$$

$$\mathbf{Cu_2O} + 3H_2O = 2HCuO_2^{-} + 4H^{+} + 2e^{-}$$
$$E_0 = +1.783 - 0.1182pH + 0.0591 \log[HCuO_2^{-}] \quad (27a)$$

$$\mathbf{Cu(OH)} + H_2O = HCuO_2^{-} + H^{+} + e^{-}$$
$$E_0 = +1.673 - 0.1182pH + 0.0591 \log[HCuO_2^{-}] \quad (27b)$$

$$\mathbf{Cu_2O} + 3H_2O = 2CuO_2^{--} + 6H^{+} + 2e^{-}$$
$$E_0 = +2.560 - 0.1773pH + 0.0591 \log[CuO_2^{--}] \quad (28a)$$

$$\mathbf{Cu(OH)} + H_2O = CuO_2^{--} + 3H^{+} + e^{-}$$
$$E_0 = +2.459 - 0.1773pH + 0.0591 \log[CuO_2^{--}] \quad (28b)$$

$$Cu^{+} + H_2O = \mathbf{CuO} + 2H^{+} + 2e^{-}$$
$$E_0 = +0.620 - 0.1182pH - 0.0591 \log[Cu^{+}] \quad (29a)$$

$$Cu^{+} + 2H_2O = \mathbf{Cu(OH)_2} + 2H^{+} + e^{-}$$
$$E_0 = +0.697 - 0.1182pH - 0.0591 \log[Cu^{+}] \quad (29b)$$

$$2Cu^{++} + 3H_2O = \mathbf{Cu_2O_3} + 6H^{+} + 2e^{-}$$
$$E_0 = +2.114 - 0.1773pH - 0.0591 \log[Cu^{++}] \quad (30)$$

$$2CuO_2^{--} + 2H^{+} = \mathbf{Cu_2O_3} + H_2O + 2e^{-}$$
$$E_0 = -0.243 + 0.0591pH - 0.0591 \log[CuO_2^{--}] \quad (31)$$

$$\mathbf{CuO} + H_2O = CuO_2^{-} + 2H^{+} + e^{-}$$
$$E_0 = +2.609 - 0.1182pH + 0.0591 \log[CuO_2^{--}] \quad (32)$$

$$\mathbf{Cu(OH)_2} = CuO_2^{-} + 2H^{+} + e^{-}$$
$$E_0 = +2.534 - 0.1182pH + 0.0591 \log[CuO_2^{--}] \quad (32b)$$

Finally, on account of the existence of the gaseous hydride ***CuH*** one should also consider the reaction *between a solid body and a gaseous body*

$$\boldsymbol{CuH} = \mathbf{Cu} + H^+ + e^- \qquad E_0 = -2.775 - 0.0591\text{pH} - 0.0591 \log p_{\text{CuH}} \tag{33}$$

4.6.1.1.3.4. *Graph of the Overall Diagram*

By plotting on the same diagram the different lines and families of lines an *overall diagram* is obtained which represents the equilibrium conditions of all the reactions considered. The resulting diagram is simplified by eliminating lines which are of no practical interest, thereby giving the diagram a more useful form.

One example of a simplification involves the line corresponding to Eq. (30) of Figure 37. This is the Cu^{++}–$\mathbf{Cu_2O_3}$ equilibrium. As the potential is increased, the region of Cu^{+++}–$\mathbf{Cu_2O_3}$ (line 20) is approached. Rather than showing the separate lines for Eqs. (20) and (30), the total solubility is considered as $Cu^{++} + Cu^{+++}$.[19] This produces a curved line as shown in the range of +2.5 V. Conversely, when line 30 intersects line 17 which corresponds to the Cu^{++}–**CuO** equilibrium, there is a sharp change in the angle of the Cu^{++} boundary. In general then, when drawing lines corresponding to a given value of the solubility of condensed phases:

1. There is a gradual curve where the composition of the system changes gradually, i.e., when the line corresponding to the predominance of another dissolved ion is intersected.
2. There is a sharp discontinuity where the composition of the system changes suddenly, i.e., when the boundary between two condensed phases is intersected. Thus, for the case of copper, the curved intersections occur near the point of intersection with the lines which separate regions of relative predominance of the dissolved bodies: Cu^+–Cu^{++} (5′), $HCuO_2^-$–CuO_2^{--} (3′), $HCuO_2^-$–CuO_2^- (10′) and CuO_2^{--}–CuO_2^- (11′). The angular change will occur at intersections with lines separating the regions of stability of condensed phases such as

$$\mathbf{Cu–Cu_2O}\ (12) \qquad \mathbf{Cu_2O–CuO}\ (14) \qquad \text{and} \qquad \mathbf{CuO–Cu_2O_3}\ (15)$$

When the given system includes solid bodies which, although thermodynamically unstable with respect to other solids (e.g., $\mathbf{Cu(OH)_2}$, which

[19] The plot of the curved portions of lines of the same solubility may be easily made by the method we outlined in Section 3.2.

tends to change by dehydration to **CuO**), have a particular practical interest, it may be useful to plot two or even more equilibrium diagrams. In this case one of these diagrams might represent only the thermodynamically stable equilibria (e.g., Figure 37, page 142, where **CuO** is included but not **$Cu(OH)_2$**) and the other only the thermodynamically unstable equilibria (e.g., Figure 34, where **$Cu(OH)_2$** is considered but not **CuO**).

It must be noted that a diagram such as Figure 7 is valid only in the presence of solutions in which the metal (copper) or metalloid exists solely in the forms which have been taken into account in establishing the diagram. When solutions contain substances which are liable to form soluble complexes with the metal or metalloid (e.g., cyanide forming the $CuCN_2^-$ ion or ammonia forming the $Cu(NH_3)_4^{++}$ ion) or even insoluble salts (e.g., carbonate forming the salt **$CuCO_3$**), it is necessary to modify the diagram taking into account the conditions of stability of these dissolved or solid substances.

4.6.1.1.3.5. *Phase Rule*

Further, we point out that the variations within a system are clearly emphasized on such a diagram without having to fall back on calculations to derive the phase rule. A *point* represents an *invariant* system; a *line* represents a *univariant* system; a family of lines each of which is relating to a value of a parameter represents a *divariant* or *trivariant* system depending on whether this parameter contains one component (one concentration) or two components (a term containing two concentrations).

For example, in Figure 37 the intersection of lines 1′, 5′, and 6′ represents the invariant system for which the activities of Cu^+, Cu^{++}, and $HCuO_2^-$ ions are equal. Line 14 represents the univariant system **Cu_2O–CuO**. The family of lines 26, represents the divariant system **Cu_2O**–Cu^{++} The family of lines 1 (only one of which has been drawn) represents the trivariant system Cu^{++}–$HCuO_2^-$.

4.6.1.2. Interpretation of Diagram. Behavior of Copper in the Presence of Aqueous Solutions

4.6.1.2.1. *General Bases for Predicting Corrosion, Immunity, and Passivation of Copper*

The potential–pH diagrams for copper as shown in Figures 34 and 37 define regions where copper is soluble notably as Cu^{++}, $HCuO_2^-$, or CuO_2^{--} ions and where it exists as condensed phases such as the pure metal or

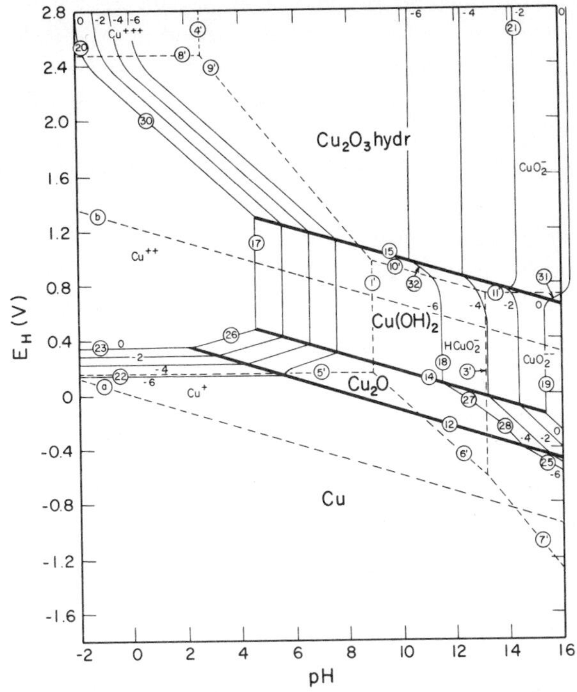

Figure 34. **Cu–H_2O** equilibrium. Overall diagram. Figure established considering the solid bodies **Cu, Cu_2O, Cu(OH)$_2$**, and **Cu_2O_3**. **CuO** was not considered.

compounds. If the pH and potential are such that the metal–environment system exists in a region where Cu^{++} is stable, then, copper may dissolve until an equilibrium Cu^{++} concentration is attained. This dissolution is simply *corrosion*. Thus, while the metal–environment system exists in the region of soluble ions, it may be expected to corrode. If, on the other hand it exists in the region of potential and pH where the metal is stable (the more negative direction of potential), then the metal will not corrode or will be *immune* from corroding. Finally, if the metal–environment system exists in a potential–pH region where a solid product, e.g., **Cu_2O** is stable, then the corrosion rate may be expected to be minimal because of a covering by this solid product. Such a regime is called a region of *passivation*, which denotes an action toward passivity. However, perfect passivity is achieved only if the solid product is perfectly protective, and this thermodynamics does not tell. Thus *passivation does not necessarily mean passivity*.

For practical purposes it is convenient to provide a definite demarcation between the regions of *corrosion and immunity* and between *corrosion and passivation*. It is convenient to take this as the line representing a solubility of *10^{-6} gram-atom of soluble ion per liter*. Thus, at higher concentrations, corrosion will be assumed to occur with more or less increased rapidity; and below this line corrosion is stifled either owing to the metal being immune or passivated.

Particular care should be taken when the metal exists in the region of passivation. If the product layer is nonporous and adherent, the underlying metal is protected. Experience shows that, at least for solutions not containing chlorides, oxide films are generally perfectly protective for aluminum, chromium, iron, and tin; they are generally not protective for copper.

Figure 35 deduced from Figure 37, illustrates on this basis the theoretical circumstances of corrosion, immunity, and passivation of copper at 25°C in the case where passivation is due to the presence of films of $\mathbf{Cu_2O}$, **CuO**, and hydrated $\mathbf{Cu_2O_3}$.

A metal may be considered more *noble* as the region of immunity lies more within the region of stability of water. Then from Figure 35 copper may be considered as a relatively noble metal (at least in the absence of complexing agents such as cyanides and ammonia which extend the regions

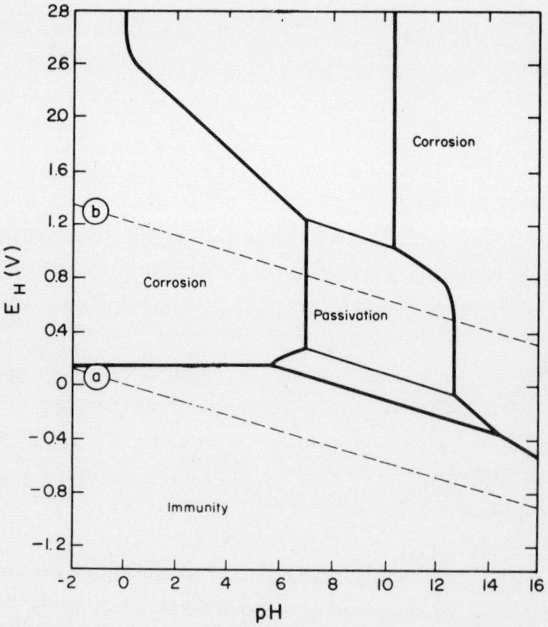

Figure 35. Theoretical circumstances of corrosion, immunity, and passivation of copper (25°C) in the case of passivation by films of $\mathbf{Cu_2O}$, **CuO**, and $\mathbf{Cu_2O_3}$.

of corrosion at the expense of the region of immunity). The two regions of corrosion in Figure 35 indicate that strongly acidic or alkaline solutions (for example, solutions of sulfuric acid or caustic soda) may vigorously corrode copper if they contain oxidants, but if oxidants are not present copper is uncorroded. Virtually neutral or mildly alkaline solutions containing oxidants passivate the metal and corrosion is greatly reduced by the formation of oxide films. Corrosion may easily be avoided even in the presence of acidic or alkaline oxidizing solutions by cathodic protection carefully maintained in such a way that the potential of the metal is less than about +0.1 V in acid solution and about −0.2 to −0.6 V, according to the pH, in highly alkaline solution.

4.6.1.2.2. *Behavior of Copper in the Presence of an Oxygen-Free Solution of Cupric Sulfate*

Consider the case of copper immersed in an *oxygen-free* 0.01 M solution of cupric sulfate (containing 2.50 g $\mathbf{CuSO_4 \cdot 5H_2O}$/liter or 635.7 mg Cu^{++}/liter). Suppose also that the activity of Cu^{++} ions in solution is equal to the concentration of ions, i.e., 0.010 g-ion/liter. In fact, the potential of copper will be that of the equilibrium between metallic copper and Cu^{++} ions; this equilibrium potential is given by Eq. (23)

$$E_0 = +0.337 + 0.0295 \log[Cu^{++}]$$

which for $[Cu^{++}] = 10^{-2}$, gives

$$E_0 = +0.337 - 0.059 = +0.278 \text{ V}$$

Figures 32, 34, and 37 show that this value of potential occurs near the region of predominance of cuprous ions, Cu^+. The equilibrium content of Cu^+ and Cu^{++} ions depends on the potential according to Eq. (5)

$$E_0 = + 0.153 + 0.0591 \log \frac{[Cu^{++}]}{[Cu^+]}$$

Thus,

$$\log \frac{[Cu^{++}]}{[Cu^+]} = \frac{0.278 - 0.153}{0.0591} = 2.11$$

and

$$\frac{[Cu^{++}]}{[Cu^+]} = 128$$

Hence, the equilibrium content of cuprous ions is 0.0100/128 = 0.000078 g-atom/liter, or 0.000078 × 63.600 = 4.95 mg Cu^+/liter.

Thus, about 5 mg Cu^{+}/liter tend to be formed. As the solution is free from oxidants other than Cu^{++}, these Cu^{+} ions may give rise only to the reactions

$$\mathbf{Cu} \rightarrow Cu^{+} + e^{-} \qquad \text{(oxidation)}$$

$$Cu^{++} + e^{-} \rightarrow Cu^{+} \qquad \text{(reduction)}$$

$$\mathbf{Cu} + Cu^{++} \rightarrow 2Cu^{+}$$

Therefore, the corrosion of copper limited to the formation of 5.0 mg Cu^{+}/liter occurs and of this amount 2.5 mg result from the corrosion of copper and 2.5 mg from the reduction of Cu^{++} ions. The final composition of the solution will be

$$635.7 - 2.5 = 633.2 \text{ mg } Cu^{++}\text{/liter}$$

$$5.0 \text{ mg } Cu^{+}\text{/liter}$$

$$638.2 \text{ mg Cu total/liter}$$

The potential of the copper remains practically unchanged at +0.278 V.

4.6.1.2.3. *Influence of pH on the Potential of Copper*

Let us examine the influence of pH on the potential of electrically insulated copper in contact with solution free from complexing agents or substances which form insoluble salts with it, e.g., pure solutions of acetic acid and caustic soda which are free from cyanide, ammonia, chloride, and carbonate.

Firstly, by way of definition consider the case of an acid solution in which the activity and concentration of dissolved copper are both equal to 10^{-4} g-atom Cu/liter (6 mg/liter). If the pH is less than about 4 the potential is near the equilibrium potential of the system **Cu**–Cu^{++} illustrated in Figures 34 and 37 by line 23 at the −4 value (+0.21 V).

The metal does or does not corrode depending on whether or not the solution contains an oxidant whose reduction potential is greater than this value of +0.21 V. If the pH is increased from about 4 to 6 the metal is covered with red oxide, $\mathbf{Cu_2O}$, and the potential virtually is that of the equilibrium $\mathbf{Cu_2O}$–Cu^{++}. The potential increases slightly, as illustrated by the line 26 marked −4, up to about +0.3 V. For values of pH greater than about 6, the metal is covered by black oxide, **CuO**, and the potential rises to nonreproducible values, which depend on the protective quality of the oxide and the content of oxidant in the solution. The potential is higher and the oxide film is more protective as the content of oxidant increases. If the pH is increased even more, precipitation in the solution of blue hydroxide, $\mathbf{Cu(OH)_2}$, as well as possibly black **CuO** is noticed from about pH = 6.6. If saturation of the solution by $\mathbf{Cu(OH)_2}$ or **CuO** is attained the content of dissolved copper in solution has a minimum value near line 1′ which separates regions predominant in Cu^{++} and $HCuO_2^-$ ions, for pH values around 8.9. If Figure 34 is correct this content of copper is about $10^{-8.3}$ (540×10^{-6} mg/liter) in the presence of hydroxide $\mathbf{Cu(OH)_2}$ *in equilibrium*; it is $10^{-9.6}$ (10×10^{-10} mg/liter) in the presence of oxide **CuO** *in equilibrium.* Values of pH greater than 8.9 produce a redissolution of the $\mathbf{Cu(OH)_2}$ and **CuO** formed; this redissolution is complete when the solubility reaches 10^{-4}, for values of pH from 13.1 to 13.9. Then the metal is only covered with $\mathbf{Cu_2O}$ and the potential falls to values represented by lines 27 and 28 and denoted −4, i.e., −0.1 to −0.5 V according to pH. This decrease of the electrode potential of copper at high values of pH is in agreement with observations made by Gatty and Spooner in 1938[20]—observations which went unexplained at that time.

4.6.1.2.4. *Electrolysis of Acidic Copper Solutions: Electrolytic Refining of Copper; Copper Voltammetry, Electrolytic Copper Plating in an Acid Bath*

Electrolysis operations of acidic cupric solutions depend on the fact that an electric current must be passed between two electrodes immersed in solution whose respective potentials must be above and below the equilibrium potential of the reaction **Cu** = $Cu^{++} + 2e^-$. This value of potential at 25°C is given by relation of $E_0 = +0.337 + 0.0295 \log[Cu^{++}]$ V. At the negative pole a deposit of copper is produced by reduction according to the reaction $Cu^{++} + 2e^- \rightarrow$ **Cu** and at the positive pole copper dissolves by oxidation with the inverse reaction **Cu** $\rightarrow Cu^{++} + 2e^-$. If these are the only two reactions occurring, they proceed with 100% current efficiency according to Faraday's law. Hence, the purification and deposition produce 1.185 g Cu/A-hr at the cathode. Copper voltammetry makes it possible to measure exactly the quantity of current used.

Generally, for *copper refining* it is important that the anodic potential, A_1, and the cathodic potential, C_1, both remain near the equilibrium potential of the reaction **Cu** = $Cu^{++} + 2e^-$ (see Figure 36). Too low a cathodic potential, C_2 (situated below the line *a* relative to the equilibrium of the

[20] Gatty and Spooner, *The Electrode Potential Behavior of Corroding Metals in Aqueous Solutions*, Clarendon Press, Oxford (1938).

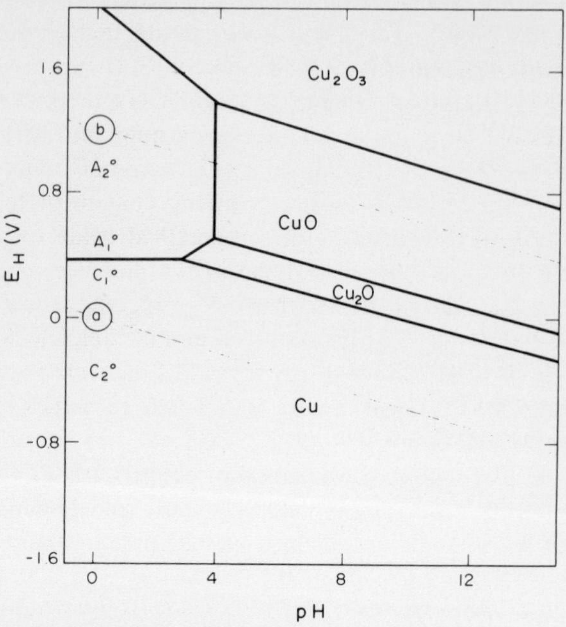

Figure 36. Electrolysis of cupric solutions.

reaction $H_2 = 2H^+ + 2e^-$ for a pressure of hydrogen of 1 atm), may cause the formation of hydrogen simultaneously with the deposition of copper on the cathode. This leads to a decrease of current efficiency and gives rise to cathodic copper containing hydrogen. Too great an anodic potential, A_2, may cause the dissolution of noble metals which subsequently are deposited on the cathode instead of being deposited as slimes during the disintegration of the anodes.

Electrolytic copper plating in acid baths is often carried out under the following conditions:

$CuSO_4 \cdot 5H_2O$	1 M, 250 g/liter
H_2SO_4	0.75 M, 74 g/liter
pH	0.75
Temperature	20° to 50°C
Current density	2 to 10 A/dm²
Stirring	by bubbling air

An increase of the copper content in solution occurs with time as well as a decrease of the acidity. This may be corrected by removing part of the copper solution and replacing it with sulfuric acid.

Since the anodic and cathodic potentials are both considerably less than the potentials indicated by line *b*, which relates to the equilibrium of the reaction $\boldsymbol{O}_2 + 4H^+ + 4e^- = 2H_2O$, a reduction of dissolved oxygen is produced mainly on the cathode according to the reaction $\boldsymbol{O}_2 + 4H^+ + 4e^- \rightarrow 2H_2O$. In addition, if the cathodic potential is less than the potentials indicated by line *a*, the formation of hydrogen by the reaction $2H^+ + 2e^- \rightarrow \boldsymbol{H}_2$ may be produced. Both these reduction reactions decrease the number of H^+ ions and therefore the acidity of the solution. On the other hand, these reactions lower to below 100% the current efficiency of the cathodic deposition of copper and increase to above 100% the current efficiency of the anodic dissolution of copper. This increase in the content of copper with diminution of acidity may be rectified by employing, simultaneously with the usual copper anodes, several lead anodes maintained at a high electrode potential and which, passivated by a protective film of lead dioxide $\mathbf{PbO_2}$, are the site of the reaction $2H_2O \rightarrow \boldsymbol{O}_2 + 4H^+ + 4e^-$. This reaction causes a decrease in the efficiency of the anodic corrosion of copper (and as a result a decrease of the copper content of the solution) and acidification of the solution.

4.6.1.2.5. *Copper Plating in Cyanide Baths*

It is well known that the acid bath procedure described above is not suitable for the copper plating of iron, ordinary steels or zinc. The copper deposit is not adherent because of corrosion of the base metal. For these metals an alkaline cyanide bath is needed. The following indicates the constituents of such a bath, taken from Blum and Hogaboom[21]:

$Cu(OH)_2 \cdot CuCO_3$	0.125 M,	27.6 g/liter
NaCN	1.15 M,	57.0 g/liter
Na_2CO_3	0.025 M,	2.6 g/liter
pH	11.5 to 12.5	
Temperature	40°C	
Current density	0.5 to 1.6 A/dm²	

[21] W. Blum and G. B. Hogaboom, *Principles of Electroplating and Electroforming*, McGraw-Hill, New York (1949), p. 215.

The possibility of copper plating steel in such a bath is due, not to the alkaline character of the bath, but rather to the presence of cyanide which has the effect of extending considerably the corrosion region of copper, by formation of copper complexes, up to potentials around -1 V (instead of about $+0.3$ V for the sulfuric acid baths). For such a cyanide bath the deposition potential of copper is nearly -1 V, while the corrosion potential of iron is about -0.7 V. Thus, cyanide baths make it possible to plate iron without corrosion of the base metal, and this leads to a more adherent deposit.

4.6.2. Theoretical Conditions of Corrosion, Immunity, and Passivation

If it is indeed reasonable to take the 10^{-6} equisolubility lines as the dividing point, a clear distinction is made between the domain where corrosion is possible (corrosion domain) and a domain where corrosion is not possible (noncorrosion domain). In the noncorrosion domain, two regions (or groups of regions) can be distinguished. In one of these regions, the solid stable form is the metal itself (called field of *immunity* following a suggestion of J. N. Agar). Although a pure metal surface is presented to the solution in this case the metal is quite uncorrodable, because the corrosion reaction is energetically impossible. In the other regions of noncorrosion, the solid stable form is not the metal but an oxide, a hydroxide, a hydride, or a salt (*passivation* domain). The metal then tends to become coated with this oxide, hydroxide, hydride, or salt which can, according to the circumstances, form on the metal either a nonporous film practically preventing all direct contact between the metal itself and the solution (in which case protection against corrosion is perfect), or a porous deposit which only partially prevents contact between the metal and the solution (in which case the protection is only imperfect). Understood in this way, *passivation thus does not necessarily imply the absence of corrosion.* Experience shows that, in the case of chloride-free solutions at least, the oxide films are generally perfect protectors for numerous metals, among which are aluminum, chromium, iron, and tin; for tantalum, niobium, and titanium, oxide films are protective even in the presence of chlorides.

In some relatively rare cases, a degradation of the metal may come about through the action, not of *dissolution*, but of *gasification* accompanied by the formation of a volatile hydride or oxide. If, as a rough guide, we accept that metal corrosion can take place in this way if the partial equilibrium pressure of the hydride or the oxide is equivalent to at least 10^{-6}

atm, then the lines of the potential–pH equilibrium diagrams which correspond to a hydride or oxide pressure equivalent to 10^{-6} atm enable us to establish a region of corrosion by gasification.

It goes without saying that the concentration (10^{-6} g-at.wt./liter) and pressure (10^{-6} atm) values adopted here for the definition of the corrosion thresholds are arbitrary and that, in practice, there may be good reasons for modifying these critical values. With this reservation, however, this delimitation can be regarded as exact as far as protection by immunity (which is one case of cathodic protection) is concerned.

To say that protection against corrosion results in the region of passivation is rather an oversimplification. Although U. R. Evans identified in 1927 the part played by oxides in protection by passivation, passivation in practice may not be uniquely due to thermodynamically stable oxides H. H. Uhlig and N. Hackerman have shown that adsorption phenomena can contribute to passivation. On the other hand, there is a great scarcity of sufficiently precise data concerning the composition and the thermodynamic properties of protective oxide films, and such data are indispensible for an exact knowledge of their stability conditions. In spite of these imperfections the conclusions provided by the theoretical "corrosion, immunity, and passivation" diagrams frequently coincide very well with actual measurements of corrosion in the various regions of potential and pH.

4.6.3. Behavior and Equilibrium Diagrams of Copper, Iron, Zinc, Aluminum, Silver, Lead, Tin, Chromium, and Arsenic

Figures 37–53[22] show how the regions of corrosion, immunity, and passivation correspond to the detailed diagrams of electrochemical equilibria for nine important metals and metalloids. Significant features of these diagrams are discussed below.

4.6.3.1. Copper (Figures 37 and 38)

As indicated in Section 4.6.1.2 above, copper is a relatively noble metal; generally it does not corrode in the presence of solutions free from oxidants (and species such as cyanides and ammonia which form soluble complexes with it), but it is corroded by acidic or alkaline solutions containing oxidants, e.g., aerated solutions of sulfuric acid or caustic soda. Generally a

[22] Most of these are taken from *Atlas of Electrochemical Equilibria*, Ref. 34.

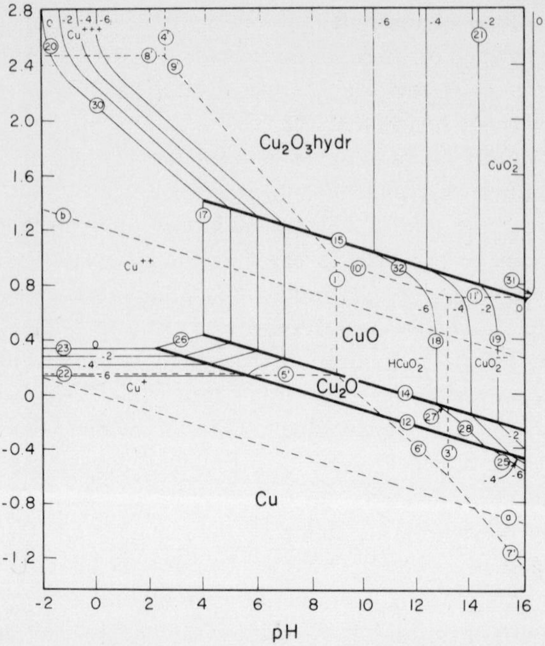

Figure 37. Potential–pH equilibrium diagram for the system copper–water at 25°C.

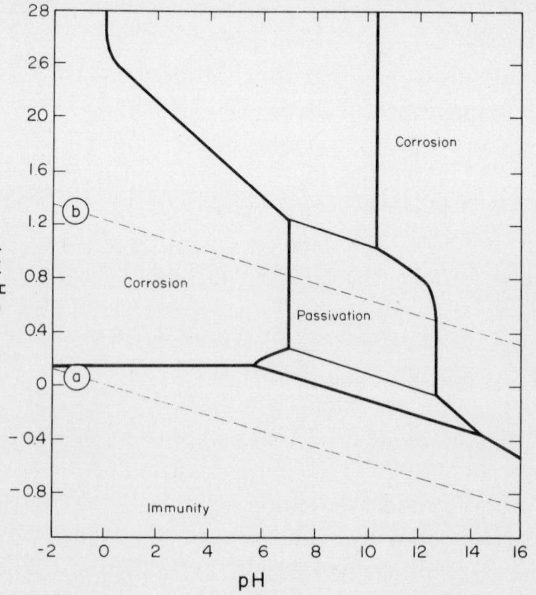

Figure 38. Theoretical conditions of corrosion, immunity, and passivation of copper at 25°C.

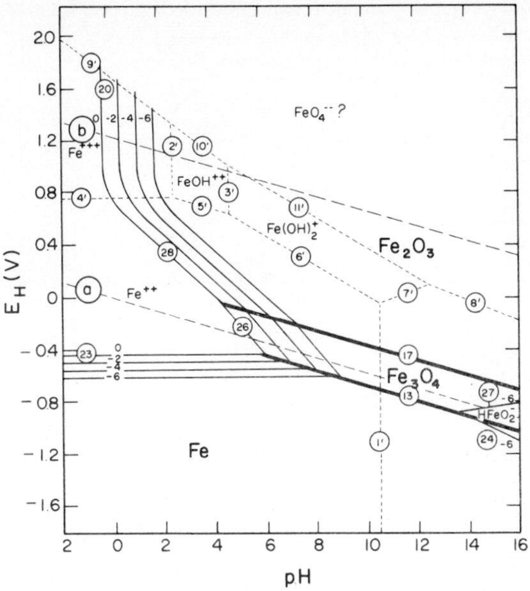

Figure 39. Potential–pH equilibrium diagram for the system iron–water at 25°C.

slight lowering of the electrode potential is sufficient to make the metal immune and thereby to cathodically protect it.

4.6.3.2. Iron (Figures 39 and 7, page 17)

Iron, which is examined in detail in Chapter 6 of these lectures is not, as is well known, a noble metal. It generally corrodes with evolution of hydrogen unless the conditions are such that it is covered with a protective oxide conferring a state of passivation. Generally this is only possible in the presence of solutions of relatively low chloride concentration. This passivation occurs more readily between pH 10 and 13 than for higher or lower values of pH.

4.6.3.3. Zinc (Figures 40 and 41)

Zinc is a non-noble metal which, nevertheless, often corrodes only very slowly due to the high overpotential for the reduction of water (i.e., $2H^+ \rightarrow H_2 + 2e^-$); passivation is possible for values of pH between 9 and 12. At pH 7 zinc does not passivate but corrodes. In the case of aerated stagnant pure water, there is local passivation of the metal due to the in-

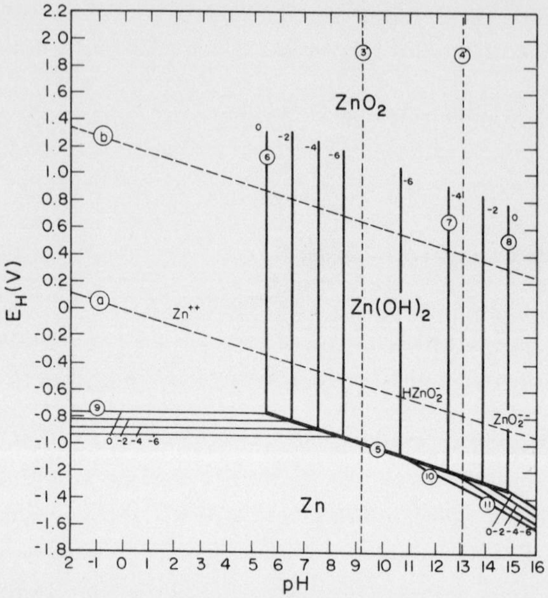

Figure 40. Potential–pH equilibrium diagram for the system zinc–water at 25°C.

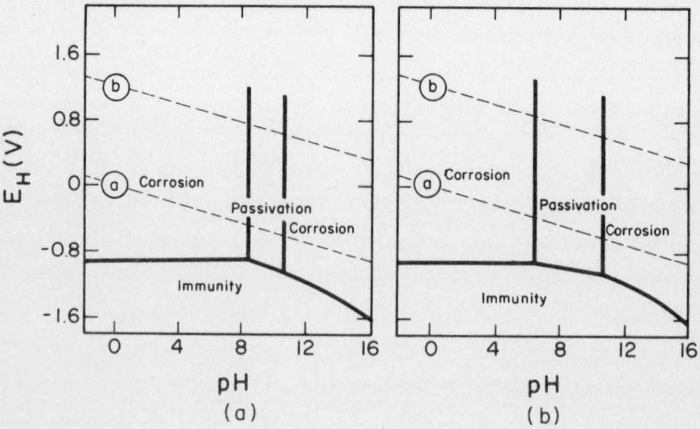

Figure 41. Theoretical conditions of corrosion, immunity, and passivation of zinc at 25°C: (a) for solutions free from CO_2; and (b) for solutions containing CO_2 (1 g-mole/liter).

crease of pH resulting from the cathodic reduction of oxygen ($O_2 + 4H^+ + 4e^- \rightarrow 2H_2O$). This local passivation leads to a localized corrosion rather than a general corrosion of the whole surface of the metal. In the presence of bicarbonated water, the corrosion resistance is increased because of the formation of insoluble carbonate (**$ZnCO_3$** or of basic carbonate 2**$ZnCO_3$** $\cdot$ 3**$Zn(OH)_2$**), which extends the region of passivation toward neutral pH values (Figure 41b).

4.6.3.4. Aluminum (Figures 42 and 43)

Although aluminum has a highly negative dissolution potential it may be used as a material for construction because it is passivated by a layer of alumina which is stable between about pH 4 and 9. The resistance of aluminum essentially is associated with the more or less protective qualities of this oxide layer. These qualities depend markedly on the circumstances in which this layer is formed: cold or hot dry oxidation, cold or hot anodic oxidation and with or without surface colmatage. The oxide layer formed on the metal may be destroyed by chlorides or dissolved by alkaline solutions.

4.6.3.5. Silver (Figures 44 and 45)

Silver, a relatively noble metal, strongly resists corrosion in solutions free from highly oxidizing (nitric acid) or complexing substances (ammonia, cyanides). The good corrosion resistance is particularly noticeable in highly alkaline solutions.

4.6.3.6. Lead (Figures 46 and 47)

Although relatively noble, lead dissolves appreciably in pure water due to the relatively high solubility of its lower oxide **PbO**: the minimal value of this solubility (along line 1′ of Figure 46, i.e., at pH 9.3) is about $10^{-5.7}$ mole/liter, i.e., 0.4 ppm Pb, which is still to high for human consumption. This solubility is $10^{-1.3}$ mole/liter, i.e., 10 g Pb/liter, at pH 7. The toxicity of lead has frequently prohibited its use as piping for water supply especially in the case of drinking water. Also, the lead content increases when the water is stagnant. Like zinc, its corrosion resistance may be increased in bicarbonated water due to the formation of an insoluble carbonate. The region of stability of carbonate extends the domain of passivation of the metal, bridging the gap to the region of immunity (Figure 47b).

Cathodic protection of lead is easily attained by depressing the potential to the region of immunity. In the presence of water and aqueous solutions

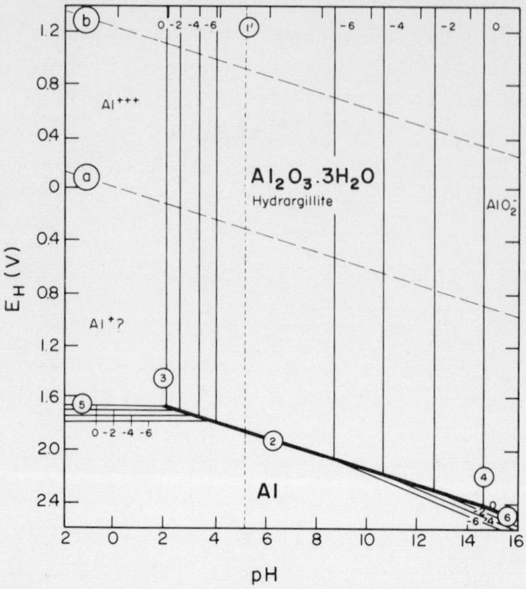

Figure 42. Potential–pH equilibrium diagram for the system aluminum–water at 25°C.

such as the sulfuric solutions of lead accumulators, however, it is essential to avoid too great a potential lowering because of "cathodic corrosion" owing, as shown by J. van Muylder (23), to the formation of gaseous lead hydride below lines 24 of Figure 46 (corrosion at low potential not to be confused with the corrosion possible in alkaline media at higher potentials). Being unstable, this gaseous hydride decomposes to gaseous hydrogen and

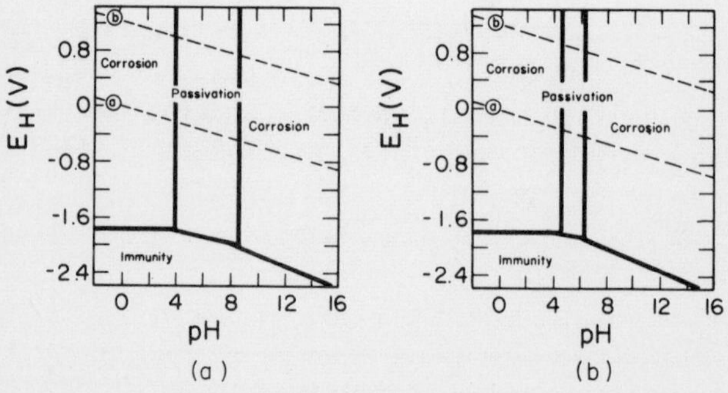

Figure 43. Theoretical conditions of corrosion, immunity, and passivation of aluminum (at 25°C): (a) passivation by a film of hydrargillite $\mathbf{Al_2O_3 \cdot 3H_2O}$; and (b) passivation by a film of böhmite $\mathbf{Al_2O_3 \cdot H_2O}$.

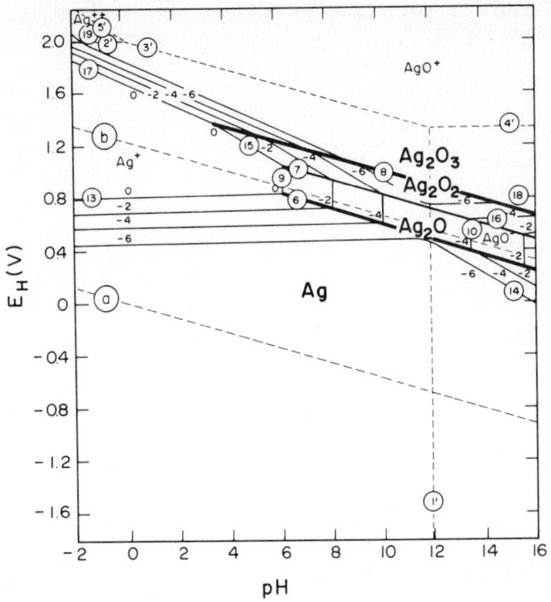

Figure 44. Potential–pH equilibrium diagram for the system silver–water at 25°C.

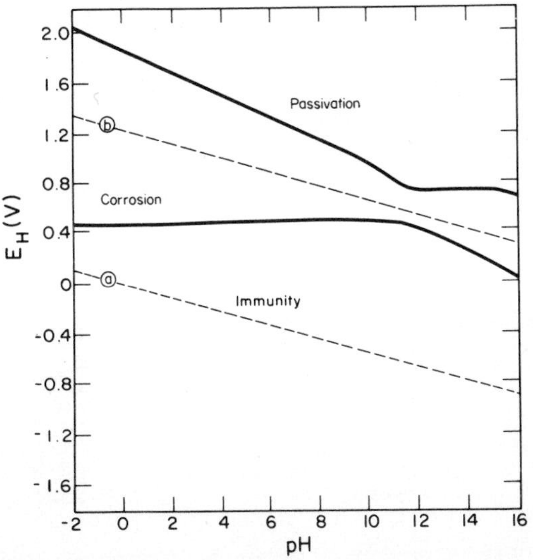

Figure 45. Theoretical conditions of corrosion, immunity, and passivation of silver at 25°C.

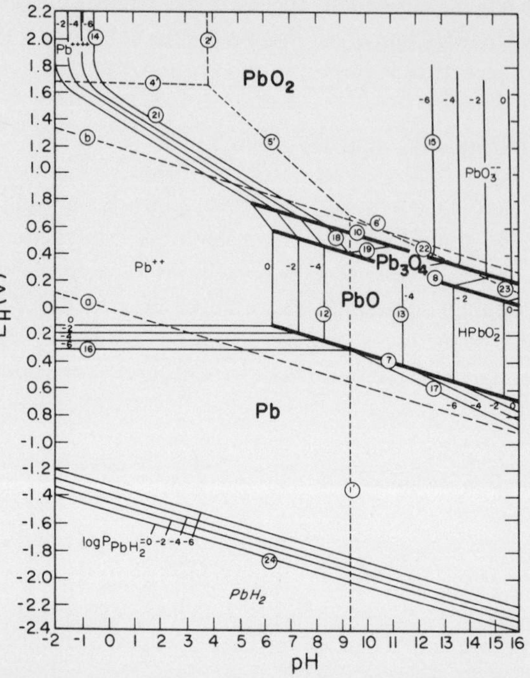

Figure 46. Potential–pH equilibrium diagram for the system lead–water at 25°C.

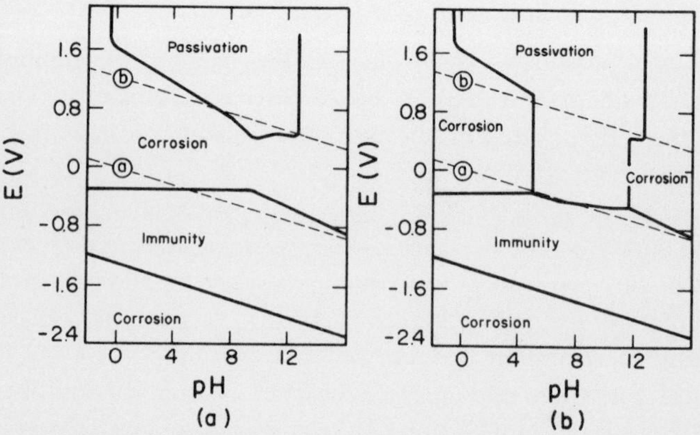

Figure 47. Theoretical conditions of corrosion, immunity, and passivation of lead (at 25°C): (a) for solutions free from CO_2; and (b) for solutions containing CO_2 (1 g-mole/liter).

metallic lead which, emitted in the form of a nonprotective cloud in the case of lead structures immersed in an aqueous solution, forms a protective screen when a structure is buried in the ground.[23]

4.6.3.7. Tin (Figures 48 and 49)

The well-known corrosion resistance of tin which enables the use of tin plate for the manufacture of "tin cans" is due to the stability of its oxide $\mathbf{SnO_2}$. Nevertheless, certain conditions of acidity, reducing power, and chloride content sometimes encountered with foodstuffs may lower the potential to one of the two triangular regions of corrosion which correspond to the reduction of $\mathbf{SnO_2}$ with formation of stannous or stannite ions where tin is divalent.

4.6.3.8. Chromium (Figures 50, 51, and 52)

Chromium is not a noble metal and may be oxidized in acid solution with the formation of blue chromous ions, Cr^{++}, green chromic ions, Cr^{+++}, and red chromic acid, H_2CrO_4, which transforms successively to orange dichromate ions, $HCrO_4^-$ (or $Cr_2O_7^{--}$), and to yellow chromate ions, CrO_4^-. When reduced this CrO_4^- gives green chromite ions, CrO_2^-, and metallic chromium. This cycle runs around a large domain of stability of solid green chromic hydroxide, $\mathbf{Cr(OH)_3}$, above which higher oxides probably exist. Chromic hydroxide, $\mathbf{Cr(OH)_3}$, (or oxide, $\mathbf{Cr_2O_3}$) which is protective in the *absence of chloride*, confers good passivity to the metal in contact with *many chloride-free* solutions.

It is now possible to point out a further method of manipulating the equilibria to obtain information about corrosion resistance. This method involves the superposition of one equilibrium diagram on another. Suppose it is appropriate to ask what inorganic species would form a stable precipitate on iron in the region of iron "corrosion" and thereby the dissolution of iron would be stifled by the presence of the insoluble film. An especially good example of this principle is available in the case of adding chromate ions to the solution for protecting iron. If the diagram of electrochemical equilibrium of chromium (Figure 50) is superimposed on the theoretical diagram of corrosion–immunity–passivation of iron (Figure 7b), it is observed that an important section of the corrosion region of iron is covered

[23] See *Rapports Techniques CEBELCOR* RT. 10 (1953), RT. 12 (1954), RT. 13 (1953), and RT. 64 (1958).

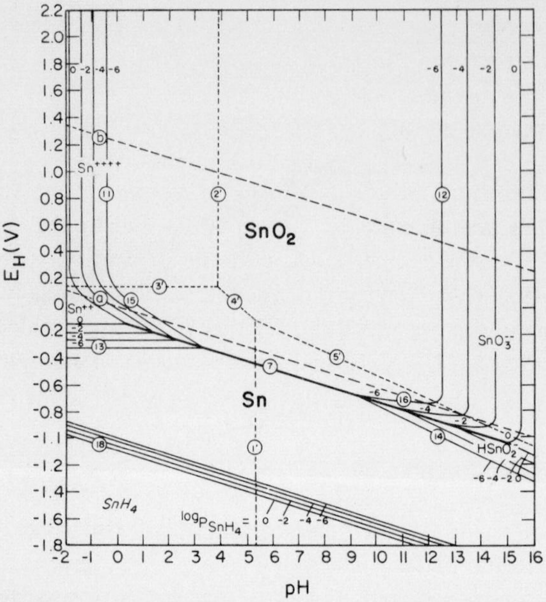

Figure 48. Potential–pH equilibrium for the system tin–water at 25°C.

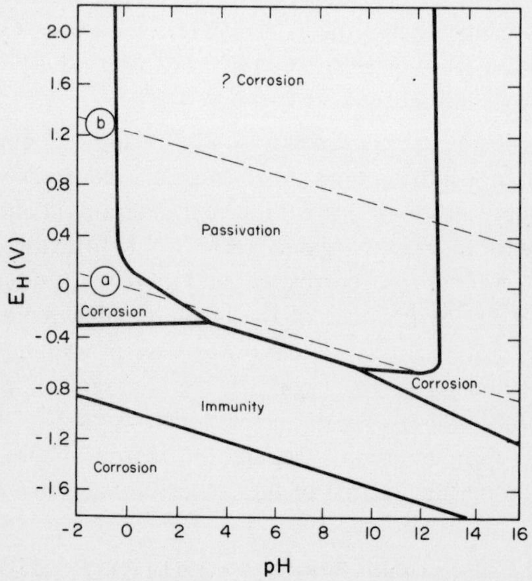

Figure 49. Theoretical conditions of corrosion, immunity, and passivation of tin at 25°C.

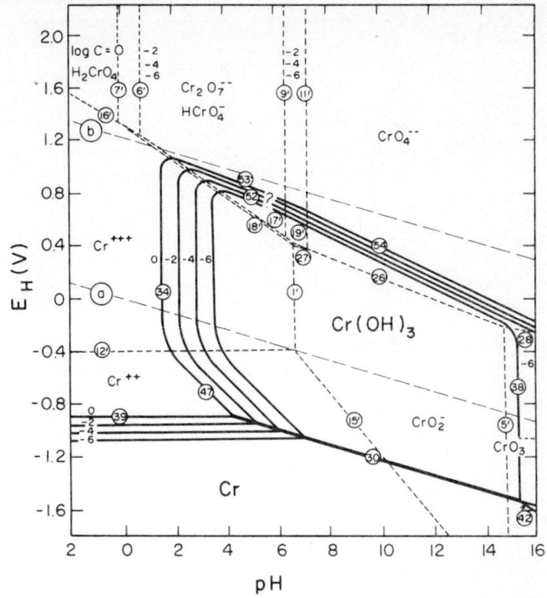

Figure 50. Potential–pH equilibrium diagram for the system chromium–water at 25°C.

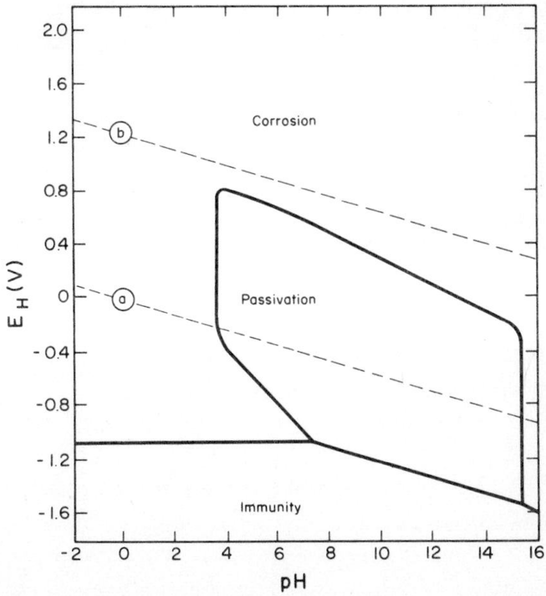

Figure 51. Theoretical conditions of corrosion, immunity, and passivation of chromium at 25°C.

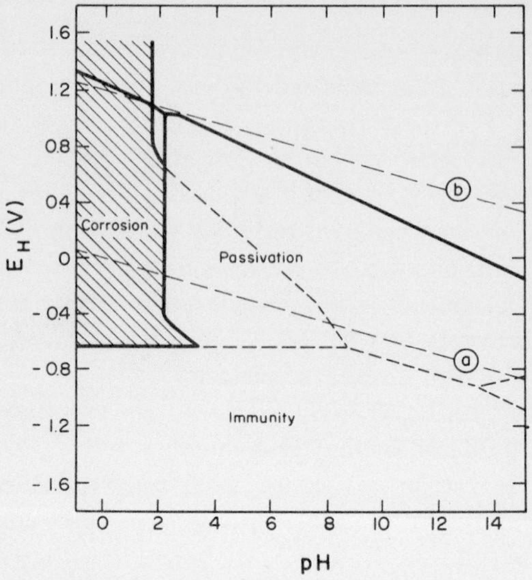

Figure 52. Theoretical domains of corrosion, immunity, and passivation of iron in the presence of solutions containing 10^{-2} g-atom Cr/liter (1.9 g K_2CrO_4/liter).

by the domain of stability of $\mathbf{Cr(OH)_3}$. This implies that corroding iron may be covered with a deposit of $\mathbf{Cr(OH)_3}$ which is passivating (if the solution is free from chloride).

Figure 52 shows the lines which delineate the various regions of corrosion, immunity, and passivation of iron. The heavy line represents the lower limit of the domain of stability of hexavalent chromium, below which reduction of chromic oxide, dichromate, and chromate is possible. The region represented in white outlines the domain where, in the presence of solutions containing 10^{-2} g-atom Cr/liter (e.g., 1.9 mg K_2CrO_2/liter), a *solid* body is stable, and this corresponds to circumstances where iron is in a state of immunity or passivation (covered by a film of iron oxide and/or chromium oxide). The grey area represents the region where no solid body is stable and this corresponds to the circumstances of corrosion of iron. It may be seen from this figure that the addition of 1.9 g K_2CrO_4/liter to a chloride-free water or solution will suppress the corrosion of iron if the pH is not below about 2 to 4. If the pH is less than these values, addition of chromate will intensify general corrosion owing to the additional reduction process afforded by reduction of the higher-valence chromate ion to the lower-valence chromic which is soluble.

4.6.3.9. Arsenic (Figure 53)

When iron corrodes with evolution of hydrogen, it has a potential slightly below the potentials indicated by line *a* of the equilibrium potential–pH diagrams. For these potentials, Figure 53 shows that the stable form of the arsenic–water system is elementary arsenic which is then practically insoluble. It follows that if arsenates or arsenites are added to a solution in which iron normally corrodes with evolution of hydrogen (for instance, sulfuric acid or hydrochloric acid) then the metal may be coated with a deposit of elementary arsenic which protects it to a certain extent against further corrosion; here we have a case of passivation, not by a film of oxide but by a film of unoxidized element.

For this reason, steel reservoirs used for storage of sulfuric acid corrode less if the acid is produced by the "lead chamber process," where nitrogen oxides are used as catalysts (if made from nonpurified roasting gases, containing arsenic), than if the acid is produced by the "contact process" on platinum (in which case the arsenic contained in the roasting gases has to be removed, because of its poisoning action on the catalyst).

We point out that such protective action of arsenic takes effect not only *vis-à-vis* iron but also other metals and alloys (e.g., certain alloy steels), the corrosion potential of which falls within the region of stability of arsenic.

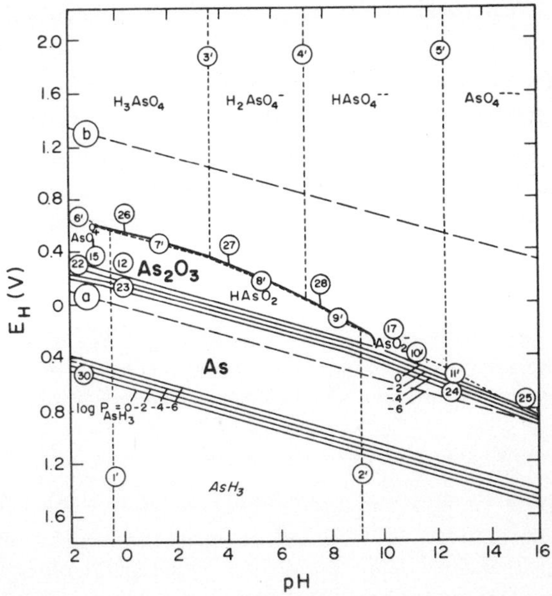

Figure 53. Potential–pH equilibrium diagram for the system arsenic–water at 25°C.

4.6.4. Nobility of Metals and Metalloids. Theoretical and Practical Bases

In Section 4.6.1.2, it was noted that an important application of the equilibrium potential–pH plots is in predicting the corrosion behavior of metals. The methods already delineated can be used now to establish diagrams of electrochemical equilibria for each element of the periodic table for which the free enthalpies of formation are known.[24]

Also in Section 4.6.1.2, it was suggested that for the purpose of considering corrosion reactions the respective equilibrium diagrams could be divided into regions of *corrosion*, *immunity*, and *passivation*. The criterion for distinguishing between corrosion and immunity or between corrosion and passivation was arbitrarily (but certainly reasonably) taken to be a concentration of the soluble species of 10^{-6} g-at.wt./liter. This corresponds to 0.06 mg/liter for iron, copper, and zinc; it corresponds to 0.03 m/liter for aluminum and 0.20 mg/liter for lead.

4.6.4.1. Thermodynamic Basis—Nobility by Immunity; Nobility by Immunity and Passivation

A metal which does not corrode is considered to be noble. In practice, this nobility is generally rated with respect to whether a metal will corrode in water. From the point of view of the potential–pH diagrams, the metal will absolutely not corrode in solutions free of oxidizing substances, such as oxygen, at these pH regions where some portion of the region of immunity lines above the line (*a*) defining the H_2–H^+ equilibrium. Thus, as the region of immunity for a given metal lies progressively higher with respect to oxidation by water, the more *noble* is the metal.

It has already been noted that, in the region of passivation the corrosion in many cases is minimal and the metal tends not to corrode. Thus,

[24] Thanks to the collaboration of many contributors among whom I wish especially to mention Jean van Muyider and Nina de Zoubov, equilibrium potential–pH diagrams have been drawn for all the elements of the periodic system, at 25°C; these diagrams have been collected in M. Pourbaix *et al.*, *Atlas of Electrochemical Equilibria in Aqueous Solutions* (Translated from the French by J. A. Franklin), Pergamon Press, Oxford, and CEBELCOR, Brussels (1966). This atlas has been prepared with the collaboration of J. Besson, J. P. Brenet, W. G. Burgers, G. Charlot, P. Delahay, E. Deltombe, R. M. Garrels, T. P. Hoar, F. Jolas, W. Kunz, M. Maraghini, M. Moussard, R. Piontelli, A. L. Pitman, J. Schmets, K. Schwabe, G. Valensi, C. Vanleugenhaghe, J. van Muylder, P. van Rysselberghe, and N. de Zoubov.

if nobility is taken effectively as including both the region of immunity and passivation then the range of nobility is extended.

Figure 54 classifies *E*–pH diagrams for 43 metals and metalloids in order of their thermodynamic nobility due to immunity. Figures 37–52 have already designated some of these areas for selected metals. Figure 54 shows the respective areas of corrosion, immunity, and passivation by

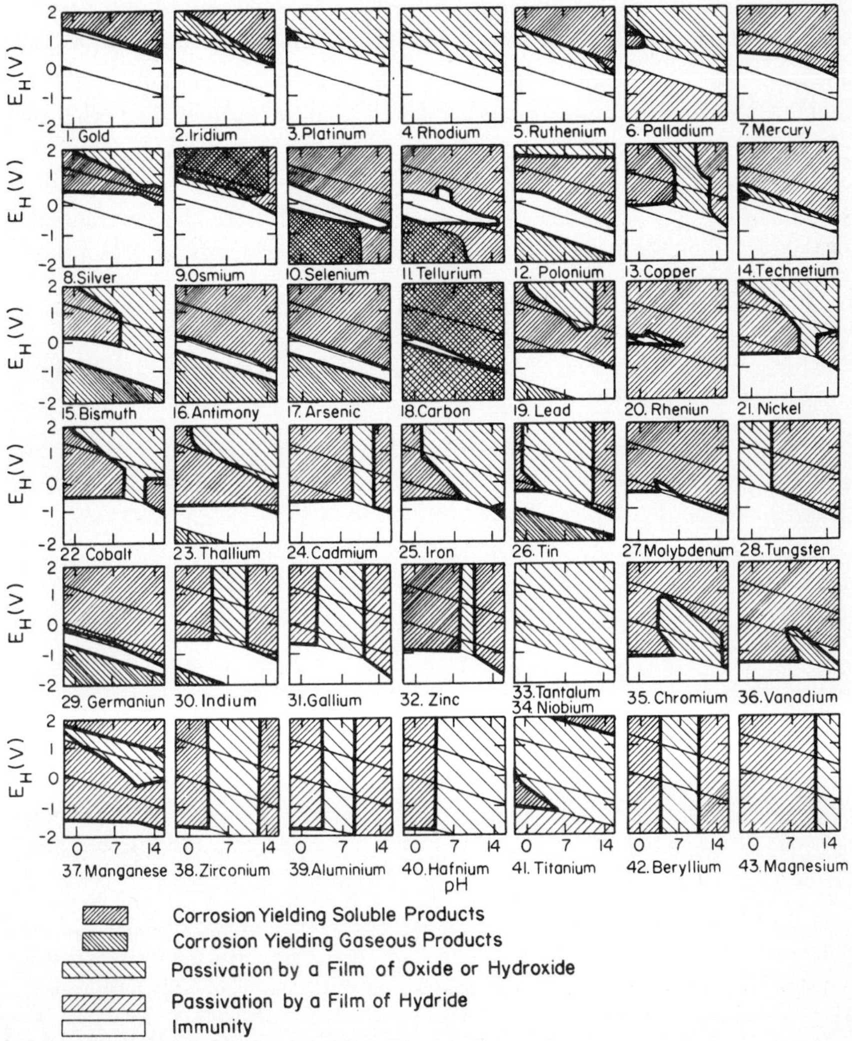

Figure 54. Corrosion, immunity, and passivation domains of metals and metalloids classified in order of nobility due to thermodynamic immunity.

various patterns of cross-hatching. The shaded areas represent the corrosion fields and were established by lines descending from right to left in the most frequent case of corrosion by dissolution and by lines descending from left to right in the relatively rare case of corrosion by gasification (e.g., by formation of OsO_4, H_2Se, H_2Te, CO_2, CH_4, AsH_3, SbH_3, BiH_3, GeH_4, SnH_4, PbH_2, InH, and TlH). The passivation fields are lightly hatched, the hatching descending from left to right in the frequent case of passivation by formation of oxide or hydroxide, and descending from right to left in the rare case of passivation by hydride formation (**Pd_2H, TiH_2**).[24] The white areas represent regions of immunity.

Table XIII shows the forty-three elements of Figure 54 arranged according to their order of nobility. Column A ranks the elements according to nobility which considers only the *region of immunity.* Column B ranks nobility based on *both* the *regions of immunity and passivation.*

According to the classification of column A in Table XIII, we may consider the first twenty-three elements (up to thallium), which have a region of immunity partially covering the domain of stability of water as having thermodynamic nobility due to immunity.

This classification given in column A, which is very similar to classical order established by W. Nernst based on values of "dissolution potentials," does not correspond well with the experimental facts; some metals like aluminum, titanium, niobium, and tantalum are much more resistant to corrosion than might be assumed both from classification A and the classification of Nernst. For the time being, leaving aside the kinetic factors and assuming as a first approximation that the passivating films considered in Figure 54 are perfectly protective, we may assume that *nobility* includes both *immunity* and *passivation.* The criterion for arranging column B of Table XIII is similar to that for column A: as the upper limit of the region of immunity plus passivation increases, the metal is more noble. This criterion is applied generally in the region of pH between 4 and 10 since these are the conditions most commonly encountered in practice.

It should be emphasized that both columns A and B of Table XIII are to be regarded as tentative subject to considerable revision, notably because the electrochemical equilibrium diagrams on which they are based are themselves approximations. In some cases drastic alterations (e.g., those for nickel and cobalt) may be required. Also, the corrosion and/or passivation reactions represented are sometimes strongly irreversible (e.g., for carbon).

In its present form, Table XIII brings out the considerable ennobling effect which passivation has on the following ten metals: niobium, tan-

TABLE XIII. Classifications of Metals and Metalloids in Order of Thermodynamic Nobility

	A (Immunity)	B (Immunity and passivation)	
	Noble metals		
1	Gold	Rhodium	1
2	Iridium	Niobium	2
3	Platinum	Tantalum	3
4	Rhodium	Gold	4
5	Ruthenium	Iridium	5
6	Palladium	Platinum	6
7	Mercury	Titanium	7
8	Silver	Palladium	8
9	Osmium	Ruthenium	9
10	Selenium	Osmium	10
11	Tellurium	Mercury	11
12	Polonium	Gallium	12
13	Copper	Zirconium	13
14	Technetium	Silver	14
15	Bismuth	Tin	15
16	Antimony	Copper	16
17	Arsenic	Hafnium	17
18	(Carbon)	Beryllium	18
19	Lead	Aluminium	19
20	Rhenium	Indium	20
21	(Nickel)	Chromium	21
22	(Cobalt)	Selenium	22
23	Thallium	Technetium	23
24	Cadmium	Tellurium	24
25	Iron	Bismuth	25
26	Tin	Polonium	26
27	Molybdenum	Tungsten	27
28	Tungsten	Iron	28
29	Germanium	(Nickel)	29
30	Indium	(Cobalt)	30
31	Gallium	Antimony	31
32	Zinc	Arsenic	32
33	Niobium	(Carbon)	33
34	Tantalum	Lead	34
35	Chromium	Rhenium	35
36	Vanadium	Cadmium	36
37	Manganese	Zinc	37
38	Zirconium	Molybdenum	38
39	Aluminium	Germanium	39
40	Hafnium	Vanadium	40
41	Titanium	Magnesium	41
42	Beryllium	Thallium	42
43	Magnesium	Manganese	43
	Non-noble metals		

TABLE XIV. Some Values of Calculated Equilibrium Potentials and of Experimental Electrode Potentials

	Calculated equilibrium potentials			Experimental electrode potentials		
	1	2	3	4	5	6
	Standard dissolution potentials[a] (W. Nernst) (pH = 0)	Thermodynamic immunity[b] (pH = 7.0)	Thermodynamic immunity and passivation[c] (pH = 7.0)	Natural sea water (J. A. Smith, R. E. Groover, T. J. Lennox, M. H. Peterson) (pH = 8.2)	Artificial sea water (J. Elze) (pH = 7.5)	Phthalate buffer (J. Elze) (pH = 6.0)
Be	−1.85	(−2.03)	B	—	—	—
Mg	−2.63	−2.45	−2.45	—	−1.355	−1.460
Al	−1.67	(−1.96)	B	−1.21	−0.667	−0.169
Ti	−1.63	(−1.72)	B	+0.45	−0.111	+0.181
Zr	−1.54	(−1.96)	B	+0.26	—	—
Hf	−1.70	(−2.09)	B	—	—	—
V	−1.17	−1.35	−0.30	+0.04	—	—
Nb	—	(−1.14)	B	+0.36	—	—
Ta	—	(−1.16)	B	+0.08	—	—
Cr	−0.91	(−1.10)	+0.55	—	—	—
Mo	−0.20	(−0.48)	−0.40	+0.08	—	—
W	—	−0.55	−0.55	+0.01	—	—
Mn	−1.18	−1.36	−1.36 — B	—	—	—
Tc	+0.40	(−0.14)	+0.10	—	—	—
Re	—	—	—	—	—	—
Fe	−0.44	−0.62	−0.62 — B	−0.50	−0.351?	−0.389
Co	−0.28	−0.46	−0.46	—	—	—

Ni	−0.25	−0.43	−0.43	+0.08 to +0.11	+0.046	+0.118
Ru	—	(+0.33)	B	—	—	—
Rh	+0.80	(+0.39)	B	—	—	—
Pd	+0.99	(+0.49)	B	+0.56	—	—
Os	—	(+0.28)	+0.46	—	—	—
Ir	—	(+0.52)	B	—	—	—
Pt	+1.19	(+0.57)	B	+0.55	—	—
Cu	+0.34	(+0.06)	+0.26	+0.11	+0.010	+0.140
Ag	+0.80	+0.43	+0.43	+0.25?	+0.149	+0.194
Au	+1.50	B	B	—	+0.243	+0.306
Zn	−0.94	−0.94	−0.94	−0.79	−0.794	−0.807
Cd	−0.40	−0.58	−0.58	−0.51	−0.519	−0.574
Hg	+0.79	+0.40	+0.40	—	—	—
Ga	−0.53	(−0.89)	B	—	—	—
In	−0.14	(−0.60)	B	−0.12	—	—
Tl	−0.34	−0.70	−0.70	—	—	—
Ge	0.00	(−0.70)	−0.66	—	—	—
Sn	−0.14	(−0.52)	B	−0.42	−0.184	−0.175
Pb	−0.13	−0.31	−0.31	—	−0.259	−0.283
As	—	−0.30	−0.30	—	—	—
Sb	—	−0.30	−0.30	—	—	—
Bi	+0.21	−0.10	−0.10	0.00	—	—
Se	—	+0.14	+0.14	—	—	—
Te	+0.57	+0.10	+0.40	—	—	—
Po	+0.65	+0.05	+0.05	—	—	—

[a] Equilibrium electrode potentials between the metal and 1 g-atom metal in solution (activity 1).

[b] Symbol B indicates that the metal is thermodynamically stable at the standard equilibrium potential for the system O_2–H_2O. This is the case for **Au**. The values of electrode potential written in parenthesis refer to equilibria between the metal and one of its oxides or hydroxides; other values refer to equilibria between the metal and 10^{-6} g-atom metal in solution.

[c] Symbol B indicates that the metal or one of its oxides or hydroxides is thermodynamically stable at the O_2–H_2O standard equilibrium potential. This is the case for **Au** (metal) and for **Be, Al, Ti, Zr, Hf, Nb, Ta, Mn, Fe, Ru, Rh, Pd, Ir, Pt, Ga, In, Sn** (oxides or hydroxides). The values of electrode-potential refer to equilibria between metal or oxides or hydroxides and 10^{-6} g-atom metal in solution.

talum, titanium, gallium, zirconium, hafnium, beryllium, aluminum, indium, and chromium. This trend in fact accounts for wide industrial application of these metals for corrosion resistance.

4.6.4.2. Actual Conditions of Corrosion and Noncorrosion of Metals

4.6.4.2.1. *Thermodynamic Nobility and Practical Nobility*

Although classification B generally conforms more closely with reality than classification A, it also must be accepted with some reservation. As Figures 37 to 54 demonstrate, the diagrams of thermodynamic stability are as different for each metal and metalloid as fingerprints are to us. It is scarcely possible to establish at one time a valid classification, for example, for acid, neutral, and alkaline media. Also for certain elements, most particularly *carbon*, the ranges of effective stability of the element are considerably extended beyond those indicated by their domains of immunity because of the great irreversibility of the oxidation and reduction reactions of the element. For other elements (*nickel*, *cobalt*) the thermodynamic data which are presently available are not sufficiently exact for the equilibrium diagrams presently derived to be considered as correct. In fact, these two metals are often rendered passive more easily than Figure 54 would lead one to predict. On the other hand as we have pointed out already on several occasions, the protectiveness of oxides of numerous metals, among which are *aluminum*, *iron*, and *chromium*, may be altered in the presence of solutions whose chloride content is greater than a certain critical value which varies according to the metal. Because of this chlorides often reduce the effectiveness of passivating oxides and thereby cause a lowering of the potential of the metal toward more negative, less "noble," more "anodic" values. Thus, the validity of classification B is only consistent for solutions relatively free from chlorides.

In Table XIV we give some values of calculated and experimental electrode potentials:

Column 1 gives the standard "dissolution potentials" after Nernst, which are the equilibrium electrode potentials between the metal and a solution containing 1 g-atom dissolved metal per liter (activity 1). These values are those used in our *Atlas*.

Column 2 gives the upper limit of the "immunity region," at pH = 7. Data in parentheses (for instance for **Be**) are the equilibrium potentials between the metal and one of its oxides. Data without parentheses (for instance for **V**) are the equilibrium potentials between the metal and a solution containing 10^{-6} g-atom/liter. Symbol B for **Au**) indicates that the metal itself is thermodynamically stable at the O_2–H_2O equilibrium potential, which is indicated by line *b* in our potential–pH charts; this potential B is +0.82 V_{SHE} at pH = 7.0.

Column 3 gives the upper limit of the "immunity + passivation region," at pH = 7.0. Symbol B indicates that an oxide (or the metal for **Au**) is stable at the O_2–H_2O equilibrium potential. When two data are given (**Mn, Fe**), there is at pH = 7.0 a corrosion region between a region of immunity and a region of passivation.

Column 4 gives experimental electrode potentials observed by Smith, Groover, Lennox, and Peterson[25] for 20 high-purity metals in natural sea water at the NRL's Marine Corrosion Research Laboratory at Key West, Florida (pH = 8.2, temperatures 19 to 24°C).

Column 5 gives experimental electrode potentials observed by J. Elze[26] in aerated artificial sea water (DIN 50.907) (pH = 7.5, temperature 25°C), for 12 technical metals.

Column 6 gives experimental electrode potentials observed by J. Elze for the same metals in an aerated artificial "chloride-free water" (phthalate buffer) (pH = 6.0, temperature 25°C).

Table XV shows the values given by J. Elze at Table XIV for the electrode potentials of 12 metals in aerated artificial sea water (column 4) and in aerated artificial "chloride-free water." According to this table, the electrode potentials of titanium and of aluminum are notably lower (less noble) in sea water than in chloride-free water.

In Table XVI, we have reproduced these two series of results of Elze (columns 4 and 6) as well as the results of Smith *et al.* (column 3), together with the upper limit shown in Table XIV for the potentials of "thermodynamic immunity" (column 2) and of "thermodynamic immunity and passivation" (column 5), at pH 7.0.

In the case of metals for which the stable form at the equilibrium potential between oxygen under 1 atm and water (lines *b* of the diagram shown at Figure 56) is the metal itself (**Au**) or a metallic oxide (**Ti, Al,** and **Sn**), this equilibrium potential between O_2 and H_2O (potential *B*) has

[25] J. A. Smith, R. E. Groover, T. J. Lennox, Jr., and M. H. Peterson, Marine corrosion studies. The electrochemical potential of high purity metals in sea water, NRL Memorandum Report 2187 (November, 1970).

[26] J. Elze and G. Oelsner, The potential series of the metals in practical corrodents, *Metalloberfläche* **12**, 129–133 (1958) (in German). See also J. Elze, *Werkstoffe und Korrosion* **10**, 737–738 (1959).

TABLE XV. Electrode Potentials of Several Metals in Stirred Solutions Saturated with Air (25°C, 1 atm)

(After J. Elze)

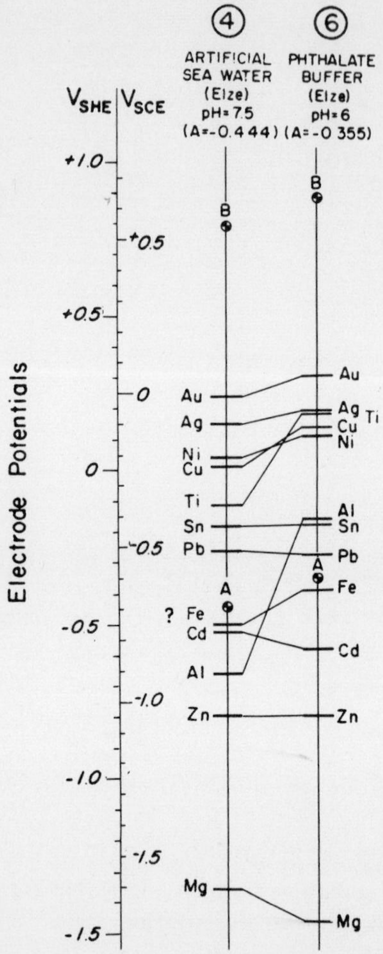

been considered as being this upper limit. This is shown in column 2 for **Au**, and in column 5 for **Au**, **Ti**, **Al**, and **Sn**. The potential *B* would be the highest possible electrode potential of a nonpolarized and uncorrodable metal or alloy in the presence of a water free from any reducing substance and saturated with oxygen under atmospheric pressure, in the ideal case where the oxygen reduction reaction $O_2 + 4H^+ + 4e^- = 2H_2O$ would be

TABLE XVI. Comparison Between the Classifications of Metals in Order of Thermodynamic Nobility and Practical Nobility

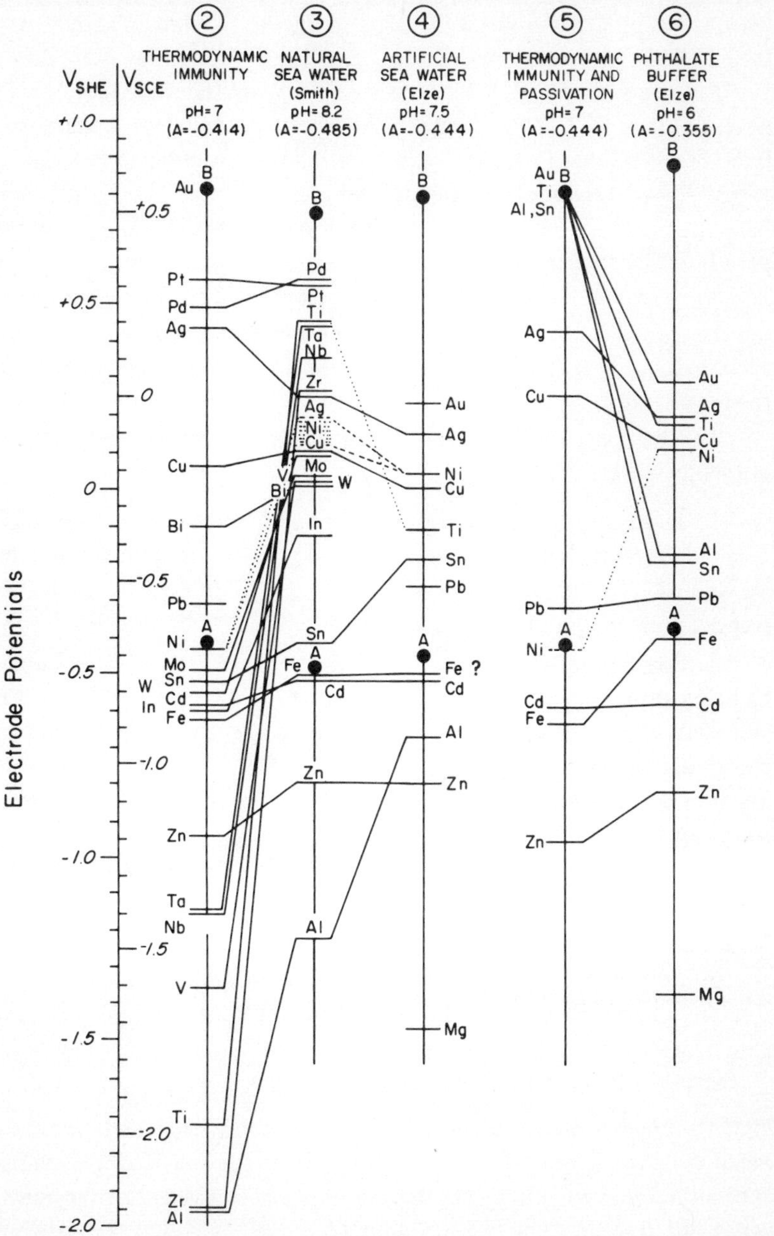

reversible. In fact, this oxygen reduction reaction is never reversible but may occur only with an overpotential of 0.20 to 0.40 V minimum. Thus the electrode potential at zero current of noncorroding metals and alloys in the presence of an aerated water free from reducing substances should be about 0.20 to 0.40 volt below the oxygen equilibrium potential B in this water. For the sea water of pH 8.2 considered by Smith *et al.* (column 3), the potential B is $+0.75\ V_{SHE}$; the zero-current potential of a metal or alloy fully uncorrodable in aerated sea water should thus be higher than about $+0.35\ V_{SHE}$; reciprocally, metals and alloys which, when nonpolarized, have an electrode potential higher than about $+0.35\ V_{SHE}$ in a water of pH about 8.2, as sea water, do generally not corrode in this water[27]; metals and alloys which have lower electrode potentials do generally corrode.

It is thus likely that, among the 19 metals considered in Smith's experiments, only **Pd**, **Pt**, **Ti**, **Ta**, and **Nb**, and perhaps also **Zr**, were corrosion resistant.

The left part of Table XVI (columns 2, 3, and 4) leads also to the following observations.

1. The results obtained by Smith *et al.* in natural sea water and by Elze in synthetic sea water are in fairly good agreement, except for **Ti**, **Sn**, and **Al**. These discrepancies seem to be due to differences in the chemical composition of the metals used in these two series of tests.

2. For nine of the metals considered in Figure 56 and in Table XIII as being nonpassivable in neutral water (**Pt**, **Pd**, **Ag**, **Cu**, **Bi**, **Pb**, **Cd**, **Fe**, and **Zn**), the experimental potential values are similar to the theoretical potentials of thermodynamic immunity. This is in agreement with the assumption that no passivating film exists on these metals under these conditions.

The experimental electrode potentials of **Fe**, **Cd**, and **Zn** are lower than the hydrogen equilibrium potential A; these three metals thus corrode with hydrogen evolution.

3. For six metals considered in Figure 54 and in Table XIII as passivable in neutral water (**Ti**, **Ta**, **Nb**, **Zr**, **In**, **Sn**), the experimental potential values in sea water (columns 3 and 4) are notably higher than the potentials of thermodynamic immunity (column 2). This denotes that a passivating film effectively exists on these metals.

[27] Let us add that such metals and alloys, which have the highest zero-current electrode potential in the presence of solutions saturated in oxygen are the best catalysts for electrochemical reduction of oxygen, or for oxidation with oxygen.

Aluminum, which has an electrode potential lower than *A*, corrodes with hydrogen evolution.

4. For **V**, **Mo**, and **W**, which have been considered in Figure 54 and in Table XIII as nonpassivable in neutral water, the electrode potentials in sea water (column 3) are higher than foreseen. It thus seems that passivating films are formed which are not shown in the equilibrium potential–pH charts.

As far as the right part of Table XVI is concerned (columns 5 and 6), the electrode potentials of the four theoretically most noble metals (**Au**, **Ti**, **Al**, and **Sn**) are less positive than would be expected (about +0.4 to +0.6 V_{SHE}) in an aerated water free of chloride and of reducing substances. This might perhaps be due to some reduction of oxygen by the phthalate.

For further information, and by kind permission of F. L. Laque, we reproduce in Table XVII a qualitative *galvanic series of metals and alloys in flowing sea water*, with approximate indication of the composition of the alloys.[28]

4.6.4.2.2. *Anodic Protection and Cathodic Protection*

In view of the above discussion, it is obvious that the notions of immunity and passivation as outlined here make it possible to define effectively the conditions of protection against corrosion.

As a general rule, the establishment of conditions of *immunity*, corresponding to the attainment of circumstances of thermodynamic stability of the metal, allows the metal to be protected *under conditions of complete safety, despite the action of chlorides*. For the first eight solid metals of classification A (Table XIII, i.e., for gold, for the six platinum metals and for silver) this state of protection generally is achieved without the aid of external action. For copper (No. 13) generally it is sufficient to avoid the presence of oxidants. For lead (No. 19) and iron (No. 25) artificial lowering of the potential by cathodic protection (protection by reduction) can bring the potential easily into the region of immunity and thereby stop corrosion.

Conversely, it has already been shown that metals often can be effectively protected if their potentials lie in the region of passivation. Thus,

[28] F. L. Laque, Private communication July 14, 1966. F. L. Laque and G. L. Cox, Some observations on the potentials of metals and alloys in sea water, *Proc. ASTM* **40**, 670–689 (1940), F. L. Laque, Marburg lecture, *Proc. ASTM* **51**, 495 (1951); F. L. Laque and H. R. Copson, Corrosion resistance of metals and alloys (2nd ed.), ACS Monograph 156, Reinhold, New York (1963), Table 4.4, p. 98. See also F. L. Laque, Materials selection for ocean engineering, Lectures at UCLA (1966) (published by John Wiley, New York, 1972).

TABLE XVII. Galvanic Series of Metals and Alloys in Flowing Sea Water
(After F. L. Laque) (with approximate compositions of alloys)

Noble metals (cathodic)	
Platinum	Pt
Carbon (graphite)	C
316 stainless steel (passive)	(18Cr, 12Ni, 3Mo)
Monel	(66Ni, 29Cu, 2.8Al, 0.9Fe, 0.4Mn)
Hastelloy C	(50Ni, 17Mo, 16.5Cr, 2.5Co, 7Fe, 1Si, 1Mn, 0.3V, 0.08C)
304 stainless steel (passive)	(18Cr, 8Ni)
Titanium	Ti
Silver	Ag
Nickel–aluminum bronze	(78/81Cu, 4.5/5.5Ni, 9/10Al, 3.5/5.5Fe, 0.5/1Mn, 0.01Pb)
Inconel	(78Ni, 13.5Cr, 6Fe)
Nickel	Ni
Cupronickel	(70Cu, 30Ni)
Cupronickel	(90Cu, 10Ni)
Admiralty brass	(70/73Cu, 26/28Zn, 0.9/1.2Sn, 0.07Pb, 0.06Fe, 0.02/0.1As, Sb or P)
M bronze	86.3Cu, 4.7Sn, 4.8Zn, 3.9Pb
G bronze	88Cu, 10Sn, 2Zn
Aluminum brass	(76/79Cu, 18/22Zn, 1.8/2.5Al, 0.07Pb, 0.06Fe, 0.02/0.1As)
Red brass	(85Cu, 15Zn)
Silicon bronze	(85Cu, 5Si, 5Zn, 2.5Fe, 1.5Al, 1.5Mn, 1Sn, 0.5Pb)
Copper	Cu
Yellow brass	(65Cu, 35Zn)
Naval brass	(60Cu, 39Zn, 1Sn)
Manganese bronze	(58/65Cu, 23/39Zn, 1.4/3Fe, 0.1/3.7Mn, 0/1Sn, 0/4.5Al)
Muntz metal	(59/63Cu, 36/40Zn, 0.3Pb, 0.07Fe)
Tin	Sn
Lead	Pb
Ni resist	(65/75Fe, 13.5/17.5Ni, 5.5/7.5Cu, 3C, 1.0/2.8Si, 1.0/1.5Mn, 1.0/2.5Cr)
316 stainless steel (active)	(18Cr, 12Ni, 3Mo)
304 stainless steel (active)	(18Cr, 8Ni)
430 chromium steel	(17Cr)
410 chromium steel	(13Cr)
Cast iron	—
Wrought iron	—
Mild steel	—

TABLE XVII (continued)

Aluminum 2117 rivet alloy	(95Al, 2.2 to 3.0Cu, 0.3Mg, <0.8Si, <1.0Fe, <0.2Mn, <0.1Cr, <0.25Zn)
Cadmium	Cd
Aluminum 3003	(95Al, 2.2/3.0Cu, 0.8Si, 0.2/0.5Mg, 0.2Mn, 1.0Fe, 0.2Zn, 0.1Cr)
Aluminum 1100	(99Al, 1Fe + Si, 0.2Cu, 0.1Zn, 0.05Mn)
Aluminum 6061	(96/97Al, 0.1/0.4Cu, 0.4/0.8Si, 0.8/1.2Mg, 0.1Mn, 0.1Zn, 0.1/0.3Cr)
Aluminum 356	(91/92Al, 6.5/7.5Si, 0.2Cu, 0.2/0.4Mg, 0.1Mn, 0.5Fe, 0.2Zn, 0.2Ti)
Aluminum 5052	(96/97Al, 0.1Cu, 0.45Fe + Si, 22/2.8Mg, 0.1Mn, 0.1Zn, 0.1/0.3Cr)
Aluminum 5086	(93/96Al, 0.1Cu, 0.4Si, 3.5/4.5Mg, 0.2/0.7Mn, 0.5Fe, 0.2Zn, 0.1/0.2Cr, 0.1Ti)
Aluminum 5456	(93/94Al, 0.1Cu, 0.4Fe + Si, 4.7/5.5Mg, 0.5/1Mn, 0.2Zn, 0.1/0.2Cr, 0.2Ti)
Aluminum 7072 (cladding alloy)	(98Al, 0.1Cu, 0.7Fe + Si, 0.1Mg, 0.1Mn, 0.8/1.3Zn)
Galvanized steel	(Zinc coating; max. 1.6Pb, 0.08Fe, 0.75Cd, 0.01Al)
B 605 aluminum anode alloy	>94.2Al, 5.5Zn, <0.15Fe, <0.15Si, <0.02Cu
Zinc	Zn
CB-75 aluminum anode alloy	>92.5Al, 7.0Zn, 0.12Sn, 0.05B, <0.01Cu, <0.10Fe, <0.10Si, <0.10Ni
Magnesium and magnesium alloys	Mg
Non-noble metals (anodic)	

if the potential normally lies in a region of corrosion, the metal can be protected by adjusting the potential artificially into the region of passivation. This procedure is generally called "anodic protection," for such a protection generally implies an anodic (or oxidizing) action. While this procedure is often effective, *it must be used with extreme caution because the film may not be entirely protective due either to the presence of chloride or the absence of complete inherent protectiveness.*

Relatively recent work[29] has shown that in the presence of chloride solutions, the protective power of passive films is often closely associated with given values of electrode potential. On account of this it will be possible to protect cathodically certain passivable metals and alloys (ordinary steels, alloy steels, and aluminum and its alloys) without it being necessary

[29] See *Rapports Techniques CEBELCOR* RT. 103, 104, 105, 120 (1962), and 157 (1959). See also *Corr. Sci.* **3**, 239–259 (1963).

to lower the potential to the region of immunity, which in the case of aluminum is not attainable.

A relatively slight lowering of the potential is then sufficient to reach a "protection potential against pitting" below which previously existing pits and crevices become harmless, mainly because they cease to be acid. The value of this "protection potential against pitting," which has a great technical importance, may be determined in every case by adequate electrochemical experiments. In this way, complete passivation is assured even in the presence of chlorides and of some other activating anions. These conditions of cathodic protection by "perfect passivation" will be considered in Section 6.6.6.1.2.

4.6.5. Resistance of Metals to Pure Water

Generally speaking, the metals showing perfect resistance to pure water at a temperature in the region of 25°C are those having an equilibrium diagram on which the perpendicular from pH = 7 crosses only the immunity or passivation field (and no corrosion field) at potentials between −0.8 V and +0.7 V (see Figure 55). These potentials are the limiting values in the normal environments to which construction metals are exposed. With some reservations which we have pointed out and apart from the six metals in the platinum group, the following thirteen metals fall into this category: beryllium, aluminum, gallium, indium, tin, silver, gold, titanium, zirconium, hafnium, niobium, tantalum, and chromium.

For the above-mentioned metals which have a region of corrosion in the neutral region at very low potentials (tin, titanium), it is particularly important to avoid all reducing action and to ensure that the surface has a good polish. Other metals (chromium and silver, and also possibly gold and titanium) have a region of corrosion at relatively high potentials and slight dissolution will be observed in the presence of a highly oxidizing environment. Chlorides should be avoided for most passivable metals, especially aluminum.

4.6.6. Metals Which Can Be Passivated and Activated

Generally speaking, the metals which at a given pH can be passivated by oxidation and activated by reduction are those which have a higher oxide less soluble than a lower oxide and thus present a triangular corrosion field (see Figure 56). The lower the apex of this triangle is in the diagram, the easier it is to passivate the metal by oxidation. This applies to the nine

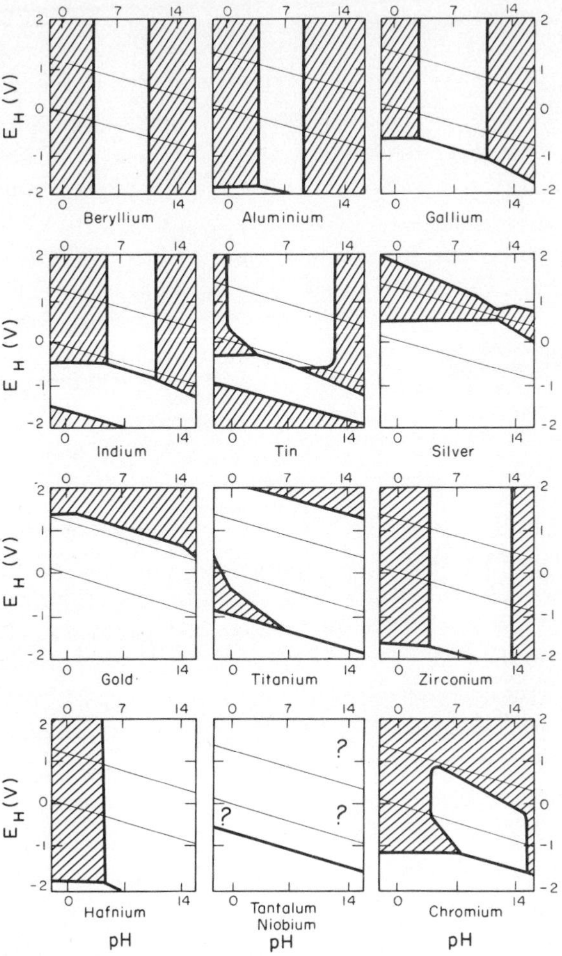

Figure 55. Metals resistant to pure water. The hatched regions indicate theoretical corrosion domains. The nonhatched regions indicate theoretical immunity and passivation domains.

following metals, in descending order of passivability: titanium, chromium, tin, iron, manganese, lead, silver, nickel, and cobalt.

For an extensive pH range, the passivation field of the first three metals (**Ti, Cr, Sn**) is situated partly below the stability field of water. For these metals passivation is very easy and occurs more often than not, spontaneously, even in the absence of an oxidizing agent. Iron passivation requires a relatively strong oxidizing action outside pH 9–13 but it tends

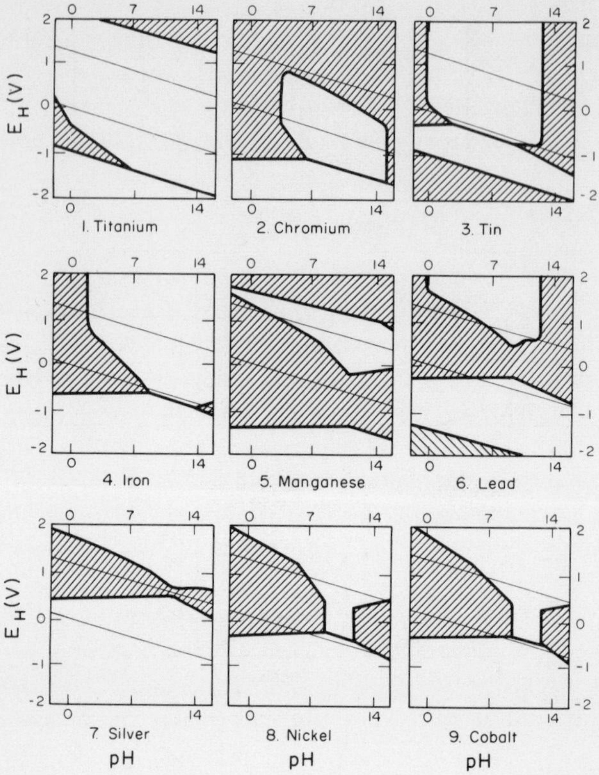

Figure 56. Passive and active metals. (See caption for Figure 55.)

to be self-protective within this range. Passivation of the five remaining metals by the formation of an oxide layer is only possible by means of a very strong oxidizing action (e.g., an anodization comparable to that used for charging lead batteries). It must, however, be understood that in spite of the difficulty of passivation, silver generally has a good corrosion resistance owing to the size of its immunity field, and that, for reasons which have as yet been only insufficiently clarified, nickel (and perhaps cobalt too) can often be passivated at lower potentials than Figure 56 leads us to anticipate.

Titanium, chromium, and tin which are normally passive can only be activated at low potentials, i.e., with relative difficulty. Activation of iron is difficult for pH values between 9 and 13 and easy outside this range. The other five metals (manganese, lead, silver, nickel, and cobalt) are very easily activated, it being nevertheless understood that, whereas activation always brings about corrosion of manganese (by no means a noble metal, the immunity field of which exists only at very low electrode potentials), activation of the other four metals only causes corrosion if there is present a certain oxidizing agent of variable intensity according to the metal (maintaining the electrode potential of the metal within its corrosion field).

4.6.7. Oxidizing Corrosion Inhibitors

Figure 42 shows that iron has a relatively wide triangular range of pH (below 9) in which corrosion can occur. Based on the discussions above, it is clear that iron can be protected if the potential is raised to the region of passivation. In Section 4.6.4.2.2 it was shown that the potential could be raised to the region of passivation by artificial means. The potential can also be raised by adding to the environment certain oxidizing species which will also raise the potential into the region of passivation.

These oxidizing species, which can raise the potential into the region of passivation, are often called "oxidizing inhibitors" because they prevent corrosion by raising the potential. Protection is particularly effective if, for the potential and pH conditions corresponding to the "corrosion fields" of the iron, the oxidizing inhibitor can be reduced together with formation of a solid. By being deposited on the "weak points" of the iron surface, this solid reduced product will improve the protective effect of the passive oxide films.

The behavior of oxidizing inhibitors can be predicted approximately by superimposing their potential–pH diagrams on the theoretical "corrosion–immunity–passivation diagram" for iron (in the case of Figure 57) to be protected. The equilibrium diagram for the oxidizing inhibitor shows (a) the relevant equilibrium potential for the reduction reaction and (b) the range of pH in which the reduced products are soluble or solid (protective). The action of various possible oxidizing inhibitors is shown in Figure 57 where the effects have been calculated for 0.01 M solutions of the following twelve oxidizing substances: oxygen, hydrogen peroxide, permanganates, hyperosmates, pertechnetates, chromates, molybdates, tungstates, vanadates, selenates, arsenates, and antimonates.

In Figure 57 a thick line indicates the potentials below which the inhibitor can be reduced; the shaded sections represent those parts of the corrosion fields of iron for which the products of the reduction of the oxidizing agent are not solids. If, as a first approximation, we accept that the reduction reactions of the oxidizing agent are reversible and that the solids formed in this reduction constitute a protective coating on the iron,

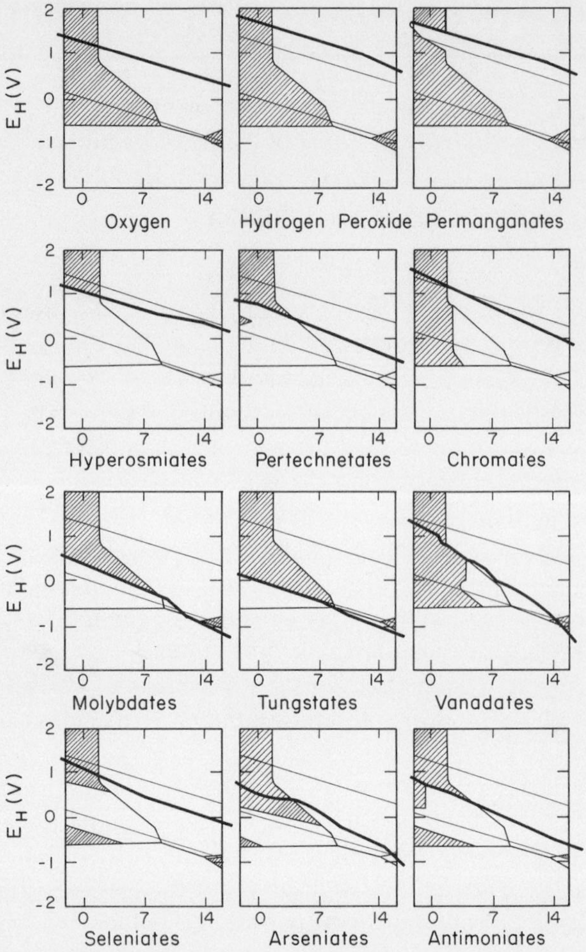

Figure 57. Oxidizing corrosion inhibitors.

then the shaded sections in Figure 57 show the theoretical conditions of corrosion, the nonshaded areas representing the theoretical areas of noncorrosion.

According to this figure, hyperosmates and pertechnetates are extremely effective inhibitors of iron corrosion, which coincides with Cartledge's observations. It is well known that chromates, although less effective, are nevertheless very good inhibitors; molybdates and tungstates are effective only to a small extent; vanadates are hardly effective at all. On the other hand, the last three diagrams of Figure 57 show that selenates,

TABLE XVIII. Equilibrium Potentials of Some Reference Electrodes

Electrode	Reaction	E_0 (25°C) (V)	$\frac{dE_0}{dt}$ (V/°C)
Standard hydrogen	$H_2 = 2H^+ + 2e^-$	0.0000	0.0×10^{-4}
0.1 M calomel	$2\mathbf{Hg} = Hg_2^{++} + 2e^-$	+0.3337	-0.7×10^{-4}
1.0 M calomel	$2\mathbf{Hg} = Hg_2^{++} + 2e^-$	+0.2800	-2.4×10^{-4}
Saturated calomel	$2\mathbf{Hg} = Hg_2^{++} + 2e^-$	+0.2415	-7.5×10^{-4}
0.1 M silver chloride	$\mathbf{Ag} = Ag^+ + e^-$	+0.2881	-6.5×10^{-4}
Saturated copper sulfate	$\mathbf{Cu} = Cu^{++} + 2e^-$	+0.30	—

arsenates, and antimonates may act as inhibitors in solutions where corrosion with hydrogen evolution is normally expected (sulfuric and hydrochloric acids). The three first diagrams provide data on the action of oxygen, hydrogen peroxide, and permanganates, which have been discussed earlier in Figures 1, 3, 9, and 10.

4.6.8. Reference Electrodes[30]

4.6.8.1. Introduction

In Section 4.3.1 a reference electrode was defined as an electrode on which the state of equilibrium of a given reversible electrochemical reaction is permanently secured under constant physicochemical conditions. Listed in Table XVIII are six commonly used reference electrodes including their electrochemical reactions, their equilibrium potential at 25°C, and their respective temperature coefficients.[31]

According to Eqs. (3.9) and (4.19) from Chapters 3 and 4 respectively, the equilibrium constant and standard equilibrium potential of the reference electrode equilibria above can be calculated from the standard free en-

[30] See M. Pourbaix, Reference electrodes, *Rapports Techniques CEBELCOR* **109**, RT. 158 (1969) (in French).

[31] A. J. de Bethune, Electrode potential–temperature coefficients, *The Encyclopedia of Electrochemistry*, Ed. J. Hampel, Reinhold, New York (1964), pp. 432–434.

[32] Standard free enthalpies of formation 25°C, *Rapport Technique CEBELCOR*, RT. 87 (1960).

TABLE XIX. Values of Molar Standard Free Enthalpies of Formation μ^0 (cal)

Liquid water and dissolved substances		Liquid substance		Solid substance		Gaseous substance	
H_2O	$-$ 56,690	*Hg*	0	Hg_2Cl_2	$-$ 50,350	H_2	0
H^+	0			KCl	$-$ 97,592		
Hg^{++}	$+$ 36,350			Ag	0		
Ag^+	$+$ 18,430			AgCl	$-$ 26,224		
K^+	$-$ 67,460			$CuSO_4 \cdot 5H_2O$	$-$ 449,300		
Cl^-	$-$ 31,350						
Cu^{++}	$+$ 15,530						
SO_4^{--}	$-$ 177,340						

thalpies of reaction. In the sections below, the equilibrium potentials of these above reference electrode equilibria will be calculated using the following free enthalpies[32] of formation already described (see Table XIX).

The first series of calculations below are performed assuming ideal solutions where the concentrations and activities are taken as being the same. In a second series of calculations we have taken into account the activity coefficients calculated using the two Debye–Hückel equations.

4.6.8.2. Calculations Without Activity Coefficients

4.6.8.2.1. *Standard Hydrogen Electrode*

As already noted in Section 4.3.3.3, the equilibrium equation for the reaction, $H_2 = 2H^+ + 2e^-$ is

$$E_0 = E_0{}^0 - 0.0591\ \text{pH} - 0.0295 \log p_{H_2} \tag{4.20}$$

where

$$E_0{}^0 = \frac{2\mu^0_{H^+} - \mu^0_{H_2}}{23{,}050 \times 2} = \frac{0 - 0}{46{,}120} = 0.000\ \text{V}$$

Under the standard conditions where pH = 0 and $p_{H_2} = 1$ atm the equilibrium potential is

$$E_0{}^0 = 0.000\ \text{V}$$

As is well known, this value is the zero for the electrochemical scale.

4.6.8.2.2. *Calomel Electrodes*

The equilibrium condition of the reaction $2\boldsymbol{Hg} = Hg_2^{++} + 2e^-$ is

$$E_0 = E_0^0 + 0.0295 \log [Hg_2^{++}] \text{ V}$$

where

$$E_0^0 = \frac{\mu^0_{Hg_2^{++}} - 2\mu^0_{Hg}}{13{,}050 \times 2} = \frac{+\ 36.350}{46{,}120} = +\ 0.788 \text{ V}$$

giving

$$E_0 = +\ 0.788 + 0.0295 \log [Hg_2^{++}] \text{ V} \tag{4.21}$$

The mercurous ion concentration is adjusted by adding mercurous chloride (calomel or Hg_2Cl_2). Here the dissociation is $\mathbf{Hg_2Cl_2} = Hg_2^{++} + 2Cl^-$ but the calomel is only slightly soluble and the condition of equilibrium is given:

$$\log [Hg_2^{++}] + 2 \log [Cl^-] = -\ \frac{\mu^0_{Hg_2^{++}} + 2\mu^0_{Cl^-} - \mu^0_{Hg_2Cl_2}}{1363}$$

$$= -\ \frac{+\ 36{,}350 - 62{,}700 + 50{,}350}{1363}$$

$$= -\ \frac{24{,}000}{1363} = -\ 17.60$$

which gives

$$\log[Hg_2^{++}] = -17.60 - 2 \log[Cl^-]$$

Combining the calomel equilibrium with the *Hg*–Hg_2^{++} equilibrium produces an equilibrium potential which is seen to depend on the chloride ion activity. On this basis there are three types of calomel reference electrodes in use according to the chloride ion activity (which is usually obtained by KCl solutions): 0.1 M, 1.0 M, and saturated KCl. The standard potentials for these three variants are calculated as follows. In the case of the two reference electrodes with 0.1 M and 1.0 M in KCl one has respectively

for 0.1 M

$$\log[Cl^-] = -1 \qquad \text{and} \qquad \log[Hg_2^{++}] = -15.60$$

for 1.0 M

$$\log[Cl^-] = \ 0 \qquad \text{and} \qquad \log[Hg_2^{++}] = -17.60$$

Using these values of $\log[Hg_2^{++}]$ in formula (4.21),

$$E_0 = +0.788 + 0.0295 \log[Hg_2^{++}] \text{ V}$$

one obtains

for 0.1 M

$$E_0 = +0.788 - 0.460 = +\mathit{0.328}\ \text{V}$$

for 1.0 M

$$E_0 = +0.788 - 0.519 = +\mathit{0.296}\ \text{V}$$

In the third case, a saturated KCl solution, the activities of K^+ and Cl^- in the solution must satisfy the equilibrium conditions of the dissociation **KCl** $= K^+ + Cl^-$. This condition is

$$\log [K^+] + \log [Cl^-] = -\frac{\overset{0}{\mu}_{K^+} + \overset{0}{\mu}_{Cl^-} - \overset{0}{\mu}_{KCl}}{1363}$$

$$= -\frac{-67{,}460 - 31{,}350 + 97{,}592}{1363}$$

$$= +\frac{1218}{1363} = +0.893$$

For a solution prepared by dissolution of pure **KCl** in pure water, the activities in both K^+ and Cl^- are equal and their common value is

$$\log[K^+] = \log[Cl^-] = 0.893 \times \tfrac{1}{2} = 0.4465$$

which corresponds to an activity of 2.80 moles KCl (209 g)/kg H_2O.

This leads, for a calomel electrode saturated in KCl to the following value of the activity in Hg_2^{++}

$$\log[Hg_2^{++}] = -17.60 - 2\log[Cl^-] = -17.60 - (0.4465 \times 2) = -18.49$$

and to the following values of the electrode potential

$$E_0 = +0.788 + 0.0295 \log[Hg_2^{++}] = +0.788 - (0.0295 \times 18.49)$$
$$= +0.788 - 0.546 = +\mathit{0.242}\ \text{V}$$

4.6.8.2.3. *Silver Chloride Electrodes*

The silver–silver chloride electrode is similar in many ways to the calomel electrode. The equilibrium condition of reaction **Ag** $= Ag^+ + e^-$ is

$$E_0 = E_0^0 + 0.0591 \log [Ag^+]$$

where

$$E_0^0 = \frac{\overset{0}{\mu}_{Ag^+} - \overset{0}{\mu}_{Ag}}{23{,}060} = \frac{+18{,}430}{23{,}060} = +0.799\ \text{V}$$

This leads to

$$E_0 = +0.799 + 0.0591 \log[Ag^+] \tag{4.22}$$

The activity of the Ag^+ is defined by using silver chloride which has a relatively low solubility and where the equilibrium condition is calculated as follows:

$$\log [Ag^+] + \log [Cl^-] = -\frac{\overset{0}{\mu}_{Ag^+} + \overset{0}{\mu}_{Cl^-} - \overset{0}{\mu}_{AgCl}}{1363}$$

$$= -\frac{+18{,}430 - 31{,}350 + 26{,}224}{1363}$$

$$= -\frac{13{,}304}{1363} = -9.76$$

giving

$$\log[Ag^+] = -9.76 - \log[Cl^-]$$

Thus, as for the calomel electrode, the electrode potential of the silver chloride electrode may be adjusted by varying the chloride activity. Assuming that concentrations and activities are the same, the potentials corresponding to the various chloride concentrations are as follows:

for 0.001 M

$$\log[Cl^-] = -3 \quad \log[Ag^+] = -6.76 \quad E_0 = +0.799 - 0.398 = +\mathit{0.401}\text{ V}$$

for 0.010 M

$$\log[Cl^-] = -2 \quad \log[Ag^+] = -7.76 \quad E_0 = +0.799 - 0.457 = +\mathit{0.342}\text{ V}$$

for 0.100 M

$$\log[Cl^-] = -1 \quad \log[Ag^+] = -8.76 \quad E_0 = +0.799 - 0.517 = +\mathit{0.282}\text{ V}$$

4.6.8.2.4. *Copper Sulfate Electrode*

The equilibrium condition of reaction $\mathbf{Cu} = Cu^{++} + 2e^-$ is

$$E_0 = E_0^0 + 0.0295 \log [Cu^{++}]$$

where

$$E_0^0 = \frac{\overset{0}{\mu}_{Cu^{++}} - \overset{0}{\mu}_{Cu}}{23{,}060 \times 2} = \frac{+15{,}530}{46{,}120} = +0.337 \text{ V}$$

This leads to

$$E_0 = +0.337 + 0.0295 \log [Cu^{++}] \tag{4.23}$$

As the solution of a copper sulfate electrode is saturated in $\mathbf{CuSO_4 \cdot 5H_2O}$ the activities of Cu^{++} and SO_4^{--} in this solution must satisfy the equilibrium condition of the dissolution reaction $\mathbf{CuSO_4 \cdot 5H_2O} = Cu^{++} + SO_4^{--} + 5H_2O$. This condition is

$$\log [Cu^{++}] + \log [SO_4^{--}] = -\frac{\mu^0_{Cu^{++}} + \mu^0_{SO_4^{--}} + 5\mu^0_{H_2O} - \mu^0_{CuSO_4 \cdot 5H_2O}}{1363}$$

Assuming that $\mu^0_{H_2O}$ is here equal to the μ^0 for pure water (−56.690 cal), which is certainly not true in such a concentrated solution, one has

$$\log [Cu^{++}] + \log [SO_4^{--}] = -\frac{15{,}530 - 177{,}340 - 285{,}450 + 449{,}300}{1363}$$

$$= -\frac{4040}{1363} = -2.964$$

For a solution prepared by dissolution of pure $\mathbf{CuSO_4 \cdot 5H_2O}$ in pure water, if one neglects the hydrolysis reaction $Cu^{++} + 2H_2O \rightarrow \mathbf{Cu(OH_2)} + 2H^+$, which leads to the formation of $\mathbf{Cu(OH)_2}$ and of H^+ ions,[33] the activities of Cu^{++} and SO_4^{--} are approximately equal; their common value is

$$\log[Cu^{++}] = \log[SO_4^{--}] = -2.964 \times \tfrac{1}{2} = -1.482$$

This corresponds to an activity of 0.033 moles $CuSO_4$/kg water. This leads to the following value of the electrode potential of the saturated copper sulfate electrode:

$$E_0 = +0.337 - (0.0295 \times 1.482) = +\mathit{0.293}\ \mathrm{V}$$

4.6.8.3. Calculations Using Activity Coefficients

Calculations in Sections 4.6.8.2 have assumed ideal conditions wherein concentrations and activities have been interchangeable. While this assumption is reasonable for very dilute solutions, serious errors are often incurred for more concentrated solutions; and, in this case, it may be necessary to consider corrections made by activity coefficients.

[33] We have observed the following values of the pH of solutions of $\mathbf{CuSO_4 \cdot 5H_2O}$ at 23°C:

0.001 M	5.25
0.010	4.90
0.100	4.25
1.000	3.30
1.43 (sat)	3.30

The "activity," a, of a dissolved species is defined as a "corrected concentration" and is related to the concentration, c, by $a = f \times c$ where f is the activity coefficient.

In order to obtain a quantitative relationship for the activity coefficient, it is convenient to introduce the concept of "ionic strength," μ,[34] of a solution. The ionic strength is defined by the relation

$$\mu = \tfrac{1}{2} \sum_{\gamma} Z_{\gamma}^{2} c_{\gamma} \tag{4.24}$$

where Z_{γ} is the charge of each ion, γ, in the solution and c_{γ} is the concentration of these ions (in g-ion/1000 g of solvent). The use of ionic strength to calculate activity coefficients depends on the relative concentration of species and is divided into three special cases.

1. When $\mu < 0.02$, the activity coefficient of an ion may be calculated by Eq. (4.25) which is the "limit law of Debye–Hückel"[35]:

$$\log f_{\gamma} = -\, 0.509\, Z_{\gamma}^{2} \sqrt{\mu} \tag{4.25}$$

2. When $0.02 < \mu < 0.2$, the activity coefficient may be calculated by Eq. (4.26) which is the "equation of Debye–Hückel" for appreciable concentrations

$$\log f_{\gamma} = -\, \frac{0.509\, Z_{\gamma}^{2} \sqrt{\mu}}{1 + 0.33a\, 10^{8} \sqrt{\mu}} \tag{4.26}$$

where a is the mean diameter of the considered ion expressed in cm. If a is taken as 3×10^{-8} Eq. (4.26) may be expressed as

$$\log f_{\gamma} = -\, \frac{0.509\, Z_{\gamma}^{2} \sqrt{\mu}}{1 + \sqrt{\mu}} \tag{4.27}$$

3. When $\mu > 0.2$ Eqs. (4.25) and (4.27) cease to be applicable, and activity coefficients may not be calculated *a priori*.

In order to illustrate the application of Eqs. (4.25) and (4.27), we have calculated in Table XX, with respect to various chloride concentrations in the **Ag–AgCl** reference electrode, the ionic strength and corresponding values of activity coefficients and associated activities.

[34] The ionic strength, μ, is not to be confused with the chemical potential, μ.

[35] S. G. Glasstone, *Introduction to Electrochemistry* (3rd ed.), Van Nostrand, New York (1947), pp. 143–146.

TABLE XX. Calculation of the Activities of Chloride Ions Cl^- in Hg–Hg_2Cl_2, Ag–AgCl, and Cu–$CuSO_4 \cdot 5H_2O$ Reference Electrodes

Concentration (molality)	Ionic strength μ	log f		log a		Activity coefficient f		Activity a	
		Form (6)	Form (8)	Form (6)	Form (8)	Form (6)	Form (8)	Form (6)	Form (8)
Solutions of calomel and Ag–AgCl⁻ *electrodes*									
0.001	0.001	−0.0163	−0.0158	−3.015	−3.016	0.963	0.964	0.000963	0.000964
0.010	0.010	−0.0509	−0.0463	−2.051	−2.046	0.890	0.899	0.00890	0.00899
0.100	0.100	−0.163	−0.123	−1.163	−1.123	0.687	0.753	0.0687	0.0723
1.000	1.000	−0.509	−0.255	−0.509	−0.255	0.310	0.556	0.310	0.556
4.80 (sat cal)	4.80	−1.115	−0.349	−0.432	−0.332	0.077	0.448	0.37	2.15
Solution of Cu–$CuSO_4$ *electrode*									
1.43 (sat)	5.72	−4.87	−1.43	−4.70	−1.276	0.000014	0.037	0.000020	0.053

When using the two following expressions for the activities of Hg_2^{++} and Ag^+ ions (see Sections 4.8.8.2.2 and 4.8.8.2.3),

Calomel electrode

$$\log[Hg_2^{++}] = -17.60 - 2\log[Cl^-]$$

Silver chloride electrode

$$\log[Ag^+] = -\ 9.76 - \log[Cl^-]$$

the values of activities of chloride ions, Cl^-, given in Table XX lead to the values given in Table XXI for the activities of the Hg_2^{++} and Ag^+ ions. The activities of Hg_2^{++} and Ag^+ ions from Table XXI may now be used in equations for the electrode potential as follows:

Calomel electrode

$$E_0 = +0.788 + 0.0295\log[Hg_2^{++}]\ \mathrm{V} \tag{4.21}$$

Silver chloride electrode

$$E_0 = +0.799 + 0.0591\log[Ag^+]\ \mathrm{V} \tag{4.22}$$

Table XXII shows electrode potentials calculated using these two equations corresponding to a series of chloride concentrations.

TABLE XXI. Logarithms of the Activities Associated with Calomel and Silver Chloride Electrodes

Electrode	Concentration (M)	Form (6)	Form (8)
Calomel		$\log[Hg_2^{++}]$	
	0.001	−11.57	−11.57
	0.010	−13.47	−13.51
	0.100	−15.27	−15.32
	1.000	−16.58	−17.09
	4.80	−16.74	−18.26
Silver chloride		$\log[Ag^+]$	
	0.001	−6.74	−6.74
	0.010	−7.25	−7.71
	0.100	−8.60	−8.62
	1.000	−9.33	−9.51

TABLE XXII. Electrode Potentials Related to Chloride Concentrations for Calomel and Silver Chloride Reference Electrodes

Electrode	Concentration (M)	Electrode potential E_0 (V_{SHE})	
		Form (6)	Form (8)
Calomel	0.001	+0.447	+0.447
	0.010	+0.397	+0.389
	0.100	+0.338	+0.336
	1.000	+0.299	+0.284
	4.80	+0.294	+0.249
Silver chloride	0.001	+0.401	+0.401
	0.010	+0.371	+0.344
	0.100	+0.292	+0.290
	1.000	+0.248	+0.237

4.6.8.4. Comparison Between Calculated and Measured Electrode Potentials of Reference Electrodes

In Table XXIII we have summarized the electrode potentials of different calomel, silver chloride, and copper sulfate reference electrodes at 25°C:

- values obtained with Eqs. (4.21) and (4.22) without corrections due to activity coefficients.
- values obtained with that formulas, when taking into account activity coefficients calculated with the "limit law" of Debye–Hückel [Eq. (4.25)].
- values obtained with these formulas, when taking into account activity coefficients calculated with the "Debye–Hückel equation" [Eq. (4.27)].
- experimented values given by de Bethune.[36]

It may be observed that, although corrections according to the Debye–Hückel formulas may be helpful, results obtained without these corrections may be often sufficiently accurate.

[36] A. J. de Bethune, Electrode potentials, temperature coefficients, *The Encyclopedia of Electrochemistry*, Ed. C. A. Hampel, Reinhold, New York (1964), pp. 432–434.

TABLE XXIII. Comparison Between Experimental and Calculated Values of Electrode Potentials for Various Reference Electrodes

Electrode	Concentration (M)	Calculated			Experimental
		Without correction	Debye–Hückel (6)	Debye–Hückel (8)	
Calomel	0.001	+0.446	+0.447	+0.477	—
	0.010	+0.387	+0.391	+0.391	—
	0.100	+0.328	+0.338	+0.336	+0.333
	1.000	+0.269	+0.299	+0.284	+0.280
	4.80 (sat)	+0.242	+0.294	+0.249	+0.241
Silver chloride	0.001	+0.401	+0.401	+0.401	+0.401
	0.010	+0.342	+0.371	+0.344	+0.343
	0.100	+0.282	+0.292	+0.290	+0.288
	1.000	+0.222	+0.248	+0.237	+0.235
Copper sulfate	(sat)	+0.293	+0.198	+0.299	+0.300

Chapter 5

ELECTROCHEMICAL KINETICS

5.1. GENERAL REMARKS

The diagrams of electrochemical equilibria such as those we have just examined are certainly useful for the study of different phenomena, but they generally provide only a partial resolution. Thermodynamics can tell *whether* a reaction may occur and often *what* the reaction will be, but the thermodynamic analysis does *not tell how fast* the reaction will occur. The question of *how fast* is the province of kinetics.

For a given temperature, the equilibrium diagrams indicate the conditions of pH, electrode potential, concentrations (or activities), and pressures (or fugacities) for which a given reaction is possible or impossible. However, in practice it is not sufficient to simply determine what value the electrode potential should have; it is often desirable to know what must be done to make the electrode potential attain the desired value. In actual fact it is often easier to alter the pH rather than adjust the value of electrode potential.

Further, it is not sufficient for a reaction to be thermodynamically possible for it to be feasible. The field of electrochemistry is filled with irreversible reactions, reactions which do not necessarily proceed even though they are thermodynamically possible. For example, the reduction of chromic acid and chromates and the oxidation and reduction of carbon are such reactions. Hence, it is useful to know in which circumstances a thermodynamically possible reaction is feasible and to know the rate at which the reaction takes place.

In the following discussion, we are concerned with the *direction*, *affinity*, *overpotential*, and *rate* of electrochemical reactions.

5.2. DIRECTION OF ELECTROCHEMICAL REACTIONS

Let us consider the case of metallic zinc in contact with an acidic solution containing zinc ions. The condition of equilibrium for the corrosion reaction of the metal, **Zn** $= Zn^{++} + 2e^-$, at 25°C, is $E_0 = -0.758 + 0.0295 \times \log[Zn^{++}]$ where E is expressed in volts *versus* the standard hydrogen electrode, V_{SHE}, and $[Zn^{++}]$ in g-ions/liter.[1] This equation when evaluated for various Zn^{++} activities gives

$\log[Zn^{++}]$	E_0 (V)
0	−0.758
−2	−0.817
−4	−0.876
−6	−0.935

Thus, if zinc is in equilibrium with an ideal solution containing 0.01 g-ion of zinc/liter (636 mg Zn^{++}/liter) and if the potential of the surface is the same as E_0, $-0.817\ V_{SHE}$, the reaction will not proceed in either direction. There will be neither corrosion nor electrodeposition of metal.

Suppose now that the potential is caused to deviate from the equilibrium value ($-0.817\ V_{SHE}$ for 0.01 g-ion Zn/liter) and is changed to a more positive value with the concentration of zinc ions being kept constant. Assume, for example, that the potential is raised to −0.758 V which is the equilibrium potential for an ideal solution containing 1.0 g-ion/liter of zinc ions. In order to satisfy this new equilibrium concentration, zinc will dissolve:

$$\mathbf{Zn} \rightarrow Zn^{++} + 2e^- \qquad \text{(corrosion of Zn: oxidation)}$$

Conversely, if by some means the potential of the metal is lowered below the equilibrium potential (−0.876 V), the reaction would be inclined to proceed in the reverse direction:

$$Zn^{++} + 2e^- \rightarrow \mathbf{Zn} \qquad \text{(electrolytic deposition of Zn: reduction)}$$

The same analysis can be applied in the case of iron. If iron is immersed in an acid solution containing dissolved iron, the equilibrium condition of

[1] The notation E_0 means that the potential E is relative to the equilibrium state of a reaction; the notation E_{0r} indicates the equilibrium state of a reaction r.

the reaction **Fe** = $Fe^{++}+2e^-$ is, at 25°C, $E_0 = -0.441+0.0295 \log[Fe^{++}]$. For a solution containing 10^{-6} g-ion Fe/liter, the equilibrium potential is

$$E_0 = -0.441 + 0.0295\times(-6) = -0.62 \text{ V}$$

Thus, if E is raised to a value higher than −0.62 V, and the iron ion concentration is initially unchanged, the reaction may proceed in the direction

$$\mathbf{Fe} \to Fe^{++} + 2e^- \qquad \text{(corrosion of Fe: oxidation)}$$

If E is less than −0.62 V, the reaction may proceed in the direction

$$Fe^{++} + 2e^- \to \mathbf{Fe} \qquad \text{(electrodeposition of Fe: reduction)}$$

From the above examples it may be concluded that:

1. When the potential, E, of an electrode is *greater* (more positive) than the equilibrium potential, E_0, of a reaction, the reaction may only proceed on this electrode in the direction of an *oxidation*.
2. When the potential, E, of an electrode is *less* (more negative) than the equilibrium potential, E_0, the reaction may only proceed on this electrode in the direction of a *reduction*.

In summary:

$$\text{if } E > E_0 \text{ oxidation is possible}$$
$$\text{if } E < E_0 \text{ reduction is possible}$$

5.3. AFFINITY, OVERPOTENTIAL, RATE, AND DIRECTION OF ELECTROCHEMICAL REACTIONS. THE SECOND PRINCIPLE OF ELECTROCHEMICAL THERMODYNAMICS

De Donder has shown that, in the case of a *chemical reaction*

$$\sum \nu M = 0$$

the direction of reaction as indicated by the sign of its rate (or velocity) V is related to the sign of the affinity, A, by the relation

$$A \times V \geq 0$$

where the affinity

$$A = -\sum \nu\mu$$

and the rate

$$V = \frac{d\xi}{dt} = \frac{\frac{dN}{\nu}}{dt}$$

where $d\xi$ is the "degree of completion" of the reaction, $d\xi = dN/\nu$, and N is the quantity of a reactive species, M, present in the system under consideration.

In the case of an *electrochemical reaction*,

$$\sum \nu M + ne^- = 0$$

the "degree of advancement," $d\xi$, may be expressed, not only by the variation dN/ν of the quantity of chemical species M, but also by the variation dq/n of the quantity of negative electric charge involved as free electrons, e^-,

$$d\xi = \frac{dN}{\nu} = \frac{dq}{n}$$

The rate of reaction is then

$$V = \frac{\frac{dN}{\nu}}{dt} = \frac{\frac{dq}{n}}{dt}$$

or

$$V = \frac{1}{n} \times \frac{dq}{dt}$$

Further, the extent and sign of the affinity of an electrochemical reaction may be measured as a function of the *overpotential* $E - E_0$ of the reaction[2] by the following relation:

$$A = n \times (E - E_0)$$

Thus, the expression of de Donder, $A \times V \geq 0$, related to chemical reactions may be written, for the case of electrochemical reactions, in the form

$$n(E - E_0) \times \frac{1}{n} \cdot \frac{dq}{dt} = (E - E_0) \times \frac{dq}{dt} > 0$$

By definition, the reaction velocity, dq/dt, is also the electrical reaction current, i. We will adopt the convention that i is *positive* when the reaction proceeds in the direction corresponding to a liberation of negative electric charge (*oxidation*) and that i is *negative* when the reaction proceeds in the direction corresponding to a deposition of negative electric charge (*reduction*). Then the relation $A \times V \geq 0$ may be written

$$\boxed{(E - E_0) \times i \geq 0} \qquad (5.1)^{2,3}$$

This expression relates the direction of the electrochemical reaction (defined by the sign of its reaction current i) to the sign of the difference between the reaction potential, E, and the equilibrium potential, E_0, of the reaction (i.e., its overpotential):

if $E > E_0$	$i \geq 0$	oxidation or no reaction
if $E < E_0$	$i \leq 0$	reduction or no reaction

This relation expresses the *second principle of electrochemical thermodynamics* which determines the direction of electrochemical reactions. It states that the rate of an electrochemical reaction is always, if not zero, of the same sign as its overpotential $E - E_0$.

For example:

Reaction	*If $E > E_0$ oxidation*
$\boldsymbol{H_2} = 2H^+ + 2e^-$	$\boldsymbol{H_2} \rightarrow 2H^+ + 2e^-$
$2H_2O = \boldsymbol{O_2} + 4H^+ + 4e^-$	$2H_2O \rightarrow \boldsymbol{O_2} + 4H^+ + 4e^-$
Fe $= Fe^{++} + 2e^-$	**Fe** $\rightarrow Fe^{++} + 2e^-$

If $E < E_0$ reduction

$\boldsymbol{H_2} \leftarrow 2H^+ + 2e^-$

$2H_2O \leftarrow \boldsymbol{O_2} + 4H^+ + 4e^-$

Fe $\leftarrow Fe^{++} + 2e^-$

[2] M. Pourbaix, Thermodynamique des solutions aqueuses diluées. Thesis, Delft (1945) (Refs. 2, 22). The overpotential of a reaction may be defined as being the difference $E - E_0$ between the potential E or the electrode on which the reaction is produced and the equilibrium potential E_0 of the reaction, both these values of potential being relative to the conditions of reaction rate, temperature, concentration, etc., under which the reaction is carried out.

[3] M. Pourbaix, *Applications of Thermodynamics to Electrochemistry*, C. R. Colloque de Thermodynamique, Brussels, January 1948, pp. 125–137, IUPAC, Paris (1948) (in French).

5.4. REACTION CURRENTS

The current i defined above is the reaction current of the electrochemical reaction under consideration. This current flow, considered as positive in the case of oxidation and negative in the case of reduction, is a measure of the quantity and sign of the rate of the electrochemical reaction.

If iron or zinc dissolves electrochemically according to the reaction,

$$\mathbf{Fe} \rightarrow Fe^{++} + 2e^{-}$$

or

$$\mathbf{Zn} \rightarrow Zn^{++} + 2e^{-}$$

the reaction is produced with the liberation of 1 faraday per gram-equivalent of metal (in this case, there are 2 faradays per gram-atom of metal).

Now

$$\begin{aligned} 1 \text{ faraday} &= 96{,}484 \text{ coulombs} \\ &= 96{,}484 \text{ A} \cdot \text{sec} \\ &= 96{,}484/3600 = 26.8 \text{ A} \cdot \text{hr} \end{aligned}$$

so that a current of 1 A corresponds to

$$\begin{aligned} \frac{1}{96{,}484} &= 0.00001036 \text{ g-eq/sec} \\ &= 0.01036 \text{ mg-eq/sec} \end{aligned}$$

or to

$$\begin{aligned} \frac{1}{26.8} &= 0.0373 \text{ g-eq/hr} \\ &= 37.3 \text{ mg-eq/hr} \\ &= 37.3 \times 10^{-3} \times 24 \times 365 = 326 \text{ g-eq/year} \end{aligned}$$

The relationship of electrochemical current and corrosion rates can be illustrated by the following examples.

1. If iron corrodes according to the reaction, $\mathbf{Fe} \rightarrow Fe^{++} + 2e^{-}$, under the action of a current of 1 mA, the rate of corrosion per hour is

$$37.3 \times 10^{-3} \text{ mg-eq/hr}$$

$$= \frac{55.84}{2} \times 37.3 \times 10^{-3} = 1.040 \text{ mg Fe/hr}$$

or per year[4]:

$$\frac{55.84}{2} \times 326 \times 10^{-3} = 9.1 \text{ g Fe/year}$$

A current of 1 A corresponds to the dissolution of 9.1 kg of iron per year.

2. If iron corrodes in sulfuric acid according to the overall reaction, $\mathbf{Fe} + 2H^+ \rightarrow Fe^{++} + \boldsymbol{H_2}$, there will be a definite relationship between the amount of hydrogen liberated and the amount of iron dissolved because of the requirement for electrical neutrality. For every 55.8 mg of **Fe** dissolved per hour, 2.0 mg of $\boldsymbol{H_2}$ will be liberated per hour according to the following reactions:

$$\text{for } \mathbf{Fe} \rightarrow Fe^{++} + 2e^-, \qquad i_1 = 26.8 \times 2 = +53.6 \text{ mA}$$

$$\text{for } 2H^+ + 2e^- \rightarrow \boldsymbol{H_2}, \qquad i_2 = -26.8 \times 2 = -53.6 \text{ mA}$$

These two reaction currents are equal and of opposite sign; the resulting current sum, $I = i_1 + i_2$, is zero. Thus, all the electrons produced on the surface by the oxidation are consumed in the reduction.

5.5. POLARIZATION CURVES

The relation $(E - E_0) \times i \geq 0$ is an extremely useful aid for the experimental study of electrochemical kinetics. It relates three quantities E, E_0, and i of which two are electrode potentials and the third a current flow. It states (Figure 58) that if the potential of a reacting electrode, or the "reaction potential" E, is plotted as ordinate on a graph highest potentials at the top, and the flow of the "reaction current" i as abscissa (increasing currents to the right) then the influence of this electrode potential on the rate of any electrochemical reaction is represented by a curve ascending from left to right. This curve passes through the origin for $E = E_0$, that is where the potential of the electrode is equal to the equilibrium potential of the reaction. Such a curve is a *polarization curve*.[5]

[4] When there is general corrosion, a corrosion current density of 1 mA/cm² corresponds to an annual depth of attack of iron equal to 9.1/7.9 = 1.1 cm; a current density of 1 μA/cm² corresponds to a depth of attack of 0.011 mm/year or 11 μm (or microns) per year or 0.44 mpy (mils per year).

[5] When several reactions *a*, *b*, *c*, ... are considered, it is useful, in order to differentiate the various equilibrium potentials and reaction currents concerned with each reaction, to use the notations E_{0a}, E_{0b}, E_{0c}, ... and i_a, i_b, i_c, ... The fundamental relation then has the form

$$(E - E_{0a}) \times i_a \geq 0 \qquad (E - E_{0b}) \times i_b \geq 0 \qquad (E - E_{0c}) \times i_c \geq 0 \ldots$$

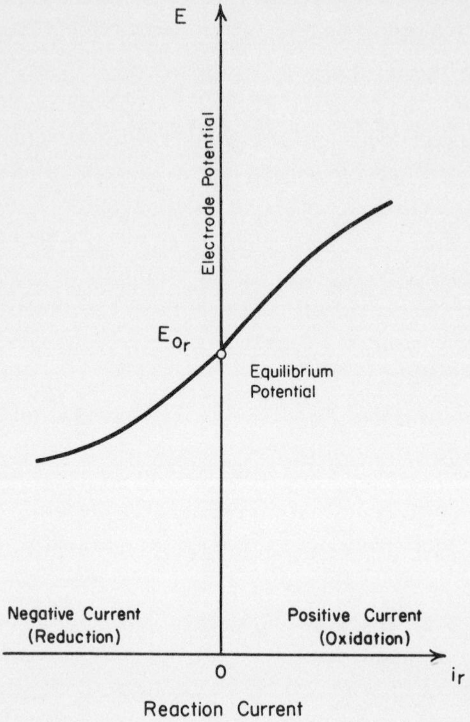

Figure 58. Polarization curve of an electrochemical reaction.

5.5.1. Reversible and Irreversible Reactions: Oxidation Potential, Reduction Potential, and Oxidation–Reduction Potential

For all chemical reactions whose equilibrium potential is E_0, the difference, $E - E_0$ (the overpotential), measures, in magnitude and sign, the *affinity* of the reaction. In the case of appreciably *reversible* reactions (Figure 59a) for which there need only be a small value of affinity to give a measurable value of the rate i, the polarization curve cuts the ordinate axis at a point where the electrode potential is equal to the equilibrium potential of the reaction. In the case of *irreversible* reactions (Figure 59) for which the rate remains almost equal to zero for measurable values of affinity, the curve is practically superimposed on the ordinate axis for a relatively wider range of electrode potential both above and below the equilibrium potential, E_0. The curve does not depart significantly from the ordinate axis until the overpotential $E - E_0$ exceeds a certain critical value above which the reaction proceeds at an appreciable rate.

According to the terminology proposed by P. Van Rysselberghe, the *oxidation potential* is the potential of the electrode above which the reaction proceeds at an appreciable rate (for example, 10^{-8} amp/cm²) in the direction of an oxidation; the *reduction potential* is the potential of the electrode below which the reaction proceeds at an appreciable rate in the direction of a reduction. In the case of reversible reactions for which the rate becomes appreciable for small deviations from the equilibrium potential, the oxidation potential and the reduction potential have practically the same value. This is the *oxidation–reduction potential* which is identical to the *equilibrium potential* of the reaction.

It must be pointed out here that a strict application of theory to the interpretation of polarization curves involves the term $E - E_0$. Note the possibility that E_0 may change as corrosion proceeds due to its dependence on activity $[M]$ of the ($E_0 = E_0{}^0 + (RT/nF)\ln[M]$) reacting species, M. Especially in the case of unstirred solutions, the value of $[M]$ may alter significantly as $E - E_0$, and hence i, increases. This effect due to local increases in $[M]$ is called "concentration polarization." Such polarization curves are thus not experimentally valid when there is any concentration polarization in the neighborhood of the electrode. This being so, and since diffusion phenomena are generally slow, *the experimental determination of*

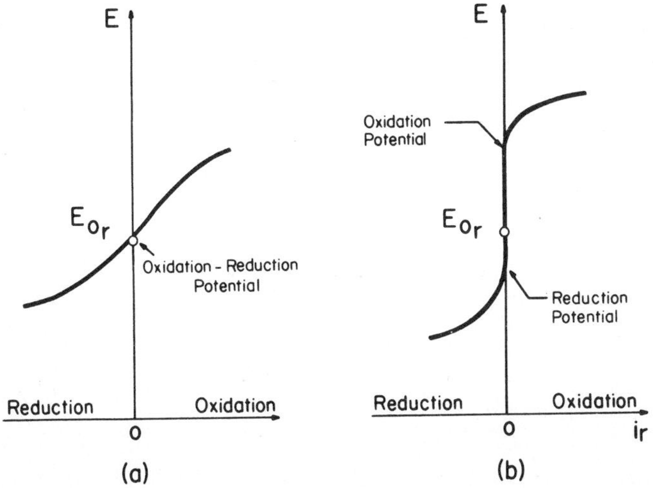

Figure 59. Polarization curve of electrochemical reactions: (a) reversible reaction; and (b) irreversible reaction.

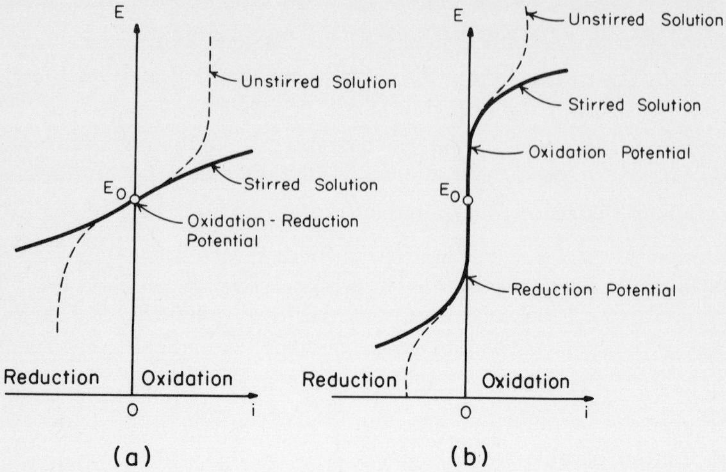

Figure 60. Influence of stirring on the shape of polarization curves: (a) reversible reaction; and (b) irreversible reaction.

such curves at constant equilibrium potential generally is only possible for strongly stirred solutions. For such conditions it is reasonable to assume that the composition of the solution adjacent to the electrode is practically constant and is the same as the average composition of the bulk solution.

If this is not the case, namely if the solution is unstirred or insufficiently stirred (this generally applies to polarographic techniques), the progress of the reaction itself will cause a significant change in the composition of the fraction of solution in contact with the electrode. In the case of oxidation there will be impoverishment of reduced constituents and enrichment of oxidized constituents with a resulting increase of the equilibrium potential. In the case of reduction (this normally concerns polarography), there will be impoverishment of oxidized constituents and enrichment of reduced constituents with a resulting decrease of the equilibrium potential. For a certain critical reaction rate the supply of reactants by diffusion will be inadequate to maintain a sufficient quantity of reactant in contact with the electrode, there will then be a considerable elevation of the equilibrium potential for oxidation and a considerable depression of the equilibrium potential for reduction. Then the polarization curve takes the shape of a "wave" well known to polarographers (Figure 60). In this instance, the potential at the beginning of each wave is the reduction potential of the reaction corresponding to this wave (Figure 61).

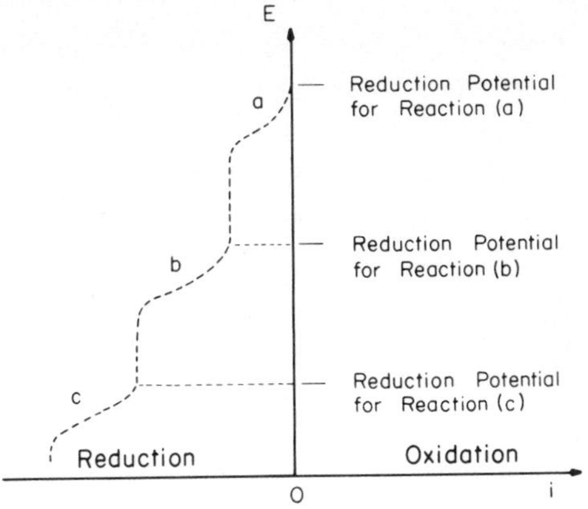

Figure 61. Polarographic curves and reduction potentials.

5.5.2. The Tafel Law. Exchange Current

For reactions carried out in solutions sufficiently stirred so that the equilibrium potential remains practically constant, it has been shown experimentally that the values of electrode potential E and reaction current i often obey the Tafel law

$$E = \alpha + \beta \log i \tag{5.2}$$

that is to say, the polarization "curve" becomes a straight line if it is plotted as a function of E *vs* $\log i$ (Figure 62). For a single reaction (such as $\boldsymbol{H_2} = 2H^+ + 2e^-$ or $\mathbf{Zn} = Zn^{++} + 2e^-$) both the oxidation and reduction

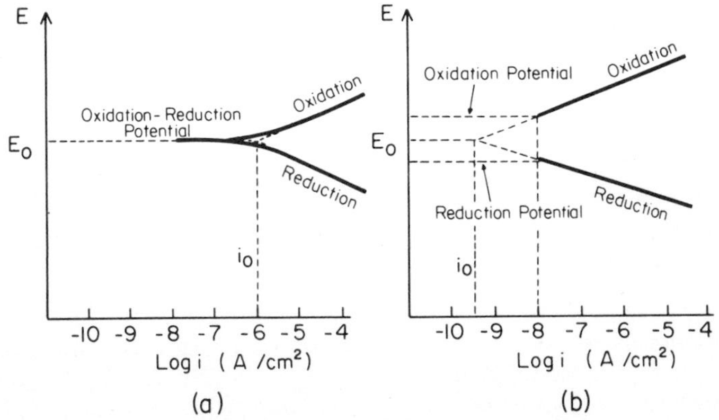

Figure 62. Tafel law—equilibrium potential and exchange currents: (a) reversible reaction; and (b) irreversible reaction.

processes can obey the semilogarithmic Tafel behavior. Thus there can be two straight line dependencies for E *vs* log i, as shown in Figure 62, although the absolute value of their slopes may not be the same. The point of intersection of the respective oxidation (i.e., **Zn** $\rightarrow$ $Zn^{++} + 2e^-$) and reduction (i.e., $Zn^{++} + 2e^- \rightarrow$ **Zn**) lines corresponds to the equilibrium potential where, by definition, the rates of oxidation and reduction reactions for a single reaction are equal. The value of current at which these lines intersect is called the *exchange current* and is designated by i_0. Thus the Tafel relationship acquires the form[6]

$$E - E_0 = \beta(\log i - \log i_0) \tag{5.3}$$

At equilibrium

$$E = E_0$$

and

$$i = i_0$$

In practice, a reaction is considered reversible when the value of its exchange current is equal to or greater than a given critical value chosen more or less arbitrarily (for example, 10^{-8} A/cm^2). A reaction is considered as irreversible when the value of its exchange current is less than this critical value. If it is assumed that this critical value is a measure of the "appreciable" rate above which a reaction may be considered to be effectively progressing, then the two potentials, which in Figure 62b correspond to these rates, are respectively the oxidation potential and the reduction potential. In the case of reversible reactions, as illustrated in Figure 62a, these two potentials are the same and yield the oxidation–reduction potential.

It must be stated that the Tafel law does not hold in every case (Figure 63); sometimes a change of slope is observed and this change may be either stable and continuous or unstable and discontinuous. Probably in the first case, the electrode may be the site of two simultaneous reactions, each in turn predominant above and below a particular electrode potential for which the rates of the two reactions are equal (Figure 63a). Probably in the second case, the electrode may exist under two different surface conditions (for example, with or without an oxide layer) according to whether or not

[6] The Tafel relationship can be formally rationalized in theoretical terms using charge dependent reaction kinetics. The formal derivation and justification of the Tafel equation has been presented by many authors. An excellent survey of the theoretical significance is given by Bockris and Reddy in Volume 1 of *Modern Electrochemistry*, Plenum Press (1970).

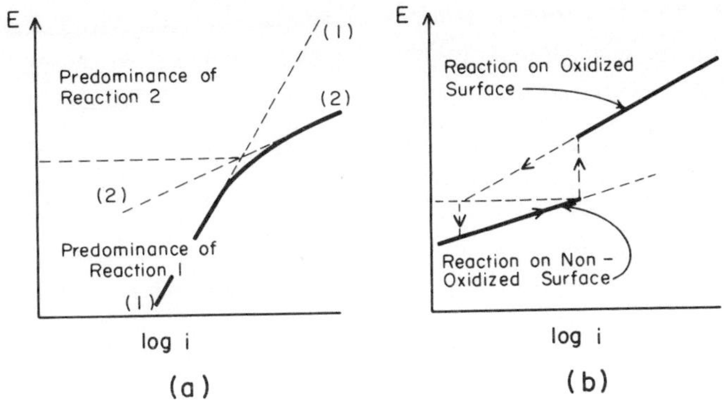

Figure 63. Apparent exceptions to Tafel law: (a) two simultaneous reactions; and (b) modifications of the surface state.

the electrode potential is greater than a certain potential which may be, for example, the equilibrium potential between the metal and one of its oxides (Figure 63b).

5.5.3. Predetermination of the Direction and Rate of Electrochemical Reactions

It follows from the preceding discussion that if the value of electrode potential E of a metallic surface in contact with an aqueous solution is known, and if the values E_{0a}, E_{0b}, and E_{0c}, ... of the equilibrium potentials of all electrochemical reactions a, b, c, ... likely to operate at this surface are known, then it is possible to undertake a thorough theoretical and experimental study of all the electrochemical phenomena which may be produced at this surface. With the relation, $(E - E_{0r}) \times i_r \geq 0$, it is possible to predetermine the direction of the various electrochemical reactions. Knowledge of the respective polarization curves, $E = f(i_r)$, will indicate the rate of these reactions. A few examples of such analyses are given in Chapter 6 where the corrosion and protection of iron and steel is discussed.

5.6. ELECTROCHEMICAL CATALYSIS OF CHEMICAL REACTIONS

The principles of electrochemical kinetics may be applied much more broadly than to only corrosion.

As outlined at the beginning of the chapter on electrochemical equilibria (see Sections 4.1 and 4.2), certain chemical reactions occurring in aqueous solutions may be considered the result of the combination of two electrochemical reactions. For example, one such reaction is the synthesis of water, $2\boldsymbol{H}_2 + \boldsymbol{O}_2 \rightarrow 2H_2O$, which may be carried out electrochemically in a fuel cell according to the following scheme:

Negative electrode	$2\boldsymbol{H}_2 \rightarrow 4H^+ + 4e^-$	(a)
Positive electrode	$\boldsymbol{O}_2 + 4H^+ + 4e^- \rightarrow 2H_2O$	(b)
	$2\boldsymbol{H}_2 + \boldsymbol{O}_2 \rightarrow 2H_2O$	

As Weydema has suggested, a metallic surface on which both of the component electrochemical reactions proceed with appreciable rapidity at a same electrode potential is likely to catalyze the overall chemical reaction. Hence, research for a catalyst for this chemical reaction would be carried out as indicated below.[7]

Using the experimental arrangement illustrated in Figure 64a to study the possible catalytic activity of a sample of metal (for instance **Pt**), the sample of the metal is immersed in an aqueous solution (such as a solution of sulfuric acid) in which the reaction is to be carried out. A stream of hydrogen is bubbled around this metal and, by means of an auxiliary cathode of incorrodible metal, the metal under study is polarized anodically with increasing current. For each value of current, the potential of the metal is measured and the current–potential curve obtained in this manner is plotted. This curve is the polarization curve of the oxidation reaction

$$\boldsymbol{H}_2 \rightarrow 2H^+ + 2e^-$$

Using the same apparatus, if the flow of hydrogen is replaced by a flow of oxygen and if the electric connections of the current supply are reversed (Figure 64b), the metal under study may be polarized with increasing current in the cathodic direction and the current–potential curve plotted. This is the polarization curve of the reaction

$$\boldsymbol{O}_2 + 4H^+ + 4e^- \rightarrow 2H_2O$$

(possibly with interference of the reduction reaction $\boldsymbol{O}_2 + 2H^+ + 2e^- \rightarrow H_2O_2$ which we do not consider here).

[7] See M. Pourbaix, Sur l'interprétation thermodynamique des courbes de polarisation, *Rapports Techniques CEBELCOR* RT. 1 (1952), The utility of thermodynamic interpretation of Polarization curves, *J. Electrochem. Soc.* **101**, 217–221c (1954).

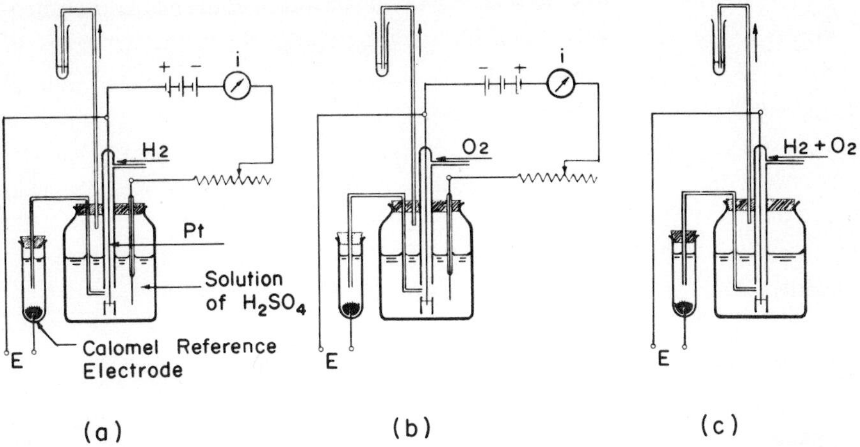

Figure 64. Study of the electrochemical catalysis of the reaction for the synthesis of water; experimental arrangement: (a) determination of the polarization curve for the oxidation of hydrogen; (b) determination of the polarization curve for the reduction of oxygen; and (c) catalysis of the reaction for the synthesis of water.

Therefore, there are two cases: either the two polarization curves diverge sufficiently from the ordinate axis, so that for a certain range of values of electrode potential, the reactions may both proceed at an appreciable rate (Figure 65a); or the polarization curves are superimposed with the ordinate axis such that for no value of electrode potential are the rates of two reactions appreciable. This leads immediately to the conclusion that the metal will catalyze the reaction $2\boldsymbol{H}_2 + \boldsymbol{O}_2 \rightarrow 2H_2O$ in the first

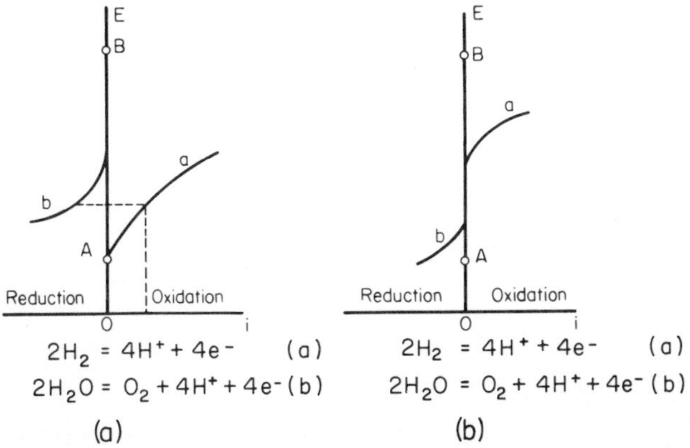

Figure 65. Polarization curves and catalysis of chemical reaction: (a) catalysis; and (b) no catalysis.

case; it will not catalyze it in the second case. This assumes that the simultaneous presence of both gases do not inhibit the other reaction.

In the case of catalysis, the electrode potential of the metal is approximately that for which the two reaction currents, i_a and i_b, are equal and of opposite sign, giving no net overall current. The rate of the overall chemical reaction, $2\boldsymbol{H}_2 + \boldsymbol{O}_2 \rightarrow 2H_2O$, is the rate common to the two electrochemical reactions.

Using the apparatus illustrated in Figure 64c, these facts may be easily verified by passing a mixture of hydrogen and oxygen around some electrochemically isolated metal.

The procedures described above are useful for many technical applications, notably in the field of industrial chemistry. This is particularly so in research on oxidation catalysts using oxygen gas with substances for which oxidation in solution is irreversible.

Chapter 6

CORROSION AND PROTECTION OF IRON AND STEEL

In this chapter, we apply the methods which have been outlined in the preceding chapters, particularly Chapters 4 (Electrochemical Equilibria) and 5 (Electrochemical Kinetics) to study the corrosion of iron and steels.

6.1. DIAGRAM OF ELECTROCHEMICAL EQUILIBRIA OF THE IRON–WATER SYSTEM AT 25°C

Figure 39 and 7, which have been described in Chapter 4, illustrate, respectively, the equilibrium potential–pH diagram of the iron–water system at 25°C and two versions of the theoretical diagram of corrosion–immunity–passivation which may be deduced by assuming that passivation results either from the formation of a film of $\mathbf{Fe_2O_3}$ (Figure 7a) or from the formation of films of $\mathbf{Fe_2O_3}$ and $\mathbf{Fe_3O_4}$ (Figure 7b).

Figure 39 shows that the region of stability of iron is entirely below the region of stability of water under atmospheric pressure. In other words, regardless of the pH of the solution, metallic iron and water are not simultaneously thermodynamically stable at 25°C under a pressure of 1 atm. The simultaneous stability of iron in the presence of water at 25°C is only attained at pressures high enough to depress the equilibrium potential of the H_2O–$\boldsymbol{H_2}$ system (represented for 1 atm by line *a*) below the equilibrium potential of the $\mathbf{Fe}$–$\mathbf{Fe_3O_4}$ system (represented by line 13). This stability is attained only at pH values between 10 and 12 for pressures greater or equal to 750 atm. At a lower pressure iron always tends, if it is not passivated, to be corroded with reduction of water and evolution of hydrogen.

This corrosion of iron gives rise to the dissolution of metal mainly with the formation of green ferrous ions Fe^{++} (reaction 1) or dihypoferrite ions FeO_2H^- (reaction 2) which are also green, depending on whether the pH is greater or less than 10.6.[1] At values of pH below about 6 to 7 these ions may change to black magnetic iron oxide, or magnetite $\mathbf{Fe_3O_4}$ (reaction 3) or into white ferrous hydroxide $\mathbf{Fe(OH)_2}$ (reaction 4). Because the area of stability of ferrous hydroxide $\mathbf{Fe(OH)_2}$ is completely covered by the region of stability of magnetite, $\mathbf{Fe_3O_4}$, $\mathbf{Fe(OH)_2}$ will be thermodynamically

[1] The chemical reactions relating to the transformations discussed in the present paragraph are outlined below. The reactions are written in the manner described in Chapter 2, p. 26. The numbers and letters associated with every constituent electrochemical reaction are those which, in Figure 39, relate to the equilibrium lines of these reactions.

Reaction 1

$$\mathbf{Fe} \rightarrow Fe^{++} + 2e^- \quad (23)$$
$$2H^+ + 2e^- \rightarrow \boldsymbol{H_2} \quad (a)$$
$$\mathbf{Fe} + 2H^+ \rightarrow Fe^{++} + \boldsymbol{H_2}$$

Reaction 2

$$\mathbf{Fe} + 2H_2O \rightarrow FeO_2H^- + 3H^+ + 2e^- \quad (24)$$
$$2H^+ + 2e^- \rightarrow \boldsymbol{H_2} \quad (a)$$
$$\mathbf{Fe} + 2H_2O \rightarrow FeO_2H^- + H^+ + \boldsymbol{H_2}$$

Reaction 3

$$3Fe^{++} + 4H_2O \rightarrow \mathbf{Fe_3O_4} + 8H^+ + 2e^- \quad (26)$$
$$2H^+ + 2e^- \rightarrow \boldsymbol{H_2} \quad (a)$$
$$3Fe^{++} + 4H_2O \rightarrow \mathbf{Fe_3O_4} + 6H^+ + \boldsymbol{H_2}$$

Reaction 4

$$Fe^{++} + 2H_2O \rightarrow \mathbf{Fe(OH)_2} + 2H^+$$

Reaction 5

$$\mathbf{3Fe(OH)_2} \rightarrow \mathbf{Fe_3O_4} + 2H_2O + 2H^+ + 2e^-$$
$$2H^+ + 2e^- \rightarrow \boldsymbol{H_2} \quad (a)$$
$$\mathbf{3Fe(OH)_2} \rightarrow \mathbf{Fe_3O_4} + 2H_2O + \boldsymbol{H_2}$$

Reaction 6

$$4Fe^{++} + 12H_2O \rightarrow \mathbf{4Fe(OH)_3} + 12H^+ + 4e^- \quad (28)$$
$$\boldsymbol{O_2} + 4H^+ + 4e^- \rightarrow 2H_2O \quad (b)$$
$$4Fe^{++} + \boldsymbol{O_2} + 10H_2O \rightarrow \mathbf{4Fe(OH)_3} + 8H^+$$

Reaction 7

$$\mathbf{4Fe_3O_4} + 2H_2O \rightarrow \mathbf{6Fe_2O_3} + 4H^+ + 4e^- \quad (17)$$
$$\boldsymbol{O_2} + 4H^+ + 4e^- \rightarrow 2H_2O \quad (b)$$
$$\mathbf{4Fe_3O_4} + \boldsymbol{O_2} \rightarrow \mathbf{6Fe_2O_3}$$

unstable with respect to $\mathbf{Fe_3O_4}$ and will tend to transform to this substance (reaction 5).

In the presence of oxygen ferrous ions and magnetite may be oxidized to ferric oxide $\mathbf{Fe_2O_3}$ (or to ferric hydroxide $\mathbf{Fe(OH)_3}$ (reactions 6 and 7).

6.2. GENERAL CONDITIONS OF CORROSION, IMMUNITY, AND PASSIVATION OF IRON

According to Figure 7, corrosion of iron is possible in two "domains" or "regions," roughly triangular in shape, which correspond to the dissolution of metal with the formation of ferrous and dihypoferrite ions, respectively. Such corrosion by dissolution will affect all the metal surface and so will be of a *general* nature.

The metal may be protected against this corrosion by lowering its electrode potential to the "region of immunity" (in which case there will be "cathodic protection," obtainable by intervention of a reducing action) or by raising its electrode potential into the "region of passivation" (in which case there will be "protection by passivation," obtained by intervention of an oxidizing action).

In the case of *protection by immunity*, the metal will be thermodynamically stable and, as a result, incorrodible; it will possess a truly metallic surface. Since water is not stable under the conditions of electrode potential and pH corresponding to this state of immunity, this protection will only be achieved by continuous consumption of external energy which causes hydrogen evolution on the iron.

In the case of *protection by passivation* the metal itself will not be stable, but will be covered by a stable oxide film ($\mathbf{Fe_3O_4}$ or $\mathbf{Fe_2O_3}$[2] according to the conditions of potential or pH). The protection will be perfect or imperfect depending on whether or not this film perfectly shields the metal from contact with the solution. In the case of imperfect protection, corrosion only affects the weak points of the passive film and therefore has a localized character. Water being stable under the conditions of electrode potential and pH which correspond to the state of passivation, this state will be attained without consumption of external energy; the solution needs only be sufficiently oxidizing for the electrode potential of the metal to be

[2] As previously said (p. 72), the protective oxide is in fact $\gamma\mathbf{Fe_2O_3}$ lattice containing hydrogen. M. C. Bloom and G. L. Goldenberg have shown that, when this lattice is stoichiometrically filled with hydrogen, its formula is HFe_5O_8.

maintained permanently in the passivation region. Remember that the theoretical Figure 7 is only valid for certain hypothetical conditions, and in the absence of substances capable of forming soluble complexes or insoluble salts with iron. The existence of soluble complexes (cyanides, tartrates) extends the theoretical domains of corrosion; the existence of insoluble salts (phosphates, sulfides) extends the domains of passivation.

6.3. POLARIZATION CURVES

As shown in Chapter 5, equilibrium diagrams such as Figure 39 (as well as Figure 57 deduced from it) are not sufficient to *solve* a given corrosion problem.

In this section, the study of techniques of protection against corrosion is approached using, besides such equilibrium diagrams, the principles of electrochemical kinetics which have been outlined in Chapter 5. In other words, in this chapter the tools afforded by both thermodynamics and kinetics will be utilized. The former will define the reactions which are possible and the latter, the rate of these reactions.

As an example, let us examine the behavior of iron, at 25°C, in the presence of 0.10 M solution of sodium bicarbonate (8.4 g $NaHCO_3$/liter), the pH being 8.4.[3] First, the behavior of iron in the presence of oxygen-free solution is examined; then its behavior in the presence of oxygen-saturated solution is studied. This is followed by a discussion of the phenomena of differential aeration.

6.3.1. Behavior of Iron in the Presence of Oxygen-Free Bicarbonate Solution

The experimental arrangement illustrated in Figure 66 is used for this study. The apparatus includes a glass flask containing 0.1 M solution of $NaHCO_3$ through which a stream of oxygen-free hydrogen is bubbled. Two freshly etched small specimens of iron *b* are immersed in this solution and each iron specimen is associated with a saturated calomel reference electrode, by means of a capillary siphon *c* of the Haber–Luggin type. A

[3] Details of these experiments are given in: Laboratory experiments for lectures in electrochemical corrosion, CEBELCOR Publication E. 65 (1967). The method used has been described in our paper, Corrosion, passivité et passivation du fer. Le rôle du pH et du potentiel. Thesis, Brussels (1945) extract, *Mémoires Société Royale Belge des Ingénieurs et des Industriels*, nr. 1, 1–40, March 1951. CEBELCOR Publication F. 21.

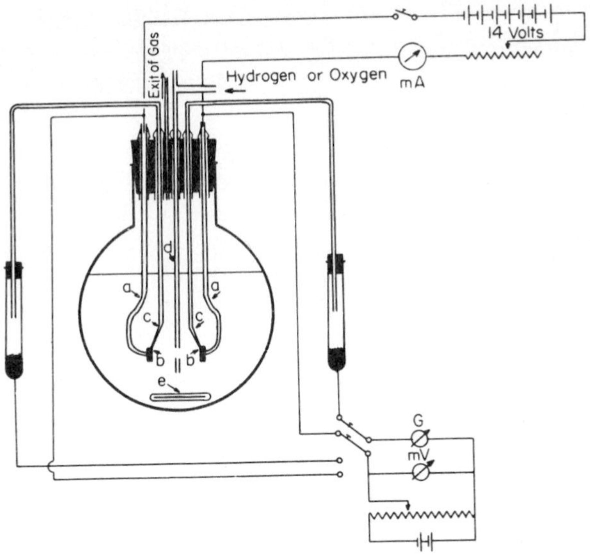

Figure 66. Experimental arrangement for the study of the electrochemical behavior of iron.

small magnetic stirrer *e* is also included. The electrical circuit, containing a 14-V battery, switch, variable resistance, and milliammeter, enables a known and variable current to be passed between the two samples of iron.

First, note the open circuit values of potential of the two electrodes (that is with no current flow) which should be the same (approximately −0.55 V with respect to the standard hydrogen electrode[4]). The switch is then closed and the resistance is gradually reduced in such a way that a progressively increasing electrolysis current flows in the cell.

For each value of electrolysis current the potentials of both electrodes are noted and then recorded graphically (see Figure 67a). It is observed that as the electrolysis current I is gradually increased the potential of the negative electrode (cathode) decreases and the potential of the positive electrode (anode) increases. For low currents, the values of current and potential are completely stable for a given value of external resistance, and are dependent on one another in a perfectly reproducible manner. When

[4] It is recalled that, at 25°C, the potential of the saturated calomel electrode is about 0.250 V greater than the potential of the standard hydrogen electrode. So this potential of −0.55 V with respect to the standard hydrogen electrode corresponds to −0.80 V with respect to the saturated calomel electrode.

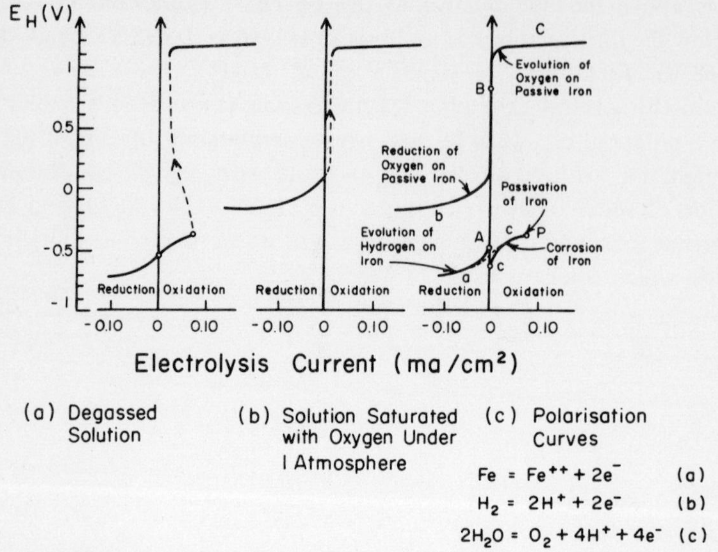

Figure 67. Behavior of mild Armco steel in 0.1 M solution of $NaHCO_3$ (pH = 8.4).

the potential of the anode reaches a certain critical value (about −0.35 V with respect to the standard hydrogen electrode) the observed current flow at once ceases to be constant and diminishes. Meanwhile, the electrode potential of the cathode varies only slightly during this fall of current and in a way that is reproducible and which conforms to the current–potential curve plotted for increasing currents. In a less reproducible manner, the electrode potential of the anode increases considerably to higher values of potential of the order of +1.15 V. Then the potential of the anode stabilizes and if the electrolysis current is again increased by further lowering the external resistance, the electrode potential of the anode again develops in a stable way, but according to a second anodic branch at higher values of potential than those of the original. As for the potential of the cathode, this decreases as a function of current and conforms to the cathodic curve plotted during the first part of the experiment.

To understand the full significance of the electrolysis curves drawn on Figure 67a, the lower branch of the electrolysis curve of this figure is redrawn in Figure 68 and the point which corresponds to the behavior of iron at zero current (for which $E = -0.55$ V and pH = 8.4) is compared with Figure 39 which represents the diagram of electrochemical equilibria of the iron–water system. Assuming that the iron content of the solution is very low (e.g., 10^{-6} g-atom/liter, or 0.056 mg/liter), it is observed that this point is above the line relative to the **Fe**–Fe^{++} equilibrium (−0.62 V), and below the lines relating to equilibria of H^+–$\boldsymbol{H_2}$ (−0.495 V), $\mathbf{Fe_3O_4}$–Fe^{++} (−0.468 V), $\mathbf{Fe_2O_3}$–Fe^{++} (−0.407 V), and $\boldsymbol{O_2}$–H_2O (+0.733 V). As a result, in the general case of a specimen iron covered with oxides and having a potential of −0.55 V in an oxygen-containing solution of pH 8.4, iron might be oxidized to the ferrous state with any of the following: reduction of water to hydrogen, reduction of the oxides $\mathbf{Fe_3O_4}$ and $\mathbf{Fe_2O_3}$ into ferrous ions, and reduction of oxygen to water (as well as to hydrogen peroxide which is not considered here).

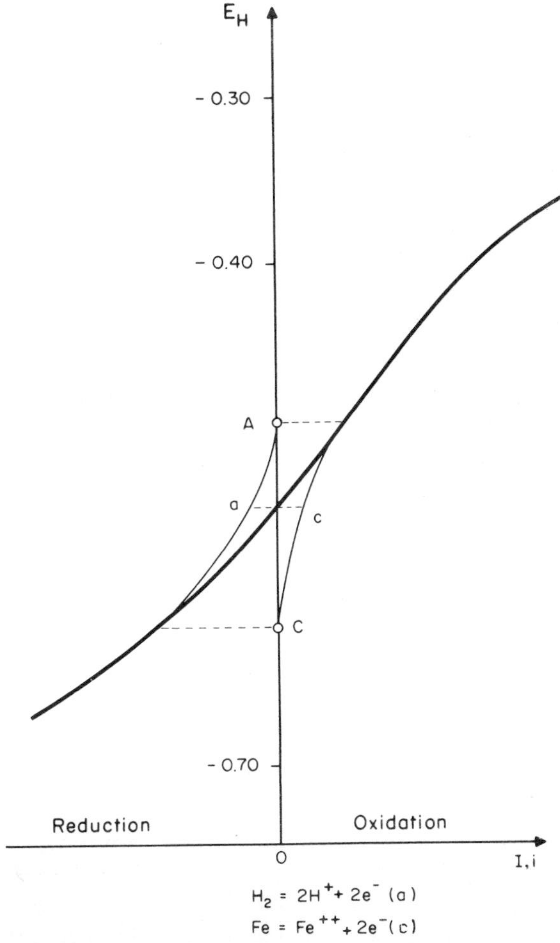

Figure 68. Interpretation of an electrolytic current–potential polarization curve. The case of iron in oxygen-free 0.1 M solution of $NaHCO_3$.

In the present case where the iron is well pickled (and consequently free from oxides) and where the solution is free from oxygen, only the following two reactions are possible:

$$2H^+ + 2e^- \rightarrow \boldsymbol{H_2} \qquad \text{(a)}$$

(with an equilibrium potential = −0.495 V)

$$\mathbf{Fe} \rightarrow Fe^{++} + 2e^- \qquad \text{(c)}$$

(with an equilibrium potential = −0.62 V)

From the theory of Wagner and Traud,[5] the current–potential electrolysis curve drawn on a large scale in Figure 68 results from the combination of two polarization curves, in the manner that has already been outlined in Chapter 5, and which are concerned respectively with the two reactions cited above. How is the electrolysis curve resolved into the two composing polarization curves?

The values, *A* and *C*, equilibrium potentials of the two reactions (a) and (c), respectively −0.495 V and −0.62 V, are plotted on the ordinate axis of Figure 68. At potential *C*, the rate of the corrosion reaction (c) is zero and the electrolysis current *I* corresponds entirely to reaction (a), the evolution of hydrogen on iron; the ordinate of point *C* of the electrolysis curve then is one of the points of the polarization curve relative to reaction (a). As the solution contains only a small quantity of dissolved iron, the rate of the deposition reaction, $Fe^{++} + 2e^- \rightarrow \mathbf{Fe}$, which may be possible at potentials lower than the potential of point *C*, is practically zero. Consequently, the whole portion of the electrolysis curve with potential values lower than the potential of point *C* is practically identical with a portion of the polarization curve relative to the hydrogen-evolution reaction on iron. An additional point on this polarization curve is point *A* which represents the equilibrium state of the $\boldsymbol{H_2}$–H^+ reaction. From here, it is easy to draw the cathodic branch of the polarization curve *a*, which represents the influence of electrode potential on the rate of the reaction $2H^+ + 2e^- \rightarrow \boldsymbol{H_2}$ (a) obtained on iron. If the ordinates are read with respect to point *A*, this curve is in fact the curve of *hydrogen overpotential* on iron.

Analogous reasoning makes it possible to draw the polarization curve *c* which defines the *E–i* behavior for the corrosion reaction of iron. At

[5] C. Wagner and W. Traud, Ueber die Deutung von Korrosionsvorgängen durch Ueberlagerung von elektrochemischen Teilvorgängen und über die Potentialbildung an Mischelektroden, *Z. Elektrochemie* **44**, 391–402 (1938).

potential A, the rate of hydrogen evolution (a) is zero and the electrolysis current corresponds entirely to reaction (c) of the corrosion of iron. Although the hydrogen oxidation is possible above the potential at A, the contribution of this oxidation reaction to that of iron oxidation is usually so small as to be neglected. Thus, the polarization curve for (c) above potential A may be taken as consisting entirely of iron oxidation. Below A, hydrogen oxidation is not possible and the oxidation curve for iron [reaction (c)] may be extended to the equilibrium potential for the **Fe**–Fe^{++} reaction at C.

For a potential which corresponds to that of electrically insulated metal (−0.55 V) the current of the oxidation reaction (c) must be equal but of opposite sign to the current of the reduction reaction (a). This permits establishing another point of curve c. Then the complete anodic branch of the polarization curve c may be drawn. This represents the influence of electrode potential on the rate of the reaction **Fe** → $Fe^{++} + 2e^-$. If the ordinates are read with respect to point C, this curve is in fact the curve of the *overpotential of corrosion of iron.*

Let us turn our attention to point P, recalling that it represents the conditions of potential and current for which, during electrolysis at increasing currents, an instability of the electrolytic system arises. At this point, there is a decrease of the electrolysis current and a discontinuous jump of potential of the anode. Comparing the characteristics of this point P ($E = -0.35$ V at pH $= 8.4$) with the diagram of electrochemical equilibria, it is observed that these characteristics lie within the region of stability of magnetite $\mathbf{Fe_3O_4}$, and at a short distance from the region of stability of the ferric oxide $\mathbf{Fe_2O_3}$ considered at this diagram. Therefore, it is quite probable that point P corresponds to the formation, on iron, of one of these two oxides[6] which, by preventing contact between the metal and the solution, suppresses all corrosion. Then, the only oxidation reaction possible after this "passivation" is the transformation into the oxides $\mathbf{Fe_3O_4}$ or $\mathbf{Fe_2O_3}$ of the small quantity of ferrous ions previously dissolved in solution during the polarization of the anode at lower potentials. This oxidation itself becomes impossible when the solution is completely denuded of ferrous ions. By maintaining a certain electrolysis current, the potential of the anode progressively increases, as Figure 67a indicates, only becoming

[6] Unpublished experimental work performed by N. de Zoubov has led to the opinion that, under the conditions outlined here, point P corresponds apparently to the formation of an iron oxide ($\mathbf{Fe_3O_4}$ or $\mathbf{Fe_2O_3}$) in the presence of a solution saturated with ferrous carbonate $\mathbf{FeCO_3}$.

stable when another oxidation reaction becomes possible. As stated before, the electrode potential stabilizes at about +1.15 V, which is higher than the value *B*, the equilibrium potential of the reaction $4H_2O \rightarrow \boldsymbol{O_2}+4H^+ + 4e^-$ (b); the electrolysis current then corresponds to an evolution of oxygen on the passive iron, according to the reaction $4H_2O \rightarrow \boldsymbol{O_2} + 4H^+ + 4e^-$.

It follows that when iron is in contact with 0.1 M solution of $NaHCO_3$ (pH = 8.4) it only corrodes if its electrode potential lies between two well-defined values, which are respectively −0.62 V and −0.35 V. Below −0.62 V, the metal is in a state of immunity and is *cathodically protected*; above −0.35 V, the metal is *protected by passivation*. The rate of corrosion possible at potentials between −0.62 and −0.35 V essentially depends on the value of the potential and is in keeping with the polarization curve c. The maximum possible rate of corrosion is given by the abscissa of point *P*. In the case of Figure 67, the maximum rate of corrosion is 0.075 mA/cm² which, is equivalent to 0.83 mm per annum, or 3.3 mpy (see p. 191).

6.3.2. Behavior of Iron in the Presence of Oxygen-Saturated Bicarbonate Solution

If the experiment which has just been described is carried out with a stream of oxygen under atmospheric pressure being passed through the 0.1 M $NaHCO_3$ solution instead of hydrogen, then it is observed that the open-circuit potentials of the samples of iron are much higher than in the absence of oxygen: about +0.1 V as opposed to −0.55 V.

Being greater than the passivation potential (−0.35 V) this value suggests that electrically insulated iron, which we have seen, corrodes if the solution does not contain oxygen, does not corrode if the solution is saturated with oxygen under atmospheric pressure: iron then is passivated by an oxide film.

Electrolysis carried out in the oxygen containing solutions with an increasing current gives rise to current–potential curves represented in Figure 67b. While the anodic curve is virtually identical to the higher of the two curves obtained in solution initially free from oxygen (which has been shown to correspond to evolution of oxygen according to the reaction $2H_2O \rightarrow \boldsymbol{O_2} + 4H^+ + 4e^-$) the cathodic curve is, at least in its initial portion, considerably above the curve obtained in the absence of oxygen. If the electrode potentials relating to this curve are compared with the electrochemical equilibrium diagram of the iron–water system (Figure 39), it may be seen that they fall within the region of stability of $\mathbf{Fe_2O_3}$ and that the only reaction with a greater value of equilibrium potential is the reaction

$2H_2O = O_2 + 4H^+ + 4e^-$ (b). This curve is none other than the cathodic branch of this reaction (b), the curve obtained at the anode being the anodic branch.[7] Therefore, during the operation of "electrolysis" the two electrodes produce merely the same reaction in opposite senses: oxidation of water to oxygen at the anode and reduction of oxygen to water at the cathode. This is almost analogous to the industrial refining of metals by electrolysis (copper, for example) where both electrodes are the site of the same reaction carried out in opposite directions ($\mathbf{Cu} \rightarrow Cu^{++} + 2e^-$ at the anode and $Cu^{++} + 2e^- \rightarrow \mathbf{Cu}$ at the cathode).

If the experiment is continued by further increasing the electrolysis current (Figure 69), the system ceases to be stable when the potential of the cathode reaches a certain critical value; the potential of the cathode decreases and gradually stabilizes on a curve about 200 mV below the curve corresponding to values existing before the appearance of the instability. Now the critical value below which the potential becomes unstable during cathodic polarization is virtually equal to −0.35 V, which is the value above which the potential becomes unstable during anodic polarization due to a passivation of the metal by formation of an oxide film. This implies that the passivation reaction is practically reversible and that the oxide film which forms by passivation above this potential is destroyed by reduction below this potential. The *activation potential* (often called the "Flade potential") is equal to the *passivation potential* and the common value of these two potentials is that of the *potential of thermodynamic equilibrium* between the passivating oxide ($\mathbf{Fe_2O_3}$) and the solution. The lower of the two cathodic curves illustrated in Figure 69 then corresponds to a reduction of oxygen on nonpassive iron ("active" iron); this reduction of oxygen will occur together with a corrosion of the metal if the potential is greater than potential *C* (equilibrium of the corrosion reaction). An evolution of hydrogen will take place if the potential is less than potential *A* (equilibrium of the hydrogen evolution reaction).

These observations confirm the conclusions which have been expressed at the bottom of Section 6.3.1 above concerning the electrochemical behavior of iron in oxygen-free solution; namely, that whether the 0.1 M $NaHCO_3$ solution contains oxygen (or another oxidant) or is free from it,

[7] As in the previous experiment, the possible formation of hydrogen peroxide by reduction of oxygen according to the equation

$$O_2 + 2H^+ + 2e^- \rightarrow H_2O_2$$

has been neglected.

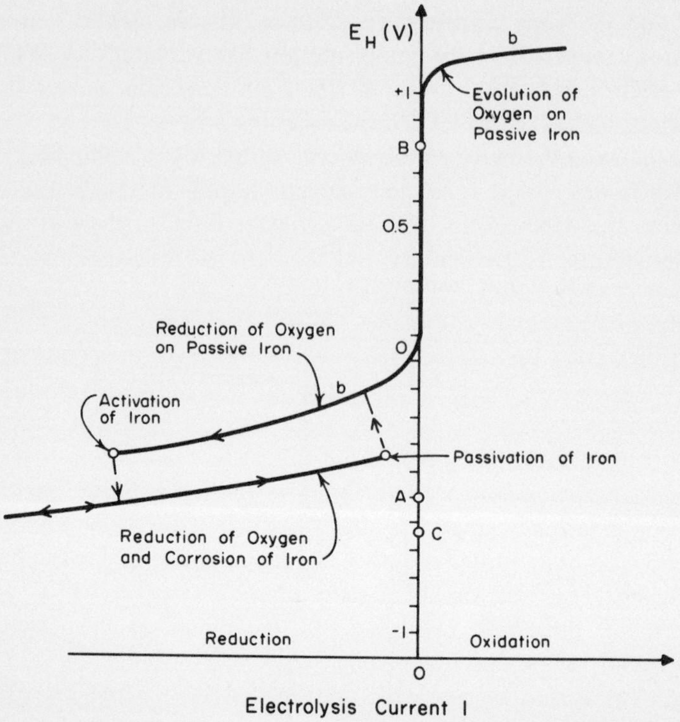

Figure 69. The behavior of iron and unalloyed steel in 0.1 M solution of $NaHCO_3$ (pH = 8.4) in the presence of oxygen. Cathodic activation of iron.

or whether the metal is electrically insulated or functions as an anode or cathode of a cell or a galvanic circuit, the metal only corrodes if its electrode potential is between about −0.62 V and −0.35 V.

6.3.3. Demonstration Experiment: Anodic Corrosion and Passivation of Iron

In Chapter 1, experiments conducted with the configuration in Figure 1d were discussed and there was noted an apparently anomalous result that iron, even though anodic, did not corrode while another iron anode in the same electrolysis circuit experienced accelerated corrosion. The bases for understanding this result is given in Figures 67a and 70. This section explains how these figures can be applied to understanding this paradoxical result. In Figure 1d[8] of Chapter 1, there are two similar electrolysis cells

[8] See *Rapports Techniques CEBELCOR* **78**, R.T. 93 (1966).

a and b and the same electric current flows between identical electrodes (etched iron) immersed in the same solution (0.1 M $NaHCO_3$). While the negative electrodes 1 and 3 are both free from corrosion, one of the two positive electrodes (2) corrodes and the other (4) does not corrode. The only difference between the two electrolysis cells is that cell a is equipped with a switch which was closed for a few seconds during the beginning of the experiment. The explanation of the experiment rests on the fact that the electromotive force of the battery and the external resistance of the elec-

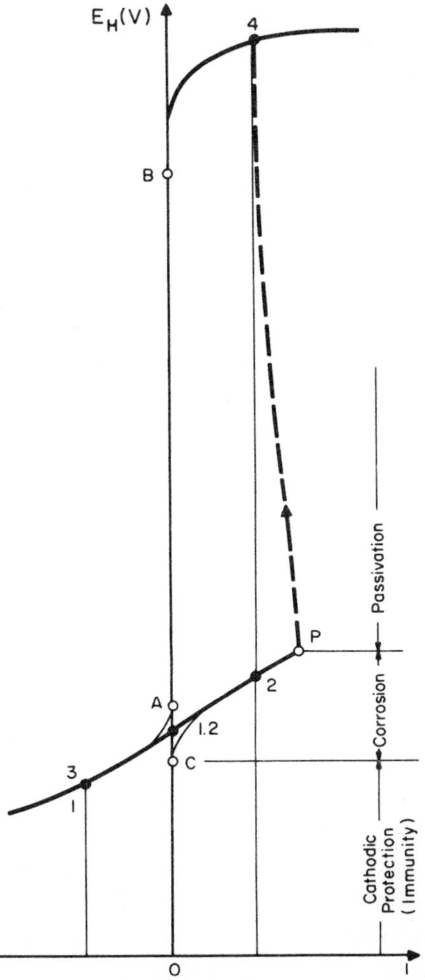

Figure 70. Explanation of experiment 1 of the experimental course.

trolysis circuit have been so chosen that, during the first phase of the experiment (with cell a short-circuited), the electrolysis current is slightly greater than the passivation current, expressed by the abscissa of point *P*. On account of this, the potential of the positive electrode 4 rapidly runs along the lower anodic branch, passes through point *P*, and stabilizes at point 4 of Figure 70 (where evolution of oxygen on the passive metal is produced). At the same time, the potential of the negative electrode 3 attains point 3 (where an evolution of hydrogen on the cathodically protected metal is produced) while the potentials of the short-circuited electrodes 1 and 2, between which no current flows, both remains at points 1 and 2 (slight corrosion with evolution of hydrogen).

During the second phase of the experiment, that is, after opening the switch, the electrolysis current which virtually maintains its previous value, expressed by the abscissa common to points 3 and 4, is insufficient to attain passivation of the positive electrode 2. Thus, the potential of this electrode runs along only a portion of the lower anodic branch to point 2′ (where corrosion without evolution of hydrogen is produced), while the potential of the negative electrode 1 goes to point 1′, identical with point 3 (where evolution of hydrogen without corrosion is produced).

6.3.4. Influence of pH and Electrode Potential on the Behavior of Iron

In 1939 and 1942–1943, some anodic polarization experiments, similar to those discussed in Section 6.3.1 above, were carried out in the presence of oxygen-free solutions of different pH values between 1.5 and 14.5 (sulfuric, acetic, bicarbonate, carbonate, and caustic solutions). For each experiment, the conditions of potential and pH were noted at which passivation appeared (i.e., noting the ordinate and abscissa of point *P* of Figure 70). These conditions of passivation are illustrated in Figures 71a and 71b. Such experiments also made it possible to establish Figures 72a and 72b which illustrate the experimental conditions of corrosion, immunity, and passivation of iron in stirred solutions together with notations of corrosion rates.

Comparison of the experimental results in Figures 71 and 72 with the theoretical potential–pH diagram for iron in Figure 39 shows that the theoretical predictions are largely verified. Effectively two regions of corrosion exist (joined by a narrow region where slight corrosion is possible) together with a region of immunity and a region of passivation of iron. The line separating the region of passivation does not correspond exactly

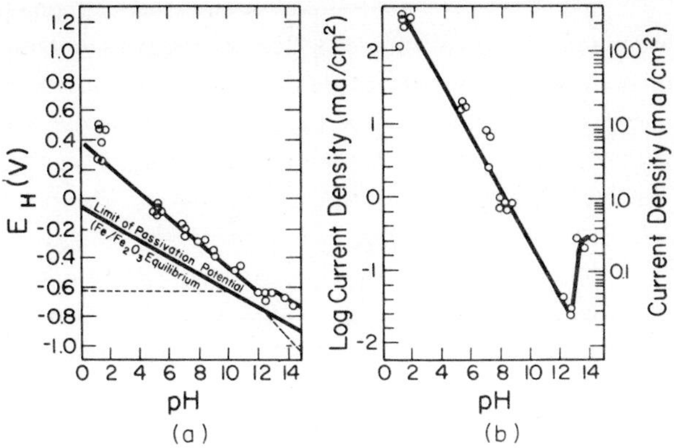

Figure 71. Influence of pH on the circumstances of anodic passivation of iron: (a) passivation potential; and (b) current densities of passivation.

with the predetermined theoretical line[9] but the general shape of the experimental figures is similar to the theoretical diagrams. Therefore, it is probable that the thermodynamic method used here to determine the general conditions of corrosion, immunity, and passivation is reasonably exact, at least in the particular case of iron. Notably, this infers the validity of the theory of passivation by oxide film which provoked much controversy when proposed in 1836 by Michael Faraday. This theory was formally vindicated in 1927 when U. R. Evans[10] isolated the film formed on iron which was passivated in chromate solution thereby making visible a film which is invisible when in contact with the metal. In 1949 and 1950, Mayne and Pryor[11] further validated the theory of passivity showing that, in the presence of an aerated caustic soda solution, iron is covered with a protective film of γFe_2O_3. Recently, the characteristics of this oxide film have been defined by M. C. Bloom and L. Goldenberg; C. L. Foley, J. Kruger, and C. J. Bechtoldt; W. T. Yolken, J. Kruger and J. P. Calvert (see p. 72).

[9] To draw the limits of the theoretical regions of immunity and passivation indicated in Figure 7, it is recalled that the solutions were arbitrarily assumed to contain 10^{-6} g-atom Fe/liter (0.06 mg/liter). Obviously in practice, this is not necessarily the case.

[10] U. R. Evans, The passivity of metals. I. The isolation of the protective film, *J. Chem. Soc.*, 1020–1040 (1927).

[11] J. E. O. Mayne and M. J. Pryor, The mechanism of inhibitors of corrosion of iron; I. By chromic acid and potassium chromate, *J. Chem. Soc.* 1831–1835 (1949); II. By sodium hydroxide solution, *J. Chem. Soc.* 229–236 (1950) (in collaboration with J. W. Menter).

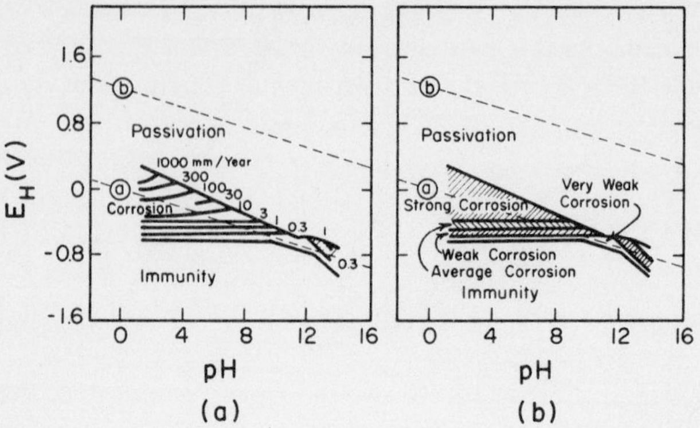

Figure 72. Experimental conditions of corrosion, immunity, and passivation of iron in the presence of stirred aqueous solutions: (a) rate of corrosion; and (b) intensity of corrosion.

Note that Figure 71b in fact indicates the critical rate of oxidation above which an oxidation is passivating and below which it is activating. From this figure, it is obvious that passivation of iron is most easily obtained for values of pH from 12.0 to 12.5. In this case, a very weak oxidizing action is sufficient to cause passivation.

Both the theoretical and experimental limits of the regions of corrosion, immunity, and passivity of iron, already illustrated respectively in Figures 7a and 71a, are reproduced in Figure 73. The corrosion behavior of iron as affected by the potential and pH obtained using various solutions are also plotted in Figure 73. These experiments were described in Chapter 1 (Figure 8). Here samples of iron were immersed in acid, neutral, or alkaline solutions and were either electrically insulated, coupled to another metal, or used as electrolytic anode or cathode. The solid symbol, ●, represents the case

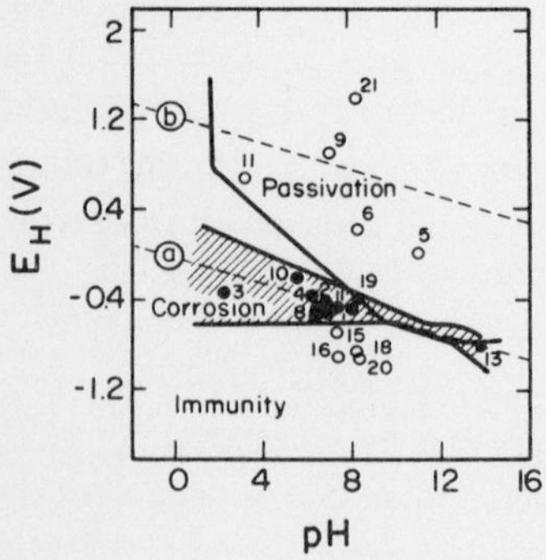

Figure 73. Theoretical and experimental conditions for corrosion or absence of corrosion of iron.

where there was corrosion and the open symbol, ○, the case where there was no corrosion. In this figure, it may be seen that there is perfect agreement between the theoretical and experimental conditions of corrosion and noncorrosion.

6.4. BEHAVIOR OF ELECTRICALLY INSULATED IRON

6.4.1. Behavior of Iron in the Absence of an Oxidant

The continuous broad line drawn in Figure 74 represents the values of electrode potential of electrically insulated iron when immersed in solutions of various pH free from oxygen and other oxidants. In such solutions, it may be seen from this figure that the potential of iron is always below line *a*, which suggests the possibility of evolution of hydrogen. At pH below about 9.5 and above about 12.5 the potential is clearly within the region of corrosion. Between pH of about 9.5 and 12.5, the potential is near the edge of the region of immunity of iron and is almost identical with the equilibrium of the system $\mathbf{Fe}$–$\mathbf{Fe_3O_4}$; under these conditions iron tends to transform to magnetite with the evolution of hydrogen. However, the rate of this transformation is small and may even become zero if the magnetite formed constitutes a protective film on the metal.

Figure 74 shows that pure water, especially free from oxygen and CO_2, is corrosive with respect to iron. As has been shown experimentally by Corey and Finnegan,[12] corrosion proceeds in such water until the solution pH is 8.3 and the content of dissolved iron is 0.2 mg/liter; this almost corresponds to the conditions of equilibrium between **Fe** and $\mathbf{Fe_3O_4}$.[13]

6.4.2. Behavior of Iron in the Presence of an Oxidant

The presence of oxygen in solution has the effect of raising the potential of iron to values indicated in Figure 74 by the dotted line. At pH values less than about 8, this elevation is insufficient to cause passivation of iron; at pH values above about 8, oxygen causes passivation of iron, with the

[12] R. O. Corey and T. J. Finnegan, *Proc. Am. Soc. Testing Materials* **39**, 1242 (1939).

[13] M. Pourbaix, Corrosion, passivity, and passivation of iron. The role of pH and potential, *Mémoires Soc. Roy. Belge. Ingénieurs Industriels* (March 1951), 30, CEBELCOR Publication F. 21 (1951) (in French).

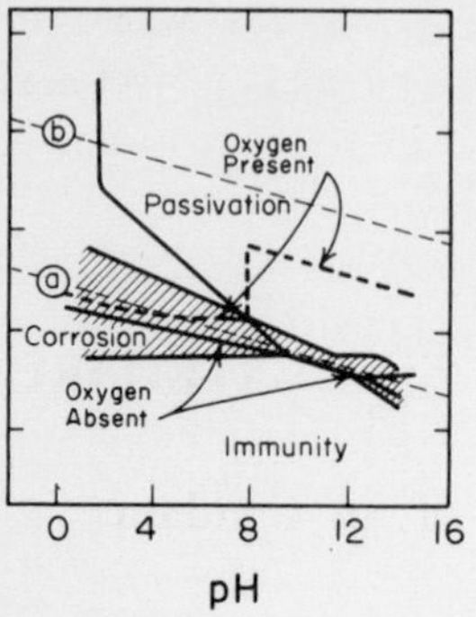

Figure 74. Electrode potential of iron in the absence and in the presence of oxygen.

formation of an oxide film which generally is protective in the case of solutions not containing chloride.

Recall that the passivating action of an oxidizing inhibitor implies some reduction of the oxidant; in the case of oxygen, this reduction takes place with the formation of water or hydrogen peroxide. The protective action of an oxidizing inhibitor will be particularly effective if it is reduced with the formation of an insoluble solid product. By being precipitated at "weak points" of the iron surface, this solid product improves the protectiveness of the passive oxide film. Therefore, the possible effectiveness of an oxidizing inhibitor may be determined by superimposing on the theoretical "corrosion–immunity–passivation" diagram of iron, the equilibrium potential–pH diagram for this inhibitor and by indicating the conditions of stability of all solid species which may be formed by oxidation of the iron and by reduction of the inhibitor. This has been done in Figure 57 where 0.01 molar solutions of eleven other oxidizing substances have also been considered together with oxygen at atmospheric pressure. Figure 57 includes hydrogen peroxide, permanganates, hyperosmiates, pertechnetates, chromates, molybdates, tungstates, vanadates, selenates, arsenates, and antimonates.

In Figure 57, the continuous broad line illustrates the potentials below which the inhibitor may be reduced. The shaded portions of the corrosion regions of iron indicate areas for which the reduction products of the oxidant are not solid bodies. To the first approximation, assuming that the reduction reactions of the oxidant are reversible and that solid bodies formed during this reduction constitute a protective coating on the iron, then the shaded zones of Figure 57 indicate the theoretical conditions of corrosion and the unshaded zones the theoretical conditions of noncorrosion. According to Figure 57, hyperosmiates and pertechnetates are extraordinarily strong corrosion inhibitors of iron; this is in agreement with the observations of

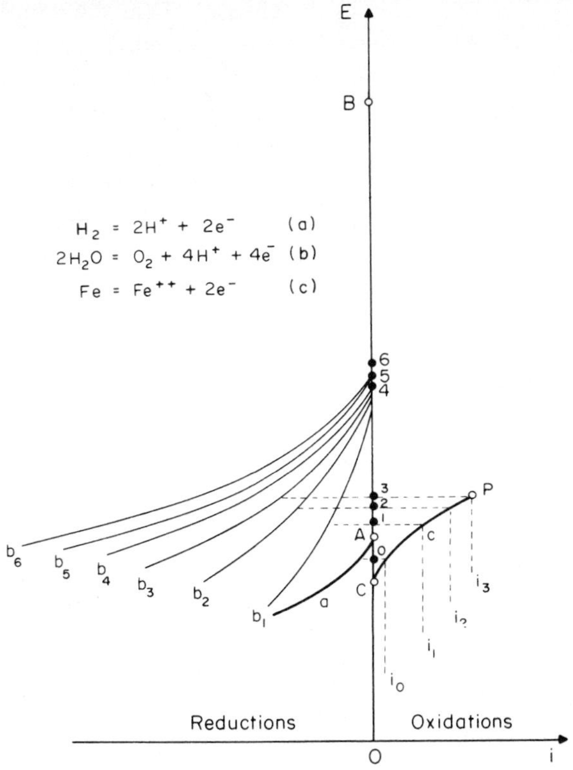

Figure 75. Behavior of iron in the presence of solutions containing different quantities of oxygen. Activating and passivating action of oxygen.

Cartledge.[14] As is well known, chromates are good inhibitors although less effective.

The polarization curves relating to the corrosion reaction of iron *c* and to the hydrogen evolution reaction on iron *a*, already represented in Figure 68, are redrawn in Figure 75 together with the polarization curves of oxygen reduction on iron for six different contents of oxygen in solution. From the theory of Wagner and Traud, according to which the electrode potential of an electrically insulated metal attains the value for which the algebraic sum of the reaction currents is nil, it is easily seen from this figure that the presence of oxygen raises the potential of the metal. As the current

[14] G. H. Cartledge, The pertechnetate ion as an inhibitor of the corrosion of iron and steel, *Corrosion* **11**, 335t–342t (1955); Action of the XO_4 inhibitors, *Corrosion* **15**, 469t–472t (1959).

of oxygen-reduction (expressed by the abscissa of curve b corresponding to the given oxygen content), is less than the passivation current of iron (expressed by the abscissa of point *P*) (points 1 and 2), the elevation of the potential is relatively small and oxygen causes an increase of the corrosion rate; when the current of oxygen reduction exceeds the passivation current (points 4, 5, and 6) the potential is significantly increased and oxygen passivates the metal; the rate of corrosion becomes practically zero and the potential stabilizes at a value near that at which curve *b* of oxygen reduction leaves the ordinate axis. The small current of oxygen reduction which is produced then corresponds to slight oxidations: oxidation of iron at weak points of the protective oxide film, oxidation of small quantities of organic material in solution, etc.

6.4.3. Differential Aeration

Let us return to the well-known experiment of U. R. Evans concerning differential aeration which has been mentioned in Chapter 1 (see the discussion associated with Figure 3). Consider the case of two homogeneous, electrically insulated samples of iron, of which one is immersed in an oxygen-free solution and the other in a solution containing oxygen in a quantity corresponding to the polarization curve b_6 of Figure 75. The situation corresponding to the b_6 curve is reproduced in Figure 76.

In the case of insulated samples, iron in oxygen-free solution has the potential denoted by 0 in Figure 76 and corrodes at the rate i_0; iron in oxygen-containing solution has the potential marked 6 and does not corrode.

If the two samples are connected by a conductor and are both immersed in an electrolyte, the potentials of the two iron samples tend to approach one another and the points representing the two samples give rise, in Figure 76, to the electrolysis curve $a + c$ (for the nonaerated sample) and the curve b_6 (for the aerated sample). These potentials stabilize at values $0'$ and $6'$ corresponding to currents which effectively flow between the "anodic" nonaerated sample and the "cathodic" aerated sample. As Evans showed, it may be seen from Figure 76 that aeration of an electrode causes increasing corrosion of the nonaerated electrode. Under the conditions represented in this figure, reduction of oxygen without corrosion is produced on the nonaerated electrode.

Figure 77a, derived from Figure 74 and 76, illustrates that the operation of "Evans cells" of differential aeration varies according to pH as

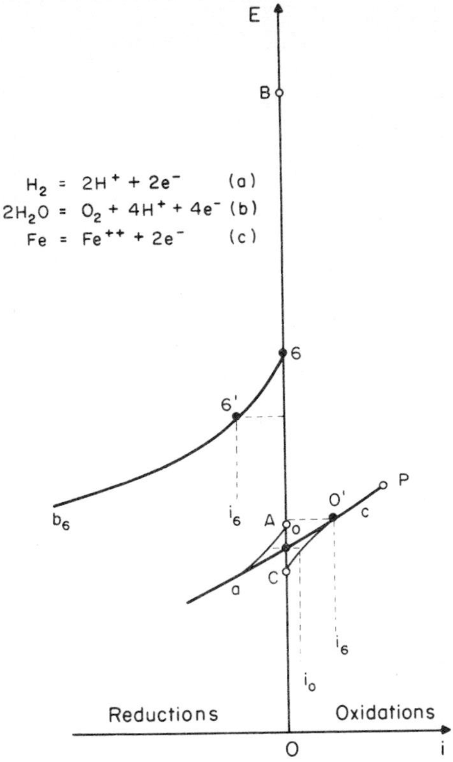

Figure 76. Differential aeration of iron. Polarization curves.

follows, *when the conditions are such that the pH is the same in both the aerated and in the nonaerated zones*:

1. pH less than 8: "abnormal operation": aeration causes a small increase of the electrode potentials of both zones, with small current flow between these zones. It causes an increase of the rate of corrosion of both zones.
2. pH between about 8 and 10: "normal operation": aeration causes an important increase of the potential of the aerated zones and a small increase of the potential of the nonaerated zones, with a strong current flow between them. It passivates the aerated zones and increases the corrosion of the nonaerated ones.
3. pH between about 10 and 13: "abnormal operation": aeration causes an important increase of the potentials of both zones, without appreciable current flow. It passivates both zones.

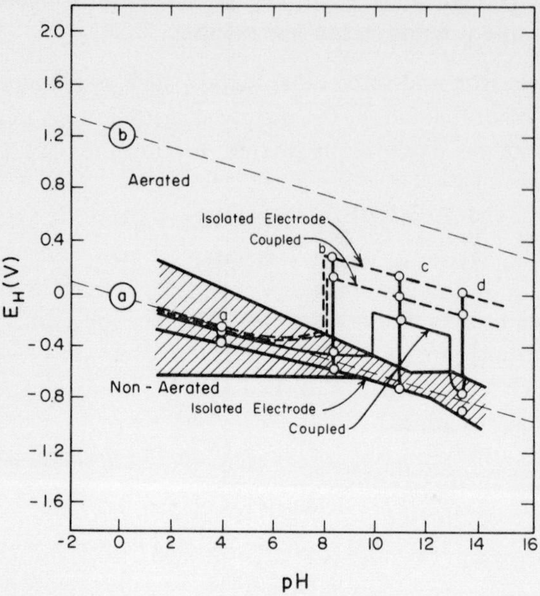

(a)

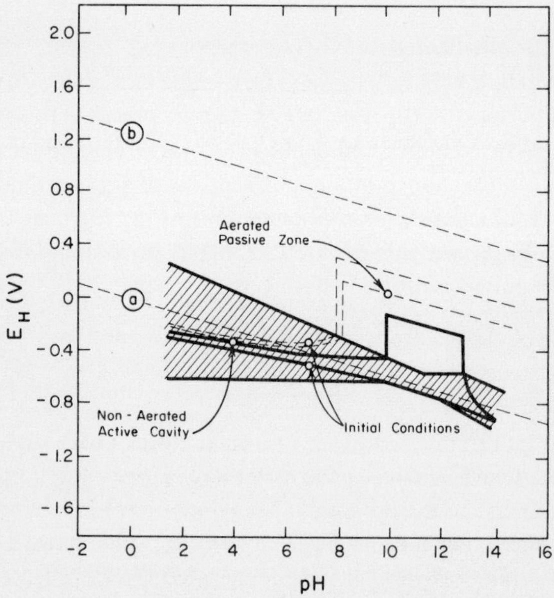

(b)

Figure 77. Differential aeration of iron. Influence of pH on the working conditions of Evans cells.

4. pH greater than about 13: "normal operation": quantitatively as between about 8 and 9, but less intense.

In fact, when iron and some other metals, such as aluminum, corrode in a somewhat neutral solution, the dissolved metallic ions may hydrolyze with formation of an oxide or hydroxide, for instance according to the reactions

$$Fe^{++} + 2H_2O \rightarrow \mathbf{Fe(OH)_2} + 2H^+$$

$$Al^{+++} + 3H_2O \rightarrow \mathbf{Al(OH)_3} + 3H^+$$

This may lead, when the corrosion occurs under conditions of restricted diffusion, for instance inside a pit or a crevice (or in an other "occluded corrosion cell" OCC, as called by B. F. Brown), to a local acidification of the nonaerated active zones.

Besides this, the reduction of oxygen on the aerated passive zones occurs according to the reaction

$$O_2 + 4H^+ \ + 4e^- \rightarrow 2H_2O$$

or

$$O_2 + 2H_2O + 4e^- \rightarrow 4OH^-$$

which leads to an alkalization of the aerated passive zones.

This case, which is very frequent, is represented by Figure 77b: the operation of the Evans cell is then "normal": a strong current is flowing between alkaline passive aerated zones and acid active nonaerated zones or cavities.

The practical importance of the phenomena of differential aeration is considerable and the reader may profitably consult the work of U. R. Evans and his school. Here, we will be content merely to point out two consequences of differential aeration, relating respectively to the formation of "corrosion pits" and to "waterline corrosion."

6.4.3.1. Corrosion Pits

Consider (Figure 78a)[15] a surface of iron in contact with slightly aerated water and covered by a passive oxide film $\mathbf{Fe_2O_3}$ containing a fissure which

[15] Figure 78b is adapted from a figure drawn up by U. R. Evans during a study of the corrosive action of drops of solution (U. R. Evans, *Metallic Corrosion, Passivity, and Protection* (2nd ed.), Arnold, London (1946), Figure 60, p. 271). Values of electrode potentials and of pH inside and outside the pit have been added (M. Pourbaix, *Electrochemical Aspects of Stress Corrosion Cracking*, NATO Research Evaluation Conference, Ericeira (March 29–April 2, 1971) Figure 3).

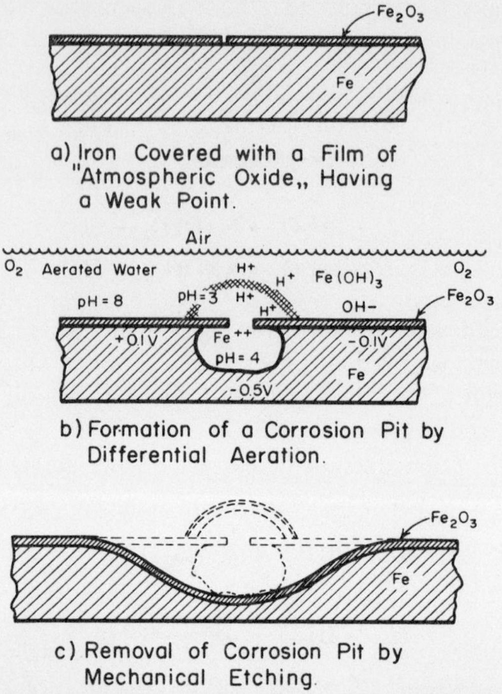

Figure 78. Corrosion pits formed on iron: (a) iron covered with a film of "atmospheric oxide" having a weak point; (b) formation of a corrosion pit by differential aeration; and (c) removal of corrosion pit by mechanical etching.

exposes bare metal. Oxygen, which has easy access to the large areas of passive oxide-covered metal, has difficulty in gaining access to the small fraction of active metal at the bottom of the fissure. A differential aeration cell will then be produced between a small "anode" at low electrode potential and a large "cathode" at high electrode potential (Figure 78b). At the "anodic" regions, for instance, where the potential is represented by the point 0′ of Figure 76, corrosion of metal occurs (with formation of ferrous ions according to the reaction $\mathbf{Fe} \rightarrow Fe^{++} + 2e^-$).

At the "cathodic" regions, the potential being represented, for example, by point 6′ of Figure 76, a reduction of oxygen occurs giving a more alkaline solution (by the formation of OH^- ions according to the reaction $\boldsymbol{O_2} + 2H_2O + 4e^- \rightarrow 4OH^-$). In the bulk of the solution between these two regions, a deposit of ferric hydroxide with acidification of the solution occurs. This results from the formation of H^+ ions according to the reaction $4Fe^{++} + \boldsymbol{O_2} + 10H_2O \rightarrow \mathbf{4Fe(OH)_3} + 8H^+$ associated with the interaction between the Fe^{++} ions formed at the anode and the oxygen existing at the cathode.

This ferric hydroxide, the essential constituent of rust, separates out into a mushroom shape above the "anodic" corrosion regions and forms a porous shield which further screens these regions from the access of oxygen and forms an "occluded corrosion cell," where the solution remains acid due to hydrolysis phenomena. In the frequent case where some chloride is present in the environment, this acid is hydrochloric acid. This intensifies considerably the corrosive action of differential aeration couples. The result is that every time such an acid "canker" of corrosion is formed under a permeable mushroom of rust, rusting proceeds unhindered under this mushroom as long as the metal remains in contact with the aerated water. The concealed character of the phenomenon is the fundamental cause of the serious nature of the corrosion of iron. By proceeding under a shield which often screens it from satisfactory inspection, corrosion is especially serious in the case of structures covered with paint which has been damaged or applied on a poorly prepared surface.

The only means of preventing such corrosion is to correctly apply cathodic protection to the structure or to remove completely, by mechanical polishing or chemical etching, all traces of rust on the surface (Figure 78c). In the latter case, having largely opened all the pits formed by corrosion, a perfect passivation of all the metal surface should be achieved for example by applying an inhibitive layer of "primer" paint on surface free from all traces of rust and moisture.

6.4.3.2. Waterline Corrosion

In the case of an ordinary steel vessel containing relatively stagnant water, it is often noticed that a particularly intense corrosion of the vessel wall occurs in the immediate neighborhood of the air–water interface (Figure 79).[16]

Corrosion initiated at a weak point of the metal surface then leads to the formation, along the meniscus where renewal of oxygen is easy, of a membrane of ferric hydroxide which shields the metallic regions underneath from access of oxygen, and causes the flow of intense differential

[16] Figure 81 is adapted from a figure drawn by U. R. Evans for the case of iron in the presence of chloride containing solutions (U. R. Evans, *Metallic Corrosion, Passivity and Protection* (2nd ed.), Arnold, London (1946), Figure 118, p. 551).

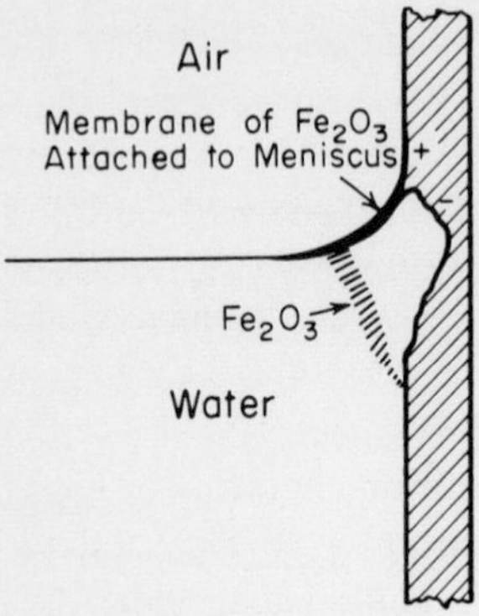

Figure 79. Waterline corrosion.

aeration currents between the less-aerated active cavities and the nearby aerated passive regions. Sometimes this may lead to rapid perforation of the wall of the water line.

6.5. BEHAVIOR OF IRON COUPLED TO ANOTHER METAL

Here, we examine the behavior of iron when it is not electrically insulated but in contact with another metal. We consider successively the cases (a) where the metal with which iron is in contact is more noble (i.e., the electrode potential of this metal, when insulated, is higher than the electrode potential of iron) and (b) where this metal is less noble than iron (i.e., the potential of the insulated metal is lower than that of iron).

6.5.1. Coupling of Iron with a More Noble Metal

When the metal is more noble than iron, the behavior of iron generally varies according to whether this metal is or is not corroded by the given solution.

6.5.1.1. More Noble Metal Not Corroded by the Solution

When the metal is not corroded by the solution, which in this case we assume to be free of oxygen or other oxidant, the presence of the metal may be expressed in Figure 68 by the addition of a supplementary polarization curve for the additional reduction of hydrogen ions on the new metal.

Besides the polarization curves relative to the corrosion of iron (c) and to the evolution of hydrogen on iron (a_{Fe}) already drawn on Figure 68,

Figure 80a shows schematically two polarization curves relative to the evolution of hydrogen, respectively, on platinum (a_{Pt}) and copper (a_{Cu}). The potential of iron, at point 1 in the case of insulated iron, increases by coupling toward value A which is the potential of insulated platinum and copper when in equilibrium with hydrogen under atmospheric pressure. The exact value of the potential of coupled iron depends mainly on the hydrogen overpotential curve of the other metal and the relative surfaces of the two metals. If, to the first approximation, it is assumed that the two metals have the same electrode potential (implying that the ohmic drops resulting from the flow of local currents are neglected) and that the polarization curves drawn on Figure 80a are relative to the real surfaces of the two metals, then the potential common to the two metals is established at a value for which the algebraic sum of the currents of oxidation and reduction is zero. According to Figure 80a, the result is that coupling of iron with platinum raises the potential of iron from value 1 to 2 and increases considerably the rate of corrosion.

By contrast, coupling of iron with copper (whose hydrogen overpotential is greater than that of iron) does not significantly modify the potential of iron nor its rate of corrosion. So in practice, copper does not affect the

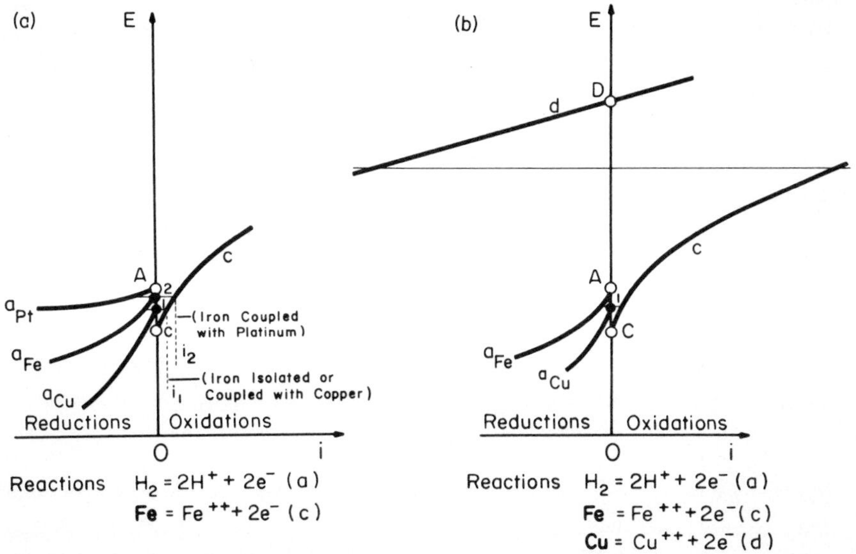

Figure 80. Coupling of iron with a more noble metal: (a) metal not corroded by the solution (platinum and copper in absence of oxidant); and (b) metal corroded by the solution (copper in the presence of oxidant).

rate of corrosion of iron *if the solution is free from oxygen and all trace of dissolved copper.*

6.5.1.2. More Noble Metal Corroded by the Solution

The action of copper coupled to iron is completely different if the solution contains dissolved copper in a form reducible to metallic copper; this is especially so in some cases for water containing oxygen or ammonia. The presence of dissolved copper is expressed by the addition to Figure 80a relative to copper-free solutions of a supplementary polarization curve, relative to the reduction of dissolved copper to metallic copper. Figure 80b gives an overall view of all the polarization curves under consideration. Obviously, the value of potential *D* below which dissolved copper is reduced to metallic copper depends greatly on the quantity of copper in solution as well as the form (complexed or not) under which it exists.

Unpublished work performed by CEBELCOR has shown that copper dissolved in water is not reducible at the usual pH of waters (about 7 to 8 where, according to Figure 37, $\mathbf{Cu_2O}$ might be stable), but is reducible at lower pH, such as those existing locally in active pits (about 3 to 5, where $\mathbf{Cu_2O}$ is not stable and where Cu^{++} ions may thus be reduced to metallic copper). This leads to the conclusion of very great practical importance that copper dissolved in water is not harmful to passive metals, as iron covered with a $\mathbf{Fe_2O_3}$ protective film, but is very harmful to metals where active pits are present.

Figure 80b shows that all traces of dissolved copper which contact nonpassive iron in an acid environment, e.g., where freshly formed $\mathbf{Fe(OH)_3}$ exists in active pits, are reduced to metallic copper which will be deposited on iron as a precipitate, with oxidation of iron to ferrous ions, Fe^{++}, or to the oxide, $\mathbf{Fe_3O_4}$ (or possibly $\mathbf{Fe_2O_3}$).[17] (See Figure 81a.) A vigorous corrosion is produced which is localized to weak points of the oxide film present on the iron and gives rise to corrosion products consisting of a mixture of metallic copper and magnetite (or possibly $\mathbf{Fe_2O_3}$). Such corrosion is particularly serious on steel tubes of certain boilers whose water circulation system incorporates copper alloy equipment. Due to the passive

[17] This will be according to the following reactions:

$$Cu^{++} + 2e^- \rightarrow \mathbf{Cu}$$
$$\mathbf{Fe} \rightarrow Fe^{++} + 2e^-$$
$$\overline{Cu^{++} + \mathbf{Fe} \rightarrow \mathbf{Cu} + Fe^{++}}$$

$$4Cu^{++} + 8e^- \rightarrow \mathbf{4Cu}$$
$$\mathbf{3Fe} + 4H_2O \rightarrow \mathbf{Fe_3O_4} + 8H^+ + 8e^-$$
$$\overline{4Cu^{++} + \mathbf{3Fe} + 4H_2O \rightarrow \mathbf{4Cu} + \mathbf{Fe_3O_4} + 8H^+}$$

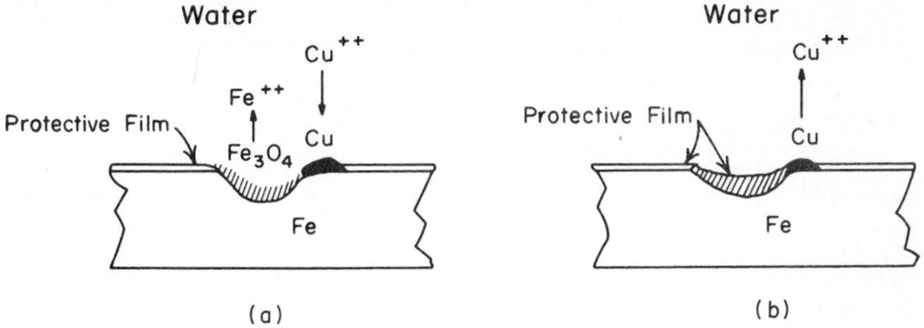

Figure 81. Formation and dissolution of copper deposits in boilers: (a) formation of copper deposit, with local corrosion of iron; and (b) dissolution of copper deposit on passive iron.

nature of iron oxide formed in this way and to variations which may occur in the nature of the water, it is possible that in such boilers a local re-dissolution of copper may be produced at certain times in the passive regions (at high electrode potential) with further deposition of copper thus dissolved in the nonpassive regions (at low electrode potential). Each deposition of copper is associated with a corrosion of iron (Figure 81b). Hence, copper liberated in the boiler may "leap-frog" around the water circulation system and each contact with iron may cause corrosion (see Fig. 82).

It is pointed out that this corroding action of solutions containing dissolved copper not only affects already corroded iron and ordinary steels but it may also affect similarly other metals less noble than copper, notably zinc and aluminum. For example, it is preferable not to collect rain water, which has run off copper roofing, in zinc pipes or containers.

6.5.2. Coupling of Iron with a Less Noble Metal: Cathodic Protection

When the metal with which iron is coupled is less noble (that is the electrode potential has a lower value than that of iron when electrically insulated) it may be considered, in the case of solution virtually free from

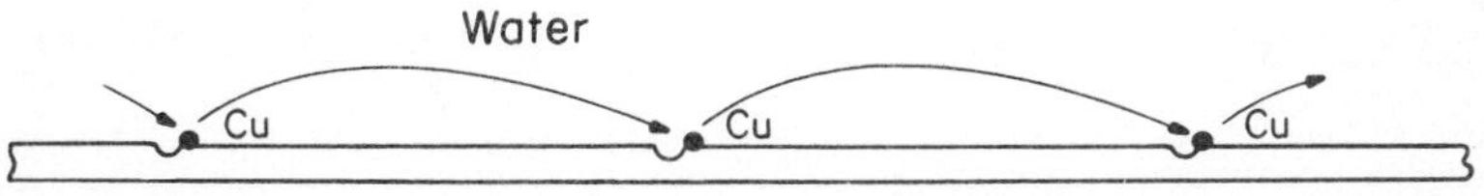

Figure 82. Local corrosion of iron, by successive deposition and dissolution of copper.

oxygen and other oxidant, that the presence of the metal may be expressed by the addition to Figure 68 of two supplementary polarization curves relating respectively to corrosion of this metal and to evolution of hydrogen on this metal.

Figure 83a shows schematically the polarization curves relative to the corrosion of iron (curve *c*) and zinc (curve *d*) as well as the polarization curves a_{Fe} and a_{Zn} relative to evolution of hydrogen on both these two metals. Points 1 and 2 in Figure 83a correspond to the corrosion of insulated specimens of iron and zinc respectively. Note that the kinetics of hydrogen reduction in zinc is much slower than on iron owing to the lower exchange current. If, as an approximation, it is assumed that, when coupled, these metals both have the same potential, Figure 83a leads to the conclusion that the common value of these potentials is point 3. Then, in the case of Figure 83a, this coupling gives rise to the four following effects:

1. Total suppression of the corrosion of iron
2. Considerable intensification of evolution of hydrogen on iron
3. Intensification of the corrosion of zinc
4. Suppression of the evolution of hydrogen on zinc

Thus, protection of iron with corrosion of zinc occurs: zinc acts as a "sacrificial anode" (or as a "reactive anode") and causes cathodic protection

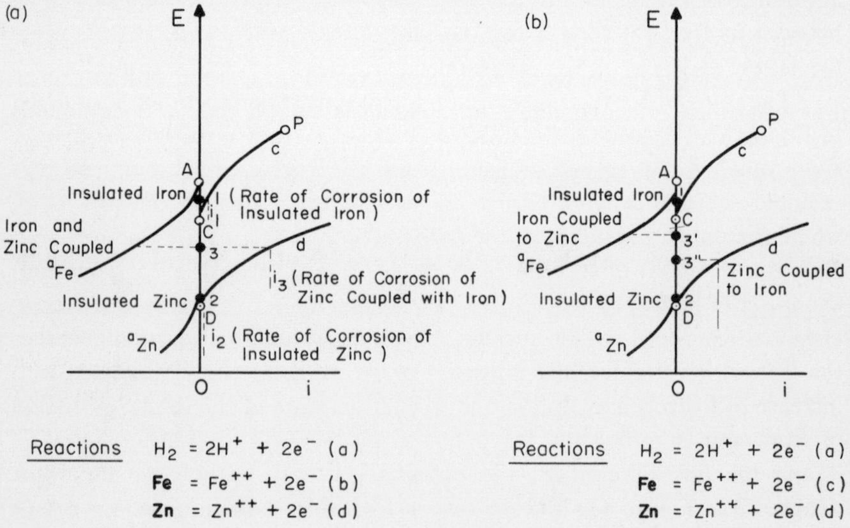

Figure 83. Coupling of iron with a less noble metal (zinc): (a) hypothetical case where the potentials of iron and zinc are equal; and (b) actual case where the potentials of iron and zinc are different.

of iron. This gives rise to a flow of current between iron and zinc. The flow is i_3, common to the rate of evolution of hydrogen on iron and the rate of corrosion of zinc. The protective action of zinc on iron results from the existence of a galvanic cell at whose electrodes the following electrochemical reactions proceed:

on iron (+ pole of the cell, or "cathode"):

$$2H^+ + 2e^- \rightarrow \boldsymbol{H_2} \qquad \text{(a)}$$

on zinc (— pole of the cell, or "anode"):

$$\mathbf{Zn} \rightarrow Zn^{++} + 2e^- \qquad \text{(d)}$$

$$\mathbf{Zn} + 2H^+ \rightarrow Zn^{++} + \boldsymbol{H_2}$$

It is recalled that to obtain Figure 83a we presumed, as an approximation, that the coupled iron and zinc both have the same electrode potential; in fact, this is only true along the line of contact of iron and zinc. In all other places, due to the existence of ohmic resistance to current flow as well as the possibility of diffusion potentials, the electrode potential of iron adopts a value between its free potential 1 and the "short-circuit potential" 3. The potential in general is then above this lattice point, for example, at point 3′ of Figure 83b. In the same way, the potential of zinc has a value between its free potential 2 and the short-circuit potential 3, that is below the latter, at 3″.

The rates with which the oxidation and reduction reactions, considered in Figure 83, proceed on the different parts of a strip of iron and a strip of zinc, attached end to end and immersed in a corrosive solution, are represented schematically in Figure 84. At the upper part of the iron strip, where the ohmic resistance of the strip curbs the effect of the coupling with zinc, the potential of the iron virtually is value 1, and iron corrodes with evolution of hydrogen 1 of Figure 83a, according to two equivalent electrochemical reactions. As the surface of iron under consideration approaches the iron–zinc interface, the potential of the iron electrode decreases with increase of hydrogen evolution and with diminution of the rate of corrosion of iron, and is zero when the potential reaches the equilibrium value C. Along the line of contact, the potential virtually establishes the value denoted 3 in Figure 83a and the rate of hydrogen evolution on iron equals the rate of corrosion of zinc. Then as we proceed away from the interface of the zinc, the potential of zinc falls with decrease of the rate of corrosion and becomes equal to the rate of free corrosion of this metal at the bottom

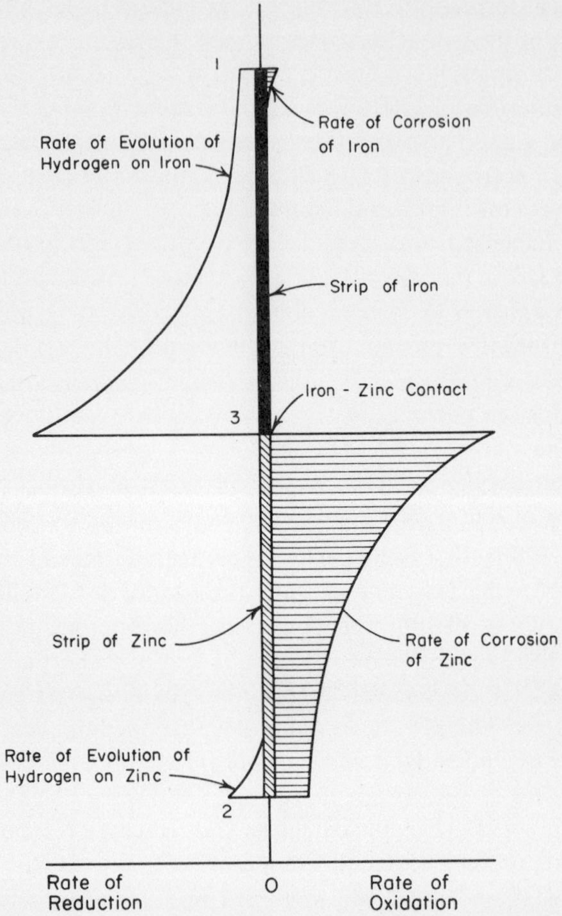

Figure 84. Distribution of the rates of corrosion and hydrogen evolution along the length of iron and zinc strips in contact with one another.

of the strip where the ohmic resistance of the zinc strip is such that the coupling with iron has no effect. At this point the rate of hydrogen evolution on zinc, which was zero in the proximity of the line of coupling with iron, is equal to the rate of corrosion of zinc. In fact, in Figure 84, it is possible to observe successively: a zone of corrosion of iron with slight hydrogen evolution; a zone of protection of iron with intense hydrogen evolution; a zone of intense corrosion of zinc without evolution of hydrogen; and a zone of slight corrosion of zinc with slight hydrogen evolution.

It is noted that, in Figure 84, the total intensity of the reduction reactions (represented by the sum of the two areas on the left of the central axis) is equal to the total intensity of the oxidation reactions (represented by the sum of the two areas to the right of this axis).

The principles illustrated in Figure 83 are used industrially in the *cathodic protection* of iron by means of "reactive anodes" or "sacrificial anodes" of zinc. This operation which is only applicable to structures immersed or surrounded by an electrolytic corrosive medium (e.g., water, wet soil, wet concrete) consists of electrically connecting the structure to some zinc immersed or surrounded by a corrosive medium. As indicated in Figure 83, it is then possible to protect iron against corrosion, due to a sacrificial corrosion of zinc which lowers the electrode potential of iron below its protection potential (generally about −0.62 V).

In the case represented in Figure 83, the four polarization curves are such that for the potential at the couple, 3, there is no corrosion of iron (the abscissa of line *c* is zero) and there is practically no evolution of hydrogen on zinc (the abscissa of the line a_{Zn} is practically zero). The rate of corrosion of zinc (abscissa of the line *d*) is then equal to the rate of evolution of hydrogen on the iron (abscissa of the line a_{Fe}), evolution which is indispensible to ensure the protection of this metal. Then the efficiency of protection due to zinc is considered as 100%.

The three diagrams of Figure 85 illustrate the effect of changing the oxidation ($\mathbf{Zn} \rightarrow Zn^{+2} + 2e^-$) and reduction ($2H^+ \rightarrow \boldsymbol{H_2} + 2e^-$) kinetics relative to the efficiency of zinc as a sacrificial anode. The hydrogen ion reduction kinetics can be accelerated by the presence on the surface of the zinc of a metal such as germanium which has a lower overpotential for this

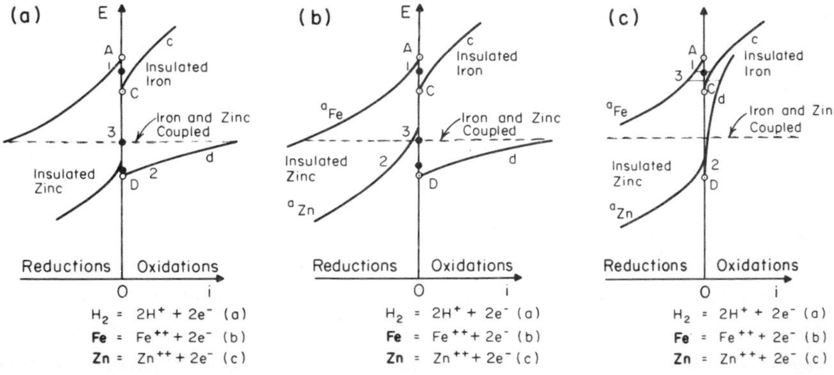

Figure 85. Coupling of iron with a less noble metal (zinc). Cathodic protection of iron by zinc under different circumstances of hydrogen overpotential and passivation of zinc: (a) zinc at high hydrogen overpotential and not passive; (b) zinc at low hydrogen overpotential and not passive; and (c) zinc at high hydrogen overpotential and passive.

reaction (Figure 85b). The oxidation kinetics of zinc can be significantly decreased by the formation of a reaction product film as shown in Figure 85c such as might form in warm water or as a result of burial in a soil which is only slightly aggressive toward zinc. It is seen that in the case only of impure zinc at low hydrogen overpotential (Figure 85b) protection of iron remains possible but at the expense of hydrogen evolution on zinc which causes corrosion of this metal thereby reducing the efficiency of the protection. The quantity of zinc needed to be dissolved to ensure protection is, in fact, equivalent to the quantity of hydrogen released on the iron (which may be considered as effective) added to the quantity of hydrogen evolved on the zinc (which is ineffective from the viewpoint of obtaining protection). In the case of passivated zinc (Figure 85c) the potential of the iron electrode will not be lowered below the equilibrium potential of corrosion *c* and there will not be significant protection of this metal.

It should be emphasized that the action of cathodic protection which has just been described for the iron–zinc couple may be obtained by coupling with any other corrodible metal having a sufficiently low electrode potential. This property is applied on a large scale in industry with the use of sacrificial anodes of magnesium, as well as of aluminum–zinc alloys.

6.6. PROTECTION OF IRON AND STEEL AGAINST CORROSION[18]

6.6.1. General Criteria

Returning to Figures 7 and 72, which express, on the basis of, respectively, theoretical and experimental conditions, the general circumstances of corrosion, immunity, and passivation of iron in the presence of aqueous solution at a temperature of 25°C, it is obvious that to protect iron against corrosion one may proceed as follows (Figure 86):

6.6.1.1. Cathodic Protection

The electrode potential can be depressed to the *region of immunity*. This is carried out by carefully employing a reducing action on the iron surface. For example, iron can be made the cathode (negative electrode) when an external source supplies *continuous electric current*. Iron may also

[18] For fuller details of this subject, the reader would be well advised to consult the *General Bibliography* under the heading *Corrosion*.

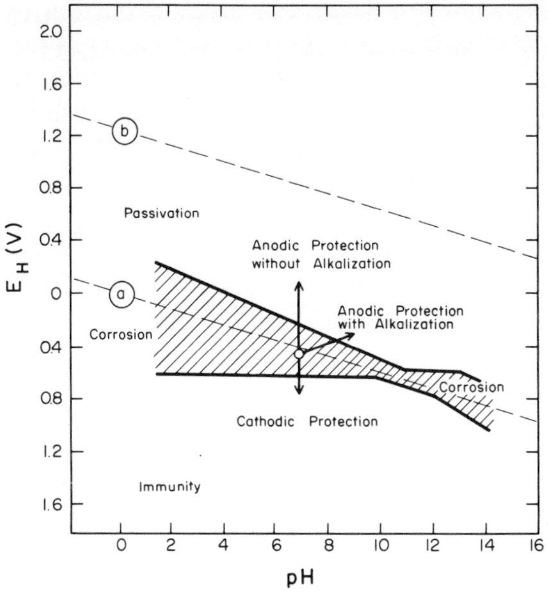

Figure 86. Protection of iron against corrosion.

be coupled with another metal corroding at a low potential, such as *sacrificial anodes* of zinc or magnesium. It is on this principle, at least in part, that the effectiveness of galvanization (hot-dip galvanization, zinc plating, zinc spraying) as well as the use of zinc-rich paints depends.

Cathodic protection will be complete or partial depending on whether or not the protection potential (generally around -0.62 V with respect to a standard hydrogen electrode) is reached. Given that the domain of stability of metallic iron is situated completely below the region of stability of water, it should be noted that the state of cathodic protection by immunity is only maintained by a permanent consumption of energy (consumption of electric current, or of zinc or magnesium) which causes evolution of hydrogen on the surfaces of protected iron.

6.6.1.2. Protection by Passivation; Oxidizing Inhibitors

The electrode potential of the metal can be raised to the region of passivity by adding oxidizing inhibitors (chromates, etc.), by anodic polarization, or simply by alkalization of the solution. The theoretical conditions of effectiveness of various oxidizing inhibitors discussed have been illustrated already within the framework of equilibrium potential–pH diagrams. However, such a passivating action is not without danger and may sometimes cause more harm than good. This is partly because, as Figure 72 shows, an insufficient elevation of potential may considerably increase the rate of corrosion and also because, if the oxide film formed by passivation is imperfectly protective and has local weak points (such as might be produced in the presence of chlorides), the general corrosion which occurred in the absence of oxidant may be replaced by an extremely vigorous *local corrosion*. Oxidizing inhibitors are completely effective only when properly used, otherwise they may be harmful. Furthermore, since the region of passivation of iron has a vast area in common with the region of stability of water, the state of protection by passivation once established may be maintained without the consumption of either energy or reactant.

6.6.1.3. Protection by Adsorption Inhibitors

A third procedure for the protection of iron against corrosion consists of using *adsorption inhibitors* (soluble oils, red lead primers) which become adsorbed to certain parts of the metal surface, separating it from contact with the aggressive medium, and in this way causing a decrease of the corrosion rate. Generally, these adsorption inhibitors act without significantly modifying the electrode potential of iron. Though they may not be as effective as oxidizing inhibitors, on the other hand, they are not dangerous.

6.6.1.4. Other Protection Procedures

Besides the three protection procedures cited above, other procedures should also be mentioned briefly. These procedures, widely used in practice, aim to prevent all contact between iron (or ordinary steel) and the corrosive medium. Such procedures include the use of impermeable coatings (paints and other coatings: metallic, plastic, etc.) which tend to cover iron with a waterproof coating. Also, use may be made of a less corrodible metal or alloy than ordinary steel: alloy steels or other special metals and alloys which passivate more readily.

6.6.2. Electrochemical Techniques for the Study of Corrosion

6.6.2.1. Intensiostatic, Potentiostatic, and Potentiokinetic Methods

How then does one establish a procedure for the protection of steel used in a particular case, and how is the effectiveness of the procedure controlled? This question has been discussed briefly in Section 6.3 (polarization curves) of this chapter, *apropos* the study of the behavior of iron in the presence of bicarbonate solutions by means of electrolysis carried out in *stirred solutions* with manual "intensiostatic" or "galvanostatic" apparatus. Since 1939, when these intensiostatic electrolyses were first carried out,[19] experiments have been undertaken using increasingly sophisticated *potentiostats*. At first, mechanical instruments (in 1952 by J. M. Bartlett and L. Stephenson) and then electronic instruments were used (in 1953 by G. Schouten and J. G. F. Doornekamp, in 1954 by J. Schoen and K. E. Staubach, and in 1955 by R. Olivier).[19]

Before proceeding, a word should be said about the inherent nature and distinguishing characteristics of "intensiostatic," "potentiostatic," and "potentiokinetic" techniques. In the intensiostat (or galvanostat), the independent variable is current. Thus, when the current exceeds point *P* of Figures 70 and 87a, there is a discontinuity in potential and the current jumps to a similar value on the oxygen evolution or pit breakdown curve. In other words, at *P*, the current is double-valued. In the case of the potentiostatic technique, the potential is the independent variable and no such double-valued situation occurs as shown in Figure 87b. In the potentiostatic technique, the potential is held constant until the current reaches a stable value. This potential and steady state current are then recorded for successive points and the curves of potential-dependent current is obtained. A "potentiokinetic" curve is obtained by continuously varying the potential and recording the current. This is the type of curve shown in Figure 87b.

For comparison, Figure 87 illustrates two current–potential curves relating to mild steel in the presence of oxygen-free 0.1 M $NaHCO_3$ solution. Figure 87a is obtained by the intensiostatic method already used in Figure 67 (but with a higher electrolysis potential and a higher external electric resistance, so that the flow of current after passivation does not appear to diminish) and during which the electrolysis current is varied manually, noting the electrode potential values corresponding to each value of current. Figure 87b is obtained by a potentiokinetic method in which the electrode potential of the metal under examination is varied linearly as a function of time and the current corresponding to each potential is automatically recorded. A comparison of the two figures shows that they both reveal a passivation, characterized by the coordinates of potential and current relating to point *P*. Also, both curves show the characteristics of general

[19] A bibliography on this subject is given in French in *Rapports Techniques CEBELCOR* RT. 123 (1964), and in English in CEBELCOR Publications E. 50 (1964) and E. 57 (1965).

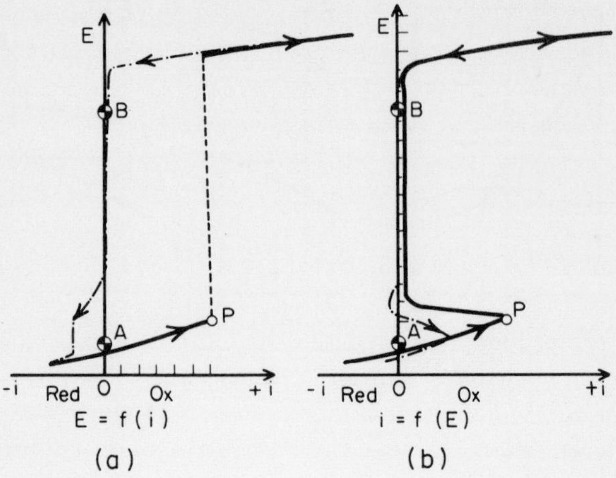

Figure 87. Polarization curves of mild steel in 0.1 M solution of $NaHCO_3$, in the absence of oxygen: (a) galvanostatic method; and (b) potentiokinetic method.

corrosion at potentials below the passivation potential, as well as characteristics of oxygen evolution at high electrode potentials. By contrast, the potentiokinetic method gives more precise indications concerning the quality of passivation (which appears when the potential increases above that of point *P*) and concerning the conditions of activation (which is shown when, during polarization at decreasing potentials, the potential becomes less than that of point *P*).

6.6.2.2. General Principles

As we have seen in Chapters 4 and 5, the conditions under which an electrochemical reaction may proceed at the interface between a metal and an electrolyte depend essentially on the relative values of the electrode potential, *E*, of the metal and the thermodynamic equilibrium potential, E_0, of the reaction. At present, these values are known for the majority of reactions occurring in aqueous corrosion, for the temperature of 25°C.[20]

When a metal is in contact with an electrolyte, the direction of each electrochemical reaction is exactly defined by the electrode potential of the metal, *E*, relative to the equilibrium potential of the reaction, E_0. However complex this medium and however numerous the electrochemical reactions,

[20] M. Pourbaix *et al.*, *Atlas d'Equilibres Electrochimiques*, Gauthier-Villars, Paris, and CEBELCOR, Brussels (1963) and *Atlas of Electrochemical Equilibria in Aqueous Solutions*, Pergamon Press, Oxford, and CEBELCOR, Brussels (1966).

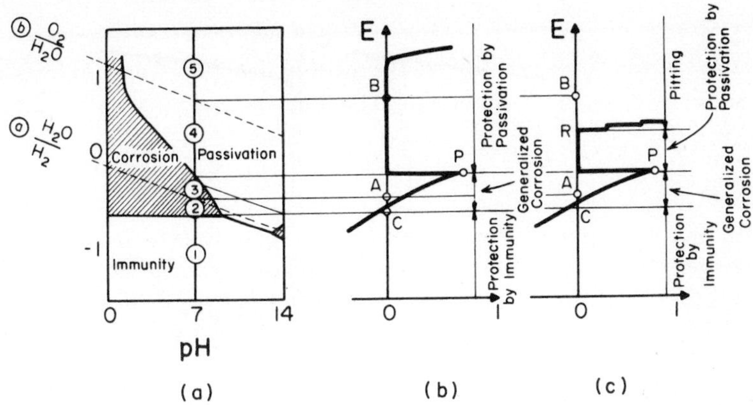

Figure 88. Theoretical and experimental predetermination of the circumstances of non-corrosion, general corrosion, and localized corrosion (pitting) in the case of iron in the presence of solution of pH = 7: (a) theoretical circumstances of corrosion, immunity, and passivation; (b) potential–current curve obtained potentiokinetically, in the case of a perfectly protective oxide film (conditions of protection and of generalized corrosion); and (c) current–potential curve determined potentiokinetically in the case of an imperfectly protective oxide film; circumstances of protection, general corrosion, and local corrosion (pitting).

if the electrode potential of the metal is *greater* than the equilibrium potential of the reaction then the reaction may only proceed in the direction of *oxidation*. Each reaction may only proceed as a *reduction* if the electrode of the metal is *less* than its equilibrium potential, i.e.,

$$\text{if } E > E_0, \quad \text{possibility of oxidation}$$
$$\text{if } E < E_0, \quad \text{possibility of reduction}$$

As an example, consider the case of iron, partially covered by "atmospheric oxide" $\mathbf{Fe_2O_3}$ immersed in a neutral aqueous solution of pH = 7. As well as the conditions of thermodynamic stability of water and its products of reduction (hydrogen) and oxidation (oxygen) under atmospheric pressure (lines *a* and *b*), Figure 88a indicates the theoretical conditions of corrosion of iron (dissolution in the form of ferrous ions, Fe^{++}, ferric ions Fe^{+++}, or dihypoferrite ions, FeO_2H^-), of immunity (stability of metallic iron), and of passivation (stability of $\mathbf{Fe_3O_4}$ or $\mathbf{Fe_2O_3}$ oxides), in the presence of solutions containing 10^{-6} g-atom Fe/liter (0.06 ppm of iron in solution).[21]

[21] Remember that this figure is only valid for solutions free of substances capable of forming soluble complexes (cyanides) or insoluble salts (phosphates, sulfides) with iron.

It is easily seen from Figure 88a that the direction (oxidation or reduction) in which the four following reactions may proceed in the presence of a solution of pH = 7 containing 0.06 ppm of iron varies according to the electrode potential of the metal as shown in Table XXIV.

As a result:

- if the electrode potential is less than −0.6 V with respect to the standard hydrogen electrode (region 1), all the reactions outlined in Table XXIV only proceed in the direction of a reduction, that is, from right to left. Iron, which then is in a state of immunity, remains uncorroded; ferrous ions possibly present in the solution and ferric oxide possibly present on the metal are reduced to the metallic state; water is reduced to gaseous hydrogen; gaseous oxygen possibly present in the solution is reduced to water (and to hydrogen peroxide not considered here). Then, by means of an external energy of reduction, *cathodic protection* of the metal occurs with evolution of hydrogen.
- if the electrode potential is between −0.6 and −0.4 V (region 2) *iron corrodes with evolution of hydrogen*, with dissolution of ferric oxide to the ferrous state, and reduction of oxygen. This and the following case correspond to normal conditions of corrosion of iron, without the influence of external energy.
- if the electrode potential is between −0.4 and −0.2 V (region 3) *corrosion takes place without evolution of hydrogen.*
- at potentials greater than about −0.2 V (regions 4 and 5), iron is covered by ferric oxide which, *if perfectly adherent and nonporous*, completely shields the metal from the solution; then the metal is *protected by passivation.* If the oxide is *not perfectly adherent* or if it is *porous*, and this may occur in solutions containing chloride, the resulting protection only affects part of the metal surface. In this case, passivation is only partial and the metal undergoes a *localized corrosion* which, because of the great affinity[22] of the corrosion reaction, may be extremely rapid.
- if the potential is between −0.2 and +0.8 V (region 4), these conditions of passivation or local corrosion may appear under the action of moderately oxidizing substances, such as oxygen, without the influence of external energy.

[22] Remember this affinity is measured by the difference $E - E_0$ between the electrode potential E of iron and the equilibrium potential E_0 of the corrosion reaction (generally about −0.6 V), that is, in Figure 88a, by the vertical distance between the point representative of the metal and the boundary between the regions of corrosion and immunity.

TABLE XXIV. Relation Between the Value of Electrode Potential and the Direction of Electrochemical Reactions, in the Case of Solutions of pH = 7 Containing 0.06 ppm of Iron

	Region	$Fe–Fe^{++}$	$H_2–H_2O$	$Fe^{++}–Fe_2O_3$	$H_2O–O_2$
		$Fe = Fe^{++} + 2e^-$	$H_2 = 2H^+ + 2e^-$	$Fe^{++} + 3H_2O = Fe_2O_3 + 6H^+ + 2e^-$	$2H_2O = O_2 + 4H^+ + 4e^-$
1	$E < -0.6$ V	reduction	reduction	reduction	reduction
2	$-0.6 < E < -0.4$	oxidation	reduction	reduction	reduction
3	$-0.4 < E < -0.2$	oxidation	oxidation	reduction	reduction
4	$-0.2 < E \quad +0.8$	oxidation	oxidation	oxidation	reduction
5	$+0.8 < E$	oxidation	oxidation	oxidation	oxidation

- potentials greater than +0.8 V (region 5) are only brought about by a particularly vigorous oxidizing action, which may cause oxidation of water to oxygen.

The existence of regions of corrosion, immunity, and passivation illustrated in Figure 88a shows that in the presence of a given aqueous solution the states of corrosion or noncorrosion of iron are closely connected with the given value of electrode potential.

Everything that has been expressed here concerning iron also applies to all metals and metalloids, and their respective diagrams of electrochemical equilibria, as already stated, are as characteristic to them as, for instance, fingerprints are to us. Figure 54, taken from the *Atlas of Electrochemical Equilibria*,[23] illustrates the theoretical diagrams of corrosion, immunity, and passivation for 43 metals and metalloids. In Figure 54, Figures 39 and 7b, which cover iron, appear as No. 25. Though still somewhat incomplete such regions of stability exist not only for metals and their oxides but also for *all* other bodies which are involved in the constitution of industrial metals and alloys and their superficial layers: carbides, sulfides, carbonates, etc. Such regions of stability exist, not only for the temperature of 25°C and atmospheric pressure but also for all temperatures and under all pressures. They exist not only when aqueous solutions are present but also in the presence of other electrolytic solutions such as organic solutions and molten salts, and even semiconducting oxides.

This particular property which the electrode potential possesses, of being connected with the conditions of stability of all the bodies which are involved in aqueous solution, is the origin of various electrochemical corrosion techniques, e.g., intensiostatic, intensiokinetic, potentiostatic, or potentiokinetic (of which polarography is a particularly important varient). In these techniques, the electrode potential is varied artificially from low to high values, and/or vice versa, thereby causing at will, corrosion and other reactions which tend to bring the system toward the state of thermodynamic stability corresponding to the value of electrode potential.

6.6.2.3. Experimental Predetermination of Circumstances of Immunity, Passivation, General Corrosion, and Localized Corrosion (Pitting)

As an example, Figures 88b and 88c illustrate the results of a potentiokinetic test for the case of iron in a solution of pH = 7 as cited above,

[23] Refs. 8 and 34, pp. 159–162.

and for which the theoretical conditions of corrosion, immunity, and passivation have been outlined in Figure 88a.

If the passivating oxide is perfectly protective, the current–potential curve determined by the potentiokinetic method at increasing potential has the general shape shown in Figure 88b. The electrolysis current I which measures the sum of the rates of all reduction reactions (considered as negative) and all oxidation reactions (considered as positive) essentially corresponds, for potentials below the equilibrium potential C of the metal, to an evolution of hydrogen on the uncorroded metal. As already explained above, there is then *cathodic protection* by placing the metal in a state of immunity. For electrode potentials between the protection potential C and the passivation potential P, the electrolysis current essentially measures the intensity of a *general corrosion* of the metal: a corrosion current density of 1 mA/cm^2 corresponds to a penetration of 30 μm(microns)/day, or 11 mm per annum or 0.44 mpy. The decrease of current which in Figure 88b occurs at the passivation potential P corresponds to the formation of an oxide film on the metal and indicates a sudden suppression of all further corrosion. This is *perfect passivation.* The oxide formed in this way is protective and remains so, even at high values of electrode potential; the oxidation which occurs at such potentials, higher than potential B, which is the equilibrium potential between water and gaseous oxygen, corresponds generally not to an oxidation of the metal but to an oxidation of water to oxygen.

In the case illustrated in Figure 88c, a premature increase of oxidation occurs at R which reveals a sudden rupture of the protective oxide film, with the appearance of *localized corrosion* (pitting). Such a curve indicates that the metal under test is protected against corrosion at potentials below C (immunity) and at potentials between P and R (passivation). The metal undergoes general corrosion between potentials C and P; it undergoes a local corrosion above R, which is called the "rupture potential," E_r.

In practice, the application of potentiokinetic methods in order to satisfactorily predetermine the circumstances of good or poor behavior of metals and alloys is generally not as simple as the above summary would lead one to believe. In certain cases, a rapid experiment carried out with a potentiostat of mediocre quality may easily produce satisfactory results. In other cases, special precautions must be taken concerning the apparatus itself, its method of use and the interpretation of the resulting curves. This is particularly so when actions of corrosion affecting only a relatively small part of the metal surface are being studied. Corrosion of this type such as pitting, intergranular corrosion, and stress corrosion, is often particularly dangerous.

As an example, let us examine here the interpretation to be given to the four polarization curves drawn in Figure 89 showing the behavior of the AISI 410 chromium steel in solutions containing increasing amounts of chloride.

Figure 89a illustrates the behavior of this steel in a chloride-free 0.10 M $NaHCO_3$ solution. Oxidation occurs only at potentials appreciably higher than the oxygen equilibrium potential *B*. If after having anodically polarized the metal at increasing potentials the metal is then polarized at decreasing

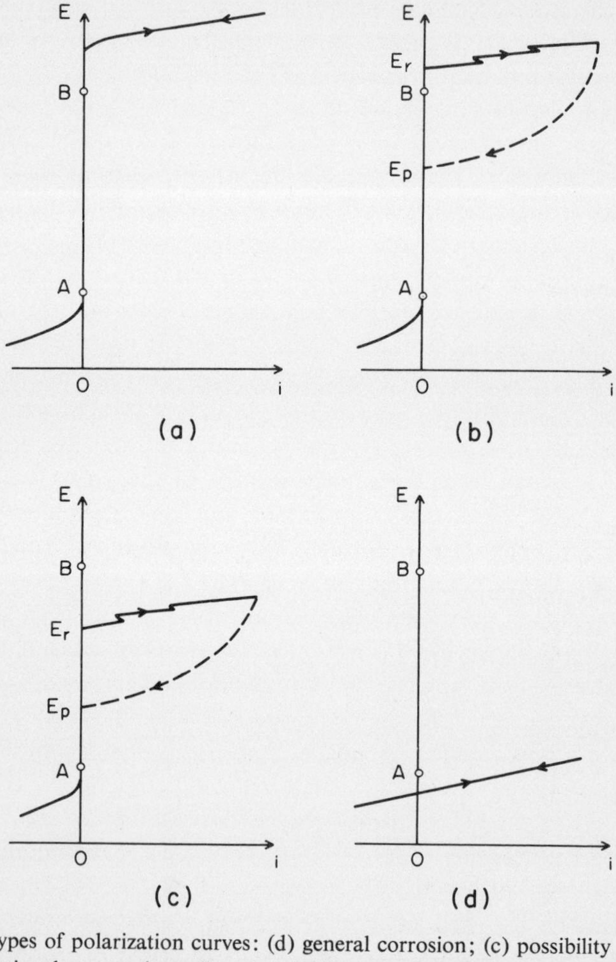

Figure 89. Types of polarization curves: (d) general corrosion; (c) possibility of pitting in cases exceptional or not depending on whether or not B–E_r and B–E_p are below or above 400 to 500 mV; (b) possibility of pitting in exceptional cases (stray currents); and (a) no possibility of corrosion.

potentials, almost identical curves are obtained when polarizing in the two directions. This means that the oxidation which has occurred at the higher potentials does not modify the surface condition of the metal, which has remained completely uncorroded. The curve shown in Figure 89a is thus characteristic of a metal or alloy which does not corrode in a particular solution over the whole possible range of potentials. Thus, it can be expected to retain a stability regardless of whether it is used without current being applied, as an anode or a cathode in an electrolysis cell, or with straight currents in the presence of the considered environment. The only reactions therefore which can occur will be at the high potentials where the evolution of oxygen is possible and at low potentials where the evolution of hydrogen is possible. However, neither of these reactions produces corrosion of the metal.

In Figures 89b and 89c, which corresponds to $NaHCO_3$ solutions with increasing amounts of NaCl, one observes the rupture potential, E_r, which is indicative of the onset of pitting. Above this rupture potential, the passivity breaks down and corrosion occurs locally producing pits. Contrary to the behavior of the type 410 in chloride-free solutions, the curves of increasing polarization and decreasing polarization do not coincide. If after having anodically polarized the metal above the rupture potential, the metal is then polarized at decreasing potentials, the curve of the decreasing potentials is lower than the curve obtained during increasing potentials and intersects the ordinate axis at a value E_p which is less than E_r. At this potential E_p, the net electrolysis current is zero, and thus, if there is no oxidizing substance in the solution, so that no reduction occurs, there is also no oxidation, and this means no corrosion. The previously formed pits thus do not grow any more, and the potential E_p may be considered a "protection potential against pit propagation," or, more simply, "protection potential against pitting." Below the value of the protection potential, E_p, the metal will be uncorroded in the same manner as in the case without chloride solutions. The curves drawn in Figures 89b and 89c show that the corrosion behavior of the alloy may be divided into three regions according to the following:

1. Below the protection potential E_p, the metal does not corrode, even after the surface is pitted.
2. Between the protection potential, E_p, and the rupture potential, E_r, the metal does not corrode if its surface is free from pitting. However, if pits are present, the metal will corrode at the site of the pits.

3. At potentials greater than or equal to E_r, the metal undergoes corrosion by pitting whether or not the surface is previously pitted.[24]

Having described the anodic polarization curves in chloride solutions as illustrated in Figure 89b and 89c, it remains to assess their significance in an alloy exposed to an environmental situation where no external polarization is applied. Since oxygen is the reducible specie of primary consideration and in general the only one which can raise the potential in the range where pitting can occur, we shall center our attention about the possibility that the presence of oxygen may raise the potential on the metal surface to the critical values of E_p or E_r. The maximum electrode potential which may be obtained on a noncorroding metal surface in a solution containing oxygen and free of any reducible specie, is the "reduction potential" of oxygen on this metal; this is the potential at which the cathodic polarization curve in the presence of oxygen leaves the ordinate axis. This potential, which is in fact, the "free potential" or the "zero-current potential" of the uncorroded metal, is represented in Figure 75, respectively, by points 4, 5, and 6, for three solutions with increasing concentrations of dissolved oxygen. In the case of solutions saturated with oxygen under atmospheric pressure, this "free potential" is generally 200 to 400 mV below the potential denoted by *B* in the potential–pH diagrams developed by CEBELCOR. This is the potential of thermodynamic equilibrium between water and oxygen at 1 atmosphere for the pH of the solution in question.

In the case of Figure 89b, the possibility of pitting is much less than in the case of Figure 89c. In fact, an oxidant such as oxygen is not capable of raising the electrode potential up to the rupture potential, E_r, and thus new pits cannot be formed. One needs only to be concerned in the instance where the protection potential, E_p, is sufficiently low for it to reach, as the result of introducing a specific oxidizing solution to the environment. In this case a certain amount of corrosion will take place at weak points of the surface where some pits may already exist. However, a vigorous

[24] These trends have been verified in research work on the behavior of type 410 stainless steel in *Rapports Techniques CEBELCOR* RT. 103 and 120 (1962) and 123 (1964) in French, and Publication E. 45 (1963) in English. These principles are applicable to all passivating metals and alloys and we believe that further studies in this area would be useful, particularly for high-strength steels, stainless steels, titanium alloys, and alloys of aluminum, nickel, and copper in the presence of sea water and other solutions containing chlorides. These ideas are applied to the problems of pitting and stress corrosion cracking in Section 6.6.6.1.2. Details are given in *Rapports Techniques CEBELCOR* RT. 157, 166, 179, and 191.

oxidation, for instance, due to the action of "stray currents," inevitably causes pitting, even on surfaces free from any defect. Finally, in the case of Figure 89d, which relates to a solution containing a very large amount of chloride, the polarization curve no longer has any range of passive currents but now appears to be a continuous line which departs from the ordinate axis below the hydrogen equilibrium potential, *A*. In this case, when in contact with an oxygen-free solution, the metal corrodes rapidly with the evolution of hydrogen; the presence of oxygen only enhances this severe corrosion and no passivation nor pitting occurs. The metal simply corrodes generally at a rapid rate.

6.6.2.4. Linear Polarization Methods

The purpose of this section is to describe a simple electrochemical technique for determining the corrosion rate of a material based on the application of principles already discussed.

As a background to describing this technique, it will be recalled from the discussion of Figures 59 and 62 that Figures 90a and 90b are the polarization curves of an electrochemical reaction wherein the electrode potential, *E*, and the reaction current, *i*, are related for the oxidation and reduction of a *single* reaction such as **Fe** $\rightleftarrows$ $Fe^{++} + 2e^-$. Figure 90a plots the *E* vs *i*

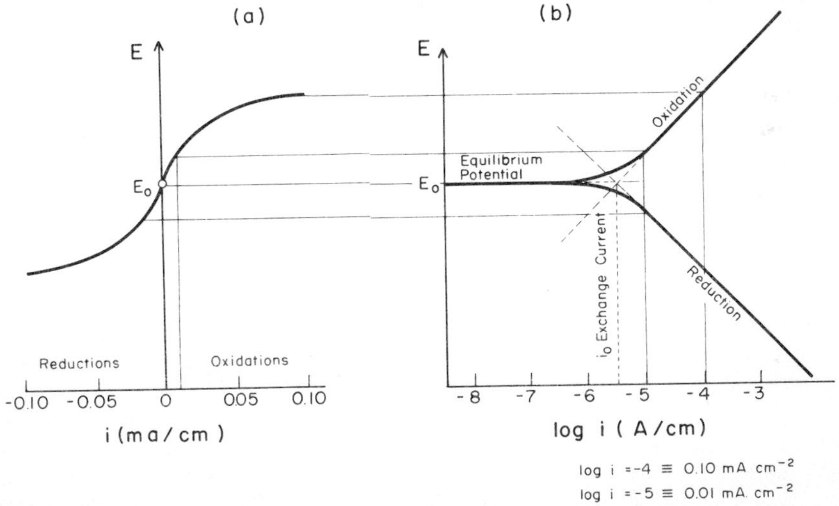

Figure 90. Polarization curves of electrochemical reaction. Equilibrium potential and exchange current: (a) arithmetic current intensity scale; and (b) logarithmic current intensity scale.

relationship using linear coordinates and Figure 90b plots the same variables semilogarithmically. The latter is usually preferred both for fundamental and practical reasons and derives initially from the experiments of Tafel already discussed. At equilibrium, where the oxidation and reduction rates for this single reaction are equal, the potential and current are those at equilibrium and are designated respectively as the "equilibrium potential," E_0, and the "exchange current," i_0.

As a simple corroding system to which we may apply this simple, relatively new technique, let us consider iron in acid solution, where the relevant reactions are

$$\mathbf{Fe} = Fe^{+2} + 2e^-$$

$$H_2 = 2H^+ + 2e^-$$

During corrosion in the acid environment, the iron reaction proceeds to the right and iron corrodes; correspondingly, the hydrogen reaction proceeds, with the absorption of electrons, to the left and hydrogen is evolved. Specifically, these reactions are depicted in Figures 91a and 91b for an acid solution at pH 0 and containing 0.010 g-ion/liter of ferrous ions. Specific parameters associated with the iron and hydrogen reactions are given in Table XXV.

The corrosion rate for iron in the acid solution is defined (in the absence of external currents) where the total oxidation current ($\mathbf{Fe} \rightarrow Fe^{++} + 2e^-$) and the total reduction current ($2H^+ \rightarrow H_2 + 2e^-$) are equal and correspond to the overall reaction

$$\mathbf{Fe} + 2H^+ \rightarrow Fe^{++} + H_2$$

TABLE XXV. Parameters for Iron and Hydrogen Reactions

	Reaction	
	$\mathbf{Fe} = Fe^{++} + 2e^-$	$H_2 = 2H^+ + 2e^-$
Equilibrium potential (V_H)	−0.500	0.000
Log exchange current (A/cm²)	−4.50	−5.85
Slopes of Tafel lines (volts per log unit) β_{ox}	+0.328	+0.123
β_{red}	−0.328	−0.123

The electrode potential corresponding to this condition of zero net current is called the "corrosion potential," E_{corr} and is $-0.250\ V_H$ for the case in Figures 91a and b. The common value of the current at the corrosion potential is called the "corrosion current," i_{corr}; and, in Figure 91, is 0.162 mA/cm² (log $i_{corr} = -3.79$). Since 1 mA/cm² corresponds to the dissolution of 9.1 g iron per year (see page 191) and to a corresponding thickness loss of 11 mm per year (or 0.44 mpy) for general corrosion, the corrosion current for Figure 91 corresponds to a loss of 1.47 g/cm² per year, or to a thickness loss of 1.8 mm per year or 0.072 mpy.

The algebraic sum of the currents (or net currents) are plotted as heavy lines in Figure 91. These currents are often called "local currents" or "electrolysis currents." Clearly, the net current at the "corrosion potential" is zero.

The analysis in the preceding paragraph suggests that the corrosion rate of a specimen can be specified if one knows the Tafel lines of the two (or more) participating reactions, i.e., **Fe** → $Fe^{++} + 2e^-$ and $2H^+ + 2e^-$ → $\boldsymbol{H_2}$. The intersection of the two lines gives the corrosion potential, E_{corr}, and the corrosion current, i_{corr}. Note that this analysis assumes

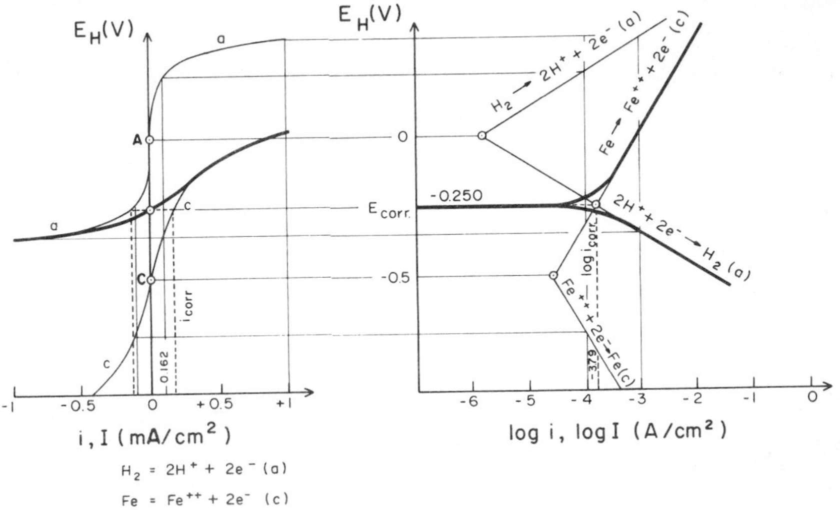

(a) (b)

Figure 91. Corrosion of iron with hydrogen evolution, in solution of pH 0 containing 0.010 g-atom Fe^{++}/liter. Anodic and cathodic polarization curves. Corrosion potential and corrosion current: (a) arithmetic current intensity scale; and (b) logarithmic current intensity scale.

implicitly that the reactions are under complete activation control (without any diffusion control) and that the specimen is electrically isolated, i.e., no anodic or cathodic currents are being applied. It will be shown shortly that the development of linear polarization methods is general enough to include any kinetic behavior which is linear on a semilogarithmic plot of E vs i. The determination of these two branches of the Tafel curves is easily performed in the laboratory by intensiostatic or by potentiokinetic experiments in a stirred solution as already described in Sections 6.3.1 and 6.6.2.

The aim of determining the corrosion current can be achieved by less laborious methods. It is possible to determine the corrosion current *in situ*, often under conditions where the metal is functioning as a part of a piece of equipment, by an elegant and simple method. This is known as the method of "linear polarization"[25] and is due to Stern, Geary, and Weisert.[26]

Figures 92a and 92b show two experimental arrangements which have already been discussed in Sections 6.3.1 and 6.6.2.1 for determining polarization curves in stirred solutions: A metallic specimen can be polarized intensiostatically (Figure 92a) or potentiokinetically (Figure 92b) and one obtains a relationship, in either case, between E and i. Figures 92c, d, e show much simplified devices for obtaining E vs i data needed for the method of polarization resistance.

The method of polarization resistance is based on the algebraic solution of the Tafel equations for the respective oxidation and reduction reactions

$$E = \alpha_a + \beta_a \log i_a$$

for anodic reaction (i.e., $\mathbf{Fe} \rightarrow Fe^{++} + 2e^-$)

$$E = \alpha_c + \beta_c \log i_c$$

for cathodic reaction ($2H^+ + 2e^- \rightarrow \boldsymbol{H_2}$). Corresponding to a potential change, $\Delta E = E - E_{corr}$, there is a change in current, $\Delta i = i - i_{corr}$. Depending whether the intensiostatic or potentiokinetic circuits are used the i or ΔE, respectively, may be the independent variables.

[25] See M. Pourbaix, On the methods for appreciating corrosion - velocity by measurement of polarization resistance, *Rapports Techniques CEBELCOR* **108**, RT. 156 (1969) (in French).

[26] See (a) M. Stern *et al.*, Electrochemical polarization, *J. Electrochem. Soc.* **104**, 56–63, 559–563, 645–650 (1957); (b) M. Stern and E. D. Weisert, Experimental observations on the relation between polarization resistance and corrosion rate, *Proc. Am. Soc. Testing Materials* **59**, 1280–1291 (1959).

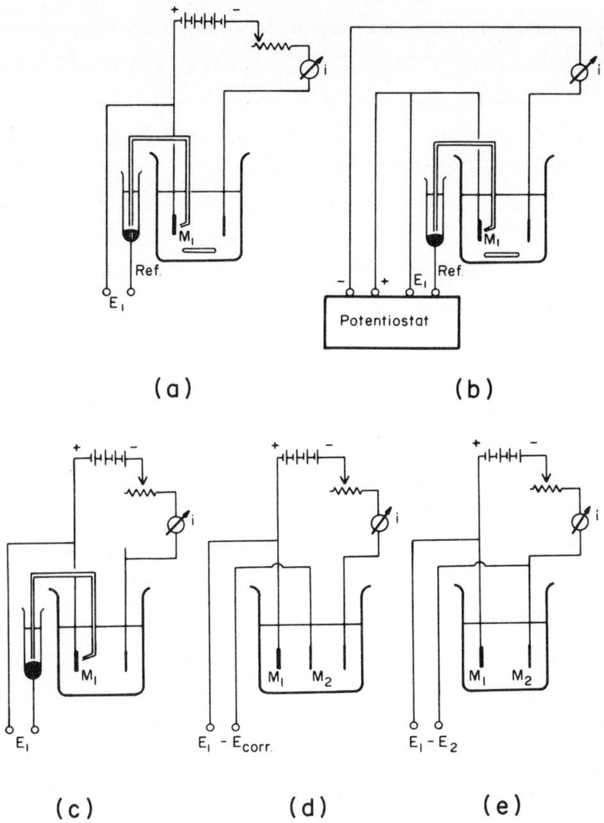

Figure 92. Experimental devices for the determination of the velocity of corrosion: (a) intensiostatic method; (b) potentiokinetic method; (c) method of polarization resistance with a reversible reference electrode; (d) method of polarization resistance with a nonreversible reference electrode; and (e) method of polarization without reference electrode.

Providing $\Delta E < 10$–20 mV, then the corrosion current may be calculated from the following equation:

$$i_{\text{corr}} = \frac{1}{2.3} \times \frac{\beta_a \cdot \beta}{\beta_a + \beta} \times \frac{I}{\Delta E}$$

According to Stern, the quotient $\Delta E/I$ is called the polarization resistance by analogy to $R = E/I$ from conventional electricity. Clerbois[27] has sug-

[27] L. Clerbois, Determination of corrosion velocity by the measurement of polarization resistance, *Rapports Techniques CEBELCOR* **108**, RT. 155 (1969) (in French).

gested that the inverse, $I/\Delta E$, be called the "corrodance" by obvious analogy to the inverse of electrical resistance which is the "conductance."

The measurement of polarization resistance is very simple. One measures the net current or electrolysis current as noted by the broad line in Figure 91. This is the current, I, actually observed in the external circuit and not the partial currents i of the oxidation or reduction processes. It is interesting that the measurement of this net current, I, together with an associated ΔE, can give the corrosion current itself.

While there is an apparently intimate dependence of the polarization resistance upon the Tafel constants, β_a and β_c, in fact, it has been shown by Stern and Weisert in Figure 93 that within a reasonable range of uncertainty the corrosion current may be predicted directly from the polarization resistance. Figure 93 shows for three sets of values for the Tafel constants the dependence of i_{corr} upon the polarization resistance $\Delta E/I$. The experimental results shown in this figure refer to different metals (nickel, iron, steel, cast iron) corroding with hydrogen evolution in different solutions (hydrochloric acid, sulfuric acid, pure water, synthetic, and natural waters). The boundary lines in this figure were selected to correspond to a reasonable range of expected values for the Tafel constants. In particular, it would be valid that one of the β values can be ∞. This is based on the

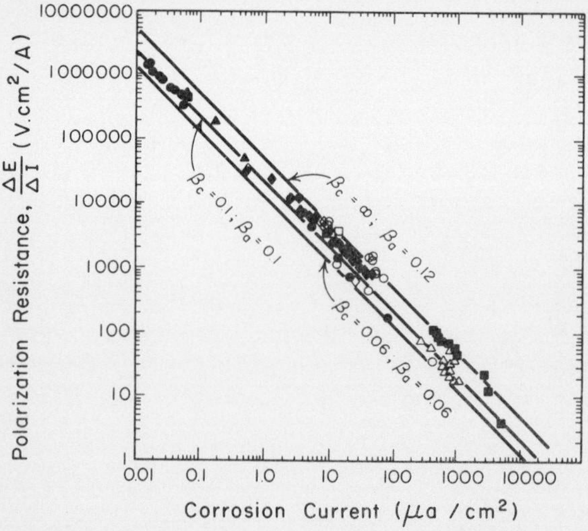

Figure 93. Experimental observations. ○ △ nickel in hydrochloric acid; ◇ iron in dilute hydrochloric acid containing sodium chloride; × various steels in sulfuric acid; □ steel and cast iron in high-conductivity waters; + cast iron in natural and synthetic waters; ● ▲ ■ ferric–ferrous exchange current on a variety of passive surfaces (after M. Stern and E. D. Weisert).

fact that cathodic reactions are often under diffusion until the limiting current is independent of potential. A similar slope of infinity may occur when the anodic reaction is stifled by a reaction product film in the surface.

It may be seen from Figure 93 that the horizontal distance between the two limiting straight lines is about 0.7 of a logarithmic unit; this means that if nothing is known about slopes β_a and β_c for a particular case and if one could assume a behavior according to the central line, the maximum possible error is 0.35 logarithmic unit. This produces an error by a factor of 2.2. Thus, for an observed value of 10 mA/cm^2 the real value may lie between 4.5 and 22 mA/cm^2.

Figures 92c, d, e show three different experimental methods for obtaining the polarization resistance. In the device depicted in Figure 92c, which is the original and best one, the polarization ΔE is the change of electrode potential of the specimen under test as measured correctly with an external and reversible reference electrode. In the arrangement shown in Figure 92d, a specimen similar to the metal under test is used as an internal, nonreversible electrode; in Figure 92e, no reference electrode is used and the potential difference between the two specimens of the metal under test is considered being equal to the polarization ΔE to be used in calculating the polarization resistance.

6.6.3. The Influence of Oxidants, Chlorides, and Orthophosphates

The study of the influence exerted by numerous factors on the corrosion of metals, the initiation of protection procedures against corrosion, and the maintainance of these procedures are based essentially on measurements of electrode potential and on the interpretation of these potentials. Current–potential curves determined in stirred solution are particularly useful in this context. As an example, we examine by means of current–potential curves obtained intensiostatically, the influence of oxygen, chloride, and orthophosphate on the behavior of mild Armco steel in 0.1 M solution of sodium bicarbonate pH = 8.4.

6.6.3.1. Influence of Oxygen

The influence exerted by oxygen has already been examined in Sections 6.3.2 and 6.4.2 as well as in Figures 69, 74, 75, and 76. Here, we are content merely to recall that oxygen renders the metal active or passive depending on whether the electrode potential which it confers on the metal is greater or less than the passivation potential. Also, the potential of passivation and generally the potential of activation (often called "the Flade potential") have the same value which is equal to the potential of thermodynamic equilibrium between the passive oxide and the solution. The oxidizing power of oxygen contained in a solution in the presence of a given metal may be measured by the value of the free potential of the metal in the presence of this solution. This potential is equal to that below which the current-potential curve leaves the ordinate axis (points 4, 5, and 6 of Figure 75) during polarization at decreasing potentials. The metals and alloys which exhibit the highest electrode potentials in the presence of a solution saturated in oxygen are the best catalysts for electrochemical reduction of oxygen in this solution.

6.6.3.2. Influence of Chlorides

Figure 94 shows current–potential curves determined intensiostatically for mild Armco steel in the presence of oxygen-free 0.1 M solution of $NaHCO_3$ free from chloride (curve *a*; this curve has already been shown in Figure 67a) and $NaHCO_3$ solutions containing increasing quantities of NaCl (curves *b* and *c*).

An examination of the three curves shows that for the $NaHCO_3$ and NaCl concentrations considered here:

1. The addition of chloride does not modify the lower branch of the current–potential curve which, as we have already seen, corresponds

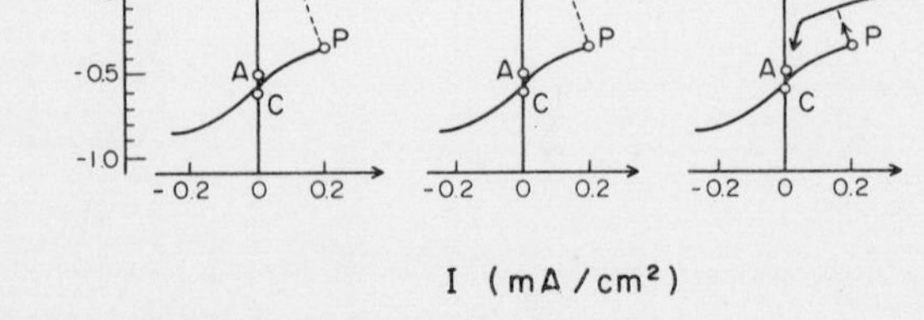

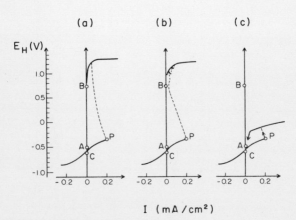

Figure 94. Influence of chloride on the behavior of mild Armco steel in 0.1 M solution of $NaHCO_3$ (pH = 8.4), in the absence of oxidant: (a) No NaCl; (b) 0.001 M NaCl; and (c) 0.01 M NaCl.

to general corrosion in nonoxidizing or feebly oxidizing media. Such contents of chloride scarcely modify the behavior of the metal under these conditions.

2. The presence of chloride exerts a most significant effect on the stability of the passive film. In Figure 94a, there is a smooth transition to the higher potentials. However in Figure 94b, there is some instability in approaching the upper portion of the curve. This results from the fact that chloride at the higher potentials prevents the film from being perfectly protective. Finally, at the higher chloride (Figure 94c), the potential stabilizes at a relatively low value and leads to pitting. In the presence of an oxidant such as dissolved oxygen, chloride may cause very accelerated corrosion; whereas in the absence of oxygen, chloride may exert relatively little effect. We remind that the three curves shown at Figure 94 have been obtained with an intensiostatic method. As will be shown later (Figure 102), more information may be obtained with potentiokinetic methods.

6.6.3.3. Influence of Orthophosphates. Oxidizing Phosphatization

The addition of a sufficient amount of disodium orthophosphate to the 0.10 M $NaHCO_3$ and 0.01 M NaCl solution of Figure 94c reverses the deleterious effect of chloride and produces a very extended range of stable passivity as shown in Figure 95c. Figure 95 shows various possible combinations of phosphate and chloride.

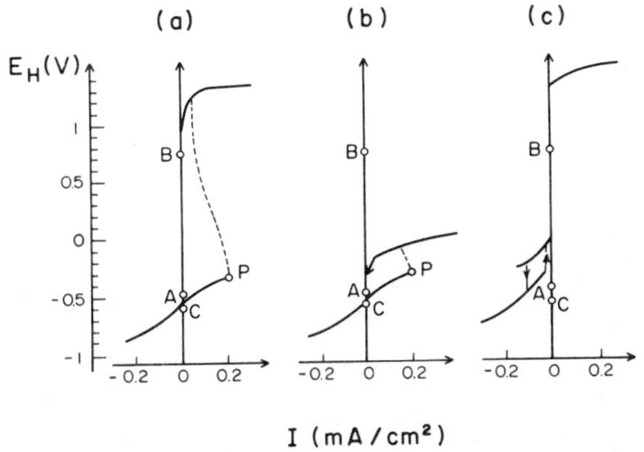

Figure 95. Influence of chloride and phosphate on the behavior of mild Armco steel in 0.1 M solution in the absence of oxidant: (a) without chloride or phosphate; (b) 0.01 M NaCl, no phosphate; and (c) 0.01 M NaCl, 0.07 M phosphate.

One observes that the addition of phosphate to the chloride-containing solutions has the effect of restoring to its normal (or nonchloride) position at high potential the second anodic branch at low potentials observed in Figure 95b. This suggests that the presence of orthophosphate gives the passive film the protective properties that it possessed in the absence of chloride. This effect appears to be produced by the deposition of insoluble iron phosphate at weak points in the film. However, as well as in the case of the bicarbonate solution free from both chloride and phosphate considered in Figure 69, the cathodic curve obtained at decreasing potentials after anodic polarization ceases to be stable when the potential reaches −0.35 V which is the common value of the activation potential and the passivation potential of iron in bicarbonate solution. Potentiostatic experiments carried out in collaboration with A. Abd el Wahed[28] have confirmed that iron then is no longer passive and corrodes.

Therefore, in the absence as well as in the presence of phosphate, it is an oxide of iron which is the essential constituent of the protective film. Phosphate exerts this synergistic effect when necessary (as in the presence of chlorides) by blocking the weak points with deposits of almost insoluble phosphate. Thus, phosphate is less effective as a corrosion inhibitor when the iron oxide does not exist (that is, at potentials less than the passivation potential) than when the iron oxide exists (at potentials higher than the passivation potential). This *oxidizing phosphatization* appears to be an essential part of the sometimes sensational effectiveness of protection procedures by phosphatization which have been successfully developed over the last few years. For example in painting, there is the use of undercoat layers of "wash primers" based on phosphoric acid and chromate. These products are a considerable improvement compared to conventional chromate "primers" which, though perfect in the absence of appreciable quantities of chloride and other aggressive substances, may fail if such substances are present. These "wash primers" also represent an improvement over original phosphoric etchants because, when properly used, their chromate content considerably improves the protective action due to the phosphate.

6.6.4. Treatment of Corrosive Water

A number of commercially important problems of corrosion can be treated by the application of electrochemical principles. Some important ones are discussed below:

[28] *Rapports Techniques CEBELCOR* RT. 19 (1954) (in French) and CEBELCOR Publication E. 22 (1955) (in English).

6.6.4.1. Drinking Water

In Chapter 3, we examined the saturation equilibria of calcium carbonate in water, as well as the treatment of aggressive water based mostly on the classical work carried out in 1912 by Tillmans and Heublein. It is recalled that the work of Tillmans and Heublein, like much other work that has been undertaken with a view to defining the aggressive characteristics of drinking water,[29] in fact tended to establish the circumstances under

[29] See especially the following publications:

1. Arbatsky, J. W., *Zeichnerische Ermittlung der Enthärtungsverhältnisse von Wassern nach dem Nomogramm von F. E. Staffeldt, und dem Kalk-Soda-Wasserbild* (Ref. 14).
2. Arbatsky, J. W., *Gas- und Wasserfach* **83**, 90, 116 (1940); Mathematische Hilfsmittel zur Enthärtungsregelung, *Wärme* **63**, 225–229, 239–241 (1940).
3. Buydens, R., L'agressivité des eaux naturelles, *Ann. Hyg. Publ. Ind. Soc.* **20**, 145–176 (1942).
4. Coin, L., Détermination de l'agressivité des eaux naturelles, *Ann. Hyg. Publ. Ind. Soc.* **20**, 145–176 (1942).
5. Demartini, F. E., La corrosion et l'indice de saturation du carbonate de calcium de Langelier, *Ref. Chim. Ind.* **40**, 59–23 (1938); *J. Am. W. W. Assoc.* **30**, 1801 (1938).
6. Franquin, J. and Marécaux, P., Etude de l'équilibre CO_3Ca–CO_2–H_2O. Théorie des eaux douces. *Chim. ind. XVIIIe Congrès Chim. Ind. Nancy*, 1938, 229–237 (1938).
7. R. Girard, Le problème de la corrosivité des eaux naturelles, *Corrosion et Anticorrosion* **12**, 347–357 (1964).
8. Guillerd, J., Courbes d'isoagressivité des eaux, *Ann. Hyg. Publ. Ind. Soc.* **19**, No. 5 (1941).
9. Langelier, W. F., The analytical control of anti-corrosion water-treatment, *J. Am. W. W. Assoc.* **28**, 1500–1508 (1936).
10. Moore, W., Graphic determination of CO_2 and the three forms of alcalinity, *J. Am. W. W. Assoc.* **31**, 51 (1939).
11. Nill, N., *Water Works Sewerage*, pp. 433, 472 (1938).
12. Pourbaix, M., Étude graphique du traitement des eaux par la chaux. Application au conditionnement des eaux et à la fabrication de magnésie hydratée, *Bull. Soc. Chim. Belg.* **54**, 10–42 (1945).
13. Riehl, M. L., Détermination graphique du pH de saturation d'après la formula de Langelier, *J. Am. W. W. Assoc.* **30**, 1801 (1938).
14. Staffeldt, F. E., Die Kontrolle der Kalk-Soda Wasserenthärtungsanlagen nach einem Nomogramm, *Gas- u. Wasserfach* **78**, 623 (1935).
15. Tillmans, J. and Heublein, D. *Gesundheitsingenieur* **35**, 669 (1912).
16. Tillmans, J., *Untersuchung von Trink- und Brauchwasser*. Berl-Lunge, *Chemisch-technische Untersuchungsmethoden* (8th ed.), Springer, Berlin (1932), pp. 289–297.
17. Tregl, J., Vereinfachte Errechnung des zugehörigen CO_2 in natürlichen Wassern, und einige praktische Anwendungen, *Gas- u. Wasserfach* **82**, 715–718 (1939).
18. Verain, M. and Franquin, J., Détermination de l'agressivité d'une eau douce, *Chim. Ind. XVIIIe Congrès Chim. Ind. Nancy, 1938*, 229–237 (1938).

which water is saturated with calcium carbonate, where any alkalization (e.g., a reduction of dissolved oxygen in the water) or heating may give rise to a deposit of calcium carbonate on the metal. Now the existence of such a deposit on the wall of a steel pipe, for example, does not necessarily imply that this wall is protected against corrosion. It appears certain that, as in the case of the phosphate, the real protective agent is not a salt (carbonate, phosphate), but an oxide of iron ($\mathbf{Fe_2O_3}$) adhering strongly to the metal. The carbonate or phosphate which may be present in the deposit covering the metal, only serves to improve the intrinsic protective qualities of the oxide by blocking possible pores and/or by covering it with a layer of variable thickness which protects it against mechanical and other degradations.

As a result, it is not sufficient that water is saturated with calcium carbonate (thereby conforming to Tillmans' curve or other analogous requirements on this point) for it not to be aggressive toward iron or ordinary steels; it is essential that it forms an adherent and protective film of oxide, $\mathbf{Fe_2O_3}$, on the metal. This implies mainly that water contains a sufficient quantity of oxidant, e.g., oxygen (without which $\mathbf{Fe_2O_3}$ will not be able to form) and does not contain too much chloride.

Therefore, the treatment of a so-called "aggressive" water in the Tillmans sense involves not only a treatment of saturation with $\mathbf{CaCO_3}$ (by filtration over $\mathbf{CaCO_3}$, by addition of lime water $\mathbf{Ca(OH)_2}$, or by filtration over semicalcined dolomite $\mathbf{CaCO_3 \cdot MgO}$, but also an aeration treatment (by bubbling or atomizing with air). Such treatments are not effective in the case of water containing much chloride such as sea water.

In other words, it is always advantageous when bringing new water supply installations into service to ensure a perfect oxidation of all the internal walls of the piping and to avoid stagnation of the water which, by causing a diminution of the oxygen content, renders the water corrosive and may give rise to deposits which generate harmful differential aeration cells. This lack of oxygen may be a major problem especially in horizontal pipes.

Because of the saturation with $\mathbf{CaCO_3}$, "nonaggressive" drinking water often leads to incrustation. The prolonged deposition of $\mathbf{CaCO_3}$ and possibly other insoluble salts ($\mathbf{MgO, SiO_2}, \ldots$) on the walls of heaters or in pipes may lead to the formation of incrustations whose annoying effects are well known. Therefore, drinking water may be either corrosive or form incrustations, and the technician will often have to reconcile two opposing requirements. We do not go into details here on precautions to be taken to reconcile the two aspects of the problem. We are content to draw attention to the influence that *physical procedures* may have in water treatment. Without appreciably modifying the composition of water and if carefully applied, these procedures cause the separation of $\mathbf{CaCO_3}$, not in the form of adherent incrustations, but in the form of sludge or slimy nonadherent deposits; this can also be accomplished by electric or magnetic procedures, the details of which are described in CEBELCOR reports.[30] Note that the aggressive nature of drinking water may often be reduced by the addition of sodium silicate whose action, as yet not fully understood, seems to depend on the formation of colloidal silicic acid which deposits in acid pits and combines with the corrosion products thereby improving their protective power.

In numerous cases really "soft" water is required; this is water free from calcium or magnesium and may be obtained by filtration of hard water over ion exchangers, such as natural zeolites regenerated by a solution of NaCl or artificial resins regenerated by a solution of NaCl or H_2SO_4. With subsequent treatment on an anionic resin, it is possible to obtain virtually pure water. Such water is corrosive for ordinary steels and must not be distributed in steel piping without some precautions. Often when bringing steel piping into service for distribution of soft water, it is advantageous to start operating the water softener only after several months during which time slightly incrusting hard water would be used. In this period, the water would deposit a coating of calcium carbonate on the walls and the presence of such a coating would effectively counter the corrosive action of the softened water subsequently passed through the pipes.

In any case, we believe it is useful to passivate a water supply system by circulation (or recirculation) of a suitable oxidizing solution during initial operation. This is true for galvanized steel pipes as well as for non-galvanized ones.

6.6.4.2. Boiler Water

It is well known that treatments applied to water supplies of boilers depend on many factors, such as the operating pressure and the rate of vaporization. Besides, these treatments must avoid not only corrosion but also scaling.

From the corrosion viewpoint, it is a question of protecting not only the parts of the installation in contact with the water in use but also those

[30] See CEBELCOR Publications F. 6, RM. 3, RT. 54, RT. 56, RT. 95, and E. 24.

portions in contact with the vapor and condensed water. As far as possible this requires complete elimination of oxygen dissolved in the water. This is achieved by a thermal degassing (by bubbling under vacuum, or in a current of water vapor) completed by a chemical "degassing" using reducing agents: sodium sulfite or hydrazine. The addition of cyclohexylamine may be useful to reduce the corrosive action of condensed water. As the theoretical diagrams of corrosion, Figures 7a and 7b, and the experimental results of Figure 71b illustrate, for values of pH around 10 to 12.5, there is minimum danger of corrosion of iron, at least at 25°C. Such results do not actually exist for boiler temperatures.

Because of the risk of priming, it is useful to maintain a pH within the range 10.0 to 10.5 preferably by the addition of trisodium orthophosphate. Furthermore, unlike alkalis (e.g., caustic soda) phosphate has the advantage of not producing local regions of very high pH (13 to 14), and probably improves the protective qualities of oxide films when chlorides are present.

These local regions of high pH are often the cause of "caustic cracking" of highly stressed regions in boilers. This adverse effect of caustic is rationalized by the region of high iron solubility at high pH as shown in Figure 39.

In addition, the precaution of using phosphates to control boiler corrosion, there are other important precautions:

1. As pointed out in Section 6.5.1.2, all dissolved copper must be avoided.
2. When boilers are out of service, precautions must be taken to avoid differential aeration cells. This can be done both by completely drying the surfaces and by excluding air by using vacuum or inert gas. Sometimes it is possible to protect a boiler not in service by refilling the system with a passivating solution such as chromate. However, care must be taken to avoid differential oxidation, which would cause local corrosion in the case of surfaces contaminated by deposits.[31]

[31] Since the present Lectures were written, R. K. Freier (*Allianz-Berichte* **16**, 8–15 (1971)) has shown that boiler corrosion may be avoided without eliminating oxygen from the water and without adding alkaline and reducing substances. By simple addition of hydrogen peroxide to pure neutral water so that the electrode potential is permanently maintained at $+400\ mV_{SHE}$, the steel tubes of a boiler may be permanently kept under conditions of perfect passivation. Brief statements about the Freier process have been published in *Rapports Techniques CEBELCOR* **120**, RT. 204, 205 (1972).

6.6.4.3. Other Industrial Waters

The fundamental principles which we have discussed earlier are applicable to cases of corrosion in all waters at all temperatures and pressures. In certain cases, the actual operating conditions are sufficiently established so that the precautions to be taken and the treatment to be applied to water may be taken into consideration in the design of the installation. In other cases, electrochemical studies help to define these precautions or treatments. In every case, electrochemical tests make it possible to estimate the effectiveness of the procedures adopted.

Regarding the particular case of central heating installations with steel radiators, serious and rapid corrosion is inevitable if definite measures are not taken to reduce the danger of differential aeration. As far as possible, it is necessary to avoid all access of oxygen resulting for example, from injection of fresh water, from aeration at an expansion tank, from leakage at a coupling, or from depression of the seating of a circulation pump. An excellent precaution, suggested by A. de Grave, consists of introducing an adequate quantity of trisodium phosphate into the water circuit.[32] The regulation and control of this quantity of phosphate may be satisfactorily achieved by electrochemical tests.[33]

6.6.5. Action of Wet Materials

The principles which have just been outlined apply not only to water but also to materials which are wet or likely to become wet: soils, porous concretes, and other porous materials used as thermic or acoustic insulators. In every case, the danger of corrosion may be evaluated by adequate electrochemical studies, carried out both on metal buried in the wet material and on metal immersed in solutions obtained by aqueous extraction of this material. In each case, such studies make it possible to initiate protection techniques and to control their efficiency.[34] Depending on the particular case, corrosion may be suppressed by modification of the chemical composition of the material (conferring passivating properties on it) or by cathodic protection. Obviously, complete protection of the material can be obtained by keeping the metal dry.

[32] A. de Grave, Corrosion and scale in central heating installations, *Chaleur et Climat*, 83–91 (1962) (in French).

[33] *Rapports Techniques CEBELCOR* RT. 134 (1965).

[34] *Rapports Techniques CEBELCOR* RT. 123 (1964).

6.6.6. Cathodic Protection

6.6.6.1. General Principles

In general terms, cathodic protection consists of lowering the electrode potential of a metal or alloy to a value for which the corrosion is zero or negligible. This lowering is brought about by applying a reducing action: electric current or a reactive "sacrificial" anode. This may be performed by placing the metal in a state of immunity; or, sometimes in the case of chloride-containing solutions, by placing the metal in a state of "perfect passivation" where previously existing pits are unable to develop.

6.6.6.1.1. *Protection by Placing the Metal in a State of Immunity*[35]

The region of electrode potential where the protection of a given metal is assured corresponds to the theoretical region of *immunity*, that is, to the circumstances of thermodynamic stability of the metal. In the case of iron and ordinary unalloyed steels, the theoretical protection potentials vary as follows as a function of pH according to Figure 39.

Recall that this figure is based on metal in the presence of solutions containing 10^{-6} g-atom Fe/liter (0.06 ppm) and free from substances capable of forming soluble complexes or insoluble salts with iron[36];

$$\begin{aligned} &\mathrm{pH} < 9.0 && E = -0.62 \ \mathrm{V} \\ &9.0 < \mathrm{pH} < 13.7 && E = -0.085 - 0.0591\mathrm{pH} \ \mathrm{V} \\ &13.7 < \mathrm{pH} && E = +0.320 - 0.0886\mathrm{pH} \ \mathrm{V} \end{aligned}$$

This leads to the following values of protection potential for iron:

pH < 9.0	$E = -0.62$ V
pH = 10.0	−0.68
pH = 11.0	−0.74
pH = 12.0	−0.80
pH = 13.0	−0.86
pH = 14.0	−0.92

[35] *Rapports Techniques CEBELCOR* RT. 85 (1960) and RT. 98 (1961).

[36] These potentials are expressed with respect to the standard hydrogen electrode. The values following are deduced from equilibrium formulas 23, 13, and 24 given on pages 308–310 of *Atlas of Electrochemical Equilibria* (Refs. 8, 33).

We remind here that these conditions of immunity relate to pure iron. In case of some steels which are sensitive to hydrogen embrittlement, immunity of iron does not mean necessarily that the steel is completely safe.

The theoretical potentials of cathodic protection for a few metals for three values of pH are:

	pH = 0	pH = 7	pH = 14
Silver	+0.44	+0.44	+0.32
Copper	+0.14	+0.04	−0.38
Lead	−0.31	−0.31	−0.74
Iron	−0.62	−0.62	−0.92

It should be recalled that these potentials are expressed with respect to the standard hydrogen electrode at 25°C. To obtain values with respect to a saturated calomel electrode, the above values must be reduced by 0.25 V (so for iron, at $pH = 7$: $-0.62 - 0.25 = -0.87$ V); to obtain values with respect to a saturated copper sulphate electrode, these values must be reduced by -0.31 V (e.g., $-0.62 - 0.31 = -0.93$ V).

6.6.6.1.2. *Protection by Placing the Metal in a State of Perfect Passivation. Protection against Pitting Corrosion, Crevice Corrosion, and Stress Corrosion Cracking*

As already mentioned in Section 6.6.2.3, cathodic protection of a passivable metal or alloy can be sometimes achieved without necessarily lowering the potential into the "immune" region, where the surface is metallic and neither dissolving nor forming an oxide. If no pits are already existing, it is sometimes sufficient that, without the metal becoming immune, its electrode potential be lower than the "rupture potential" so that pitting will not occur and corrosion is virtually nil by virtue of the passivating oxide. However, if there are pre-existing pits, then it is necessary to lower the potential somewhat more, below the "protection potential against pitting." In the first case, no pits will be formed, and in the second case, the existing pits will not grow. As already noted in Section 6.6.2.3, both the rupture potential and the protection potential against pitting may be determined by electrochemical tests. The concept of "cathodic protection by perfect passivation" (free from active pits or cracks) is best illustrated by examining Figures 96 and 97, which relate to copper, iron, and carbon steels.

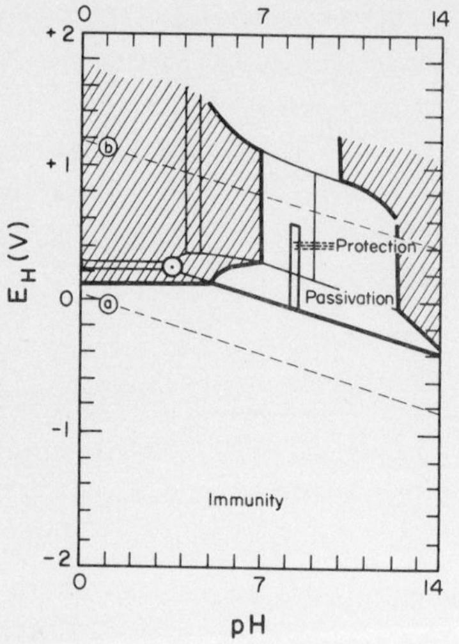

Figure 96. Copper in Brussels water. Potential and pH conditions inside pits (equilibrium conditions) and outside pits. Protection potential against pitting. Inside pits: pH = 3.5, $E = +270\ mV_{SHE}$; outside pits: pH = 8, $E = 0$ to $+650\ mV_{SHE}$.

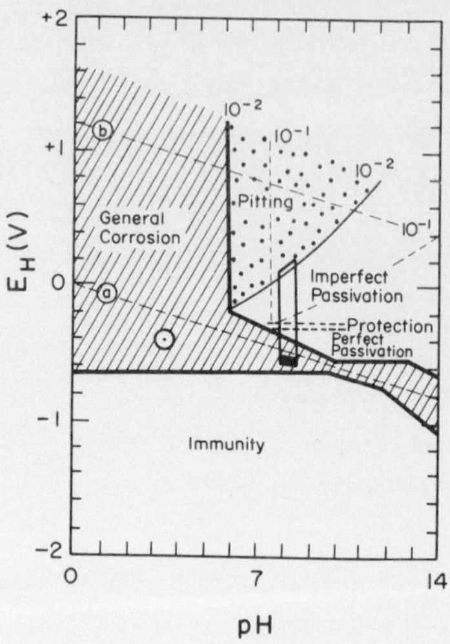

Figure 97. Armco iron in solutions containing chloride (10^{-2} and 10^{-1} g-ion/liter). Potential and pH conditions inside living pits. Rupture potentials. Protection potential against pitting.

6.6.6.1.2.1. *Pitting Corrosion of Copper and Protection*

Copper tubes used for distributing cold water sometimes experience extensive corrosion. The characteristic attack in these tubes is a type of pitting, first described in 1953 by R. May.[37] These copper pits contain a white cuprous chloride, **CuCl**, and a red cuprous oxide, **Cu_2O**. In the case of pitting in Brussels tap water, they are usually covered by a deposit containing green malachite **$CuCO_3 \cdot Cu(OH)_2$**, as shown by Figure 98.[38] These pits grow when the electrode potential of the metal, as measured outside the pit, is higher than a critical value[39] which depends upon the shape of the surface. As shown by J. van Muylder *et al.*, the critical potential is about

[37] R. May, Some observations on the mechanism of pitting corrosion, *J. Inst. Metals* **82**, pp. 65–74 (1953).

[38] *Rapports Techniques CEBELCOR* RT. 127 (1965), CEBELCOR Publication E. 61 (1967).

[39] *Rapports Techniques CEBELCOR* RT. 126 (1965).

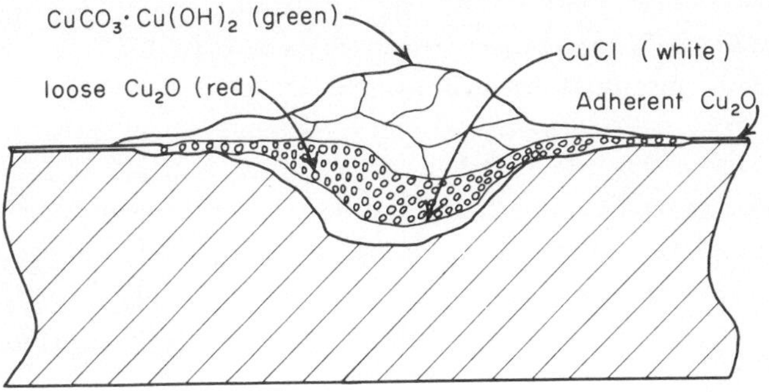

Figure 98. Copper pit in the presence of Brussels water. Cross section showing the presence of red Cu_2O and of white CuCl beneath a mushroom of green malachite.

$+170$ mV_{SCE} (i.e., $+420$ mV_{SHE}) for the concave internal surface of tubes and about $+100$ mV_{SCE} (i.e., $+350$ mV_{SHE}) for the convex surface of thin wires (0.2 mm in diameter). These observations are explained in the following paragraphs.

Figure 37 is the potential–pH diagram for the binary system Cu–H_2O at 25°C, already discussed in Section 4.6.1. Figure 99 shows the ternary system Cu–Cl–H_2O where the chloride is present in a concentration of 10^{-2} ion Cl^-/liter (355 ppm).[40] Figure 100 shows the Cu–Cl–CO_2–SO_3 system when the solutions contain 22 ppm Cl^-, 229 ppm CO_2, and 46 ppm SO_3^{--}; this corresponds to these species concentration in Brussels water.[41]

According to Figures 37, 99, and 100, especially Figure 100, the basic carbonate, $\mathbf{CuCO_3 \cdot Cu(OH)_2}$ (malachite), is the stable form of copper derivatives in the presence of Brussels water (pH about 7.9) containing oxygen. The stable form of copper in Brussels water when oxygen-free is cuprous oxide, $\mathbf{Cu_2O}$ (cuprite). Cuprous chloride, **CuCl** (nantokite), is unstable under these conditions, and thus tends to be hydrolyzed with the formation of $\mathbf{Cu_2O}$ and of hydrochloric acid according to the reaction $2\mathbf{CuCl} + H_2O \rightarrow \mathbf{Cu_2O} + 2H^+ + 2Cl^-$. This mechanism has been described elsewhere.[42] At the bottom of a copper pit, the phases **Cu**, $\mathbf{Cu_2O}$, and **CuCl**

[40] *Rapports Techniques CEBELCOR* RT. 101 (1962).

[41] CEBELCOR Publication E. 63 (1964).

[42] *Rapports Techniques CEBELCOR* RT. 126 and 127, see also M. Pourbaix, Recent applications of electrode potential measurements in the thermodynamics and kinetics of the corrosion of metals and alloys, NACE Symposium "Fundamental Corrosion Research in Progress," March 1968, Cleveland, Ohio, *Corrosion* **25**, 267–281 (1969).

coexist in contact with a solution containing about 246 ppm Cl^- and 270 ppm Cu^{++}. This situation is represented in Figure 99 by the intersection of the three lines numbered 12, 51, and 55, which is the only point on the diagram where **Cu**, **Cu_2O**, and **CuCl** may coexist as stable phases. The coordinates of this point are $E = +270\ mV_{SHE}$ and $pH = 3.5$. This point, which represents the equilibrium condition inside a copper pit, is represented in Figure 96 by a circle ⊙. The electrode potential of the surface outside the pit and directly exposed to an environment of pH about 7.8 depends on the concentration of oxygen dissolved in the water, on the degree of protection afforded by the passivating layer, and on the overpotential exerted by this layer on the reduction of oxygen. Under normal conditions, the potential outside the pits is about $+300\ mV_{SHE}$, i.e., about 70 mV higher than the equilibrium potential inside the pits; it has been observed that, under such a condition the existing pits do not grow appreciably and are not harmful. There are circumstances, however, which can lead to an

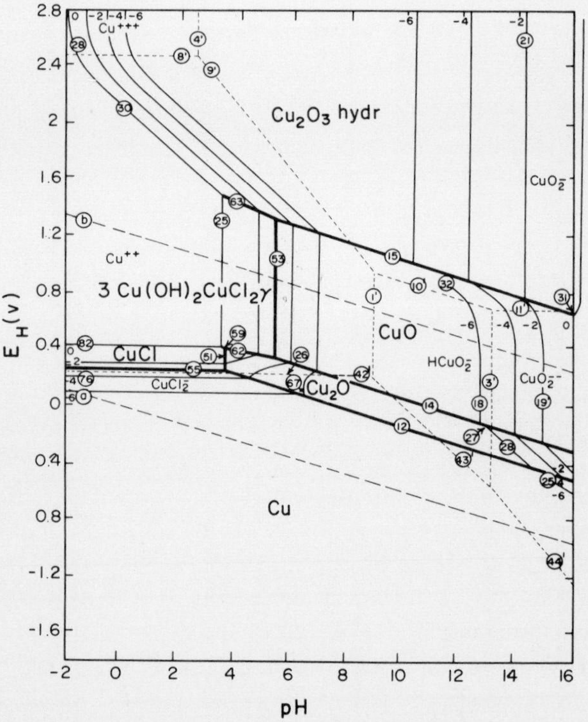

Figure 99. Potential–pH equilibrium diagram for the ternary system Cu–Cl–H_2O at 25°C (355 ppm Cl).

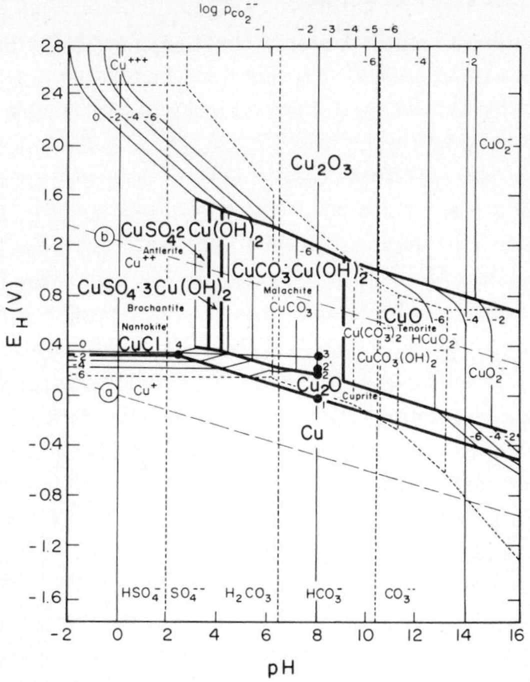

Figure 100. Potential–pH equilibrium diagram for the quinary system Cu–Cl–CO_2–SO_3–H_2 at 25°C (22 ppm Cl, 229 ppm CO_2, 46 ppm SO_3).

increase of potential on the surface. For example, it is possible that certain substances which accelerate the reduction of oxygen, i.e., have a low overpotential for oxygen reduction (such as carbon, gold, platinum, rhodium), will cause the potential to rise to higher values in the range of +350 to +420 mV_{SHE}. Such potentials are sufficiently high to overcome the diffusion potential and the ohmic drop through the layer of malachite (and other salts) covering the pits. As a result, the electrode potential inside the pits becomes higher than the equilibrium potential (+270 mV_{SHE}), and the copper dissolves very rapidly according to the reaction **Cu** → Cu^{++} + $2e^-$. This leads to severe pitting which can quickly destroy the copper tubes. This geometrical arrangement, leading to the acceleration of the pitting, causes differential aeration where the cathodic reaction $\boldsymbol{O_2} + 4H^+ + 4e^- \rightarrow 2H_2O$ occurs at about +400 to +600 mV_{SHE} on the external passivated areas, where the pH is 7.8 to 8.5, and the anodic reaction **Cu** → $Cu^{++} + 2e^-$ occurs at about +270 mV_{SHE} in the internal cavities, where the pH is about 3.5. These conditions are shown schematically in Figure 96.

If, for any reason, the electrode potential outside a pit becomes appreciably lower than $+270\ mV_{SHE}$, which is the equilibrium potential inside the pit, a reversal of the corrosion reaction inside the pit takes place. As a result, the dissolution reaction is reversed and Cu^{++} ions are reduced, according to the reaction $Cu^{++} + 2e^- \rightarrow \mathbf{Cu}$. Practically speaking, the pitting of copper as described above may be stopped by depressing the potential on the metal surface below the protection potential (about $+350$ to $+420\ mV_{SHE}$). This may be achieved by several methods. Two of these methods are the addition of iron salts to the water, and cathodic protection by the use of a sacrificial anode such as iron, zinc, brass, or aluminum–zinc alloy.

In Figure 96, we have represented the equilibrium conditions inside a copper pit ($E = +270\ mV_{SHE}$, pH $= 3.5$) and the possible potential and pH conditions outside the pits. The dotted lines represent the "protection potential" below which existing pits do not grow. Note, in accord with the effect of diffusion potential and of IR drop, that the protection potential is slightly higher than the equilibrium potential inside the pits.

To summarize the situation regarding the pitting of copper in neutral Brussels tap water we wish to emphasize the following facts:

1. As a result of the hydrolysis of the primary corrosion products, the solution existing inside a copper pit is acid even if the water outside the pit is neutral or alkaline. This solution is much more aggressive than the bulk of the water.

2. The pitting of copper may be elucidated by considering the thermodynamics and the kinetics of the electrochemical processes occurring on the passivated metal in the presence of the bulk solution and on the active metal in the presence of the *solution really existing inside the pit.*

3. Pitting of copper may be avoided by maintaining the external electrode potential of the metal below a given critical value. The critical value is the equilibrium electrode potential inside the pits (which may be predicted thermodynamically or measured by adequate experiments), to which may be added the diffusion potential and the ohmic drop through the layer of salt which is covering the pit. The diffusion potential and ohmic drop must be determined experimentally and will vary from case to case. This concept provides several means for combating the pitting of copper: These would include cathodic protection by using sacrificial materials and lowering the kinetics of the oxygen reduction reaction.

It is most likely that these concepts, although they have not yet been the subject of an extensive systematic study, may be extrapolated to every passivable material, and they may be valid not only for pitting but also in other instances where corrosion occurs under geometrical conditions of restricted diffusion: stress corrosion cracking, crevice corrosion, corrosion in recesses, corrosion under deposits, etc. Further research work along these directions should be enthusiastically supported.

6.6.6.1.2.2. *Pitting Corrosion, Crevice Corrosion, and Stress Corrosion Cracking of Iron and Steels, and Protection*

Figure 101b represents the *experimental* conditions of immunity, general corrosion, and passivity of iron in the presence of chloride-free solutions, at 25°C. Figure 101a shows, schematically, for five values of pH over the range of 5 to 13, the potentiokinetic polarization curves which were used in drawing Figure 101b. The passivation potentials, *P*, which have been determined by the potentiokinetic curves from Figure 101a have been plotted in Figure 101b, where they provide a means for separating the areas of general corrosion at lower potentials and the areas of passivity at higher potentials.

The effect of chloride ion additions to the same solution as in Figure 101 is shown in Figure 102.[43] In Figure 102, 10^{-2} g-ion Cl^-/liter (355 ppm) has been added to solutions of exactly the same values of pH as in Figure 101. These curves show the following:

1. For pH 5, increasing the potential accelerates general corrosion with no passivation.
2. In the range of pH 7–11, the rupture potential effect is observed after passivation, as described previously, with the appearance of pits at a potential *r* which increases as the pH increases. By polarizing the pitted steel at potentials decreasing toward the active direction no protection is observed at pH 7 but a protection potential is observed at pH 9–11.
3. For pH 13, no rupture potential is observed.

Polarization curves obtained in chloride solutions, as shown in Figure 102a, allow the drawing of Figure 102b which shows schematically the

[43] This work is based on the experimental observations of J. van Muylder and of N. Verhulpen–Heymans conducted in the laboratories of CEBELCOR and of Université Libre de Bruxelles.

experimental conditions of immunity, general corrosion, perfect passivity, imperfect passivity, and pitting of iron in the presence of solutions containing 355 ppm chloride ions.

Comparing the behavior of chloride-containing solutions in Figure 102b and nonchloride containing solutions in Figure 101b, one observes that the presence of chloride does not modify the behavior of iron at low potential such as might exist in the absence of oxygen or other oxidants. But in the presence of oxidants, it suppresses any passivation at pH's lower than about 6 and promotes general corrosion. Pitting occurs in the presence of oxygen for pH values from about 6 to about 10. A "protection potential" which is fairly constant in the range of −0.4 to −0.2 V, separates an area of imperfect passivity where pre-existing pits may grow from an area of perfect passivity where pre-existing pits may not grow.

Based on the information from Figure 102b, it is possible to devise conditions under which iron structures can be protected from serious corrosion. Where pitting might normally occur in aerated conditions of pH 6–10, protection might be afforded by cathodic techniques not only by depressing the potential to the area of immunity but also by depressing the potential somewhat less, only to the area of perfect passivity. Despite the substantial benefits to be obtained by the relatively small change of potential

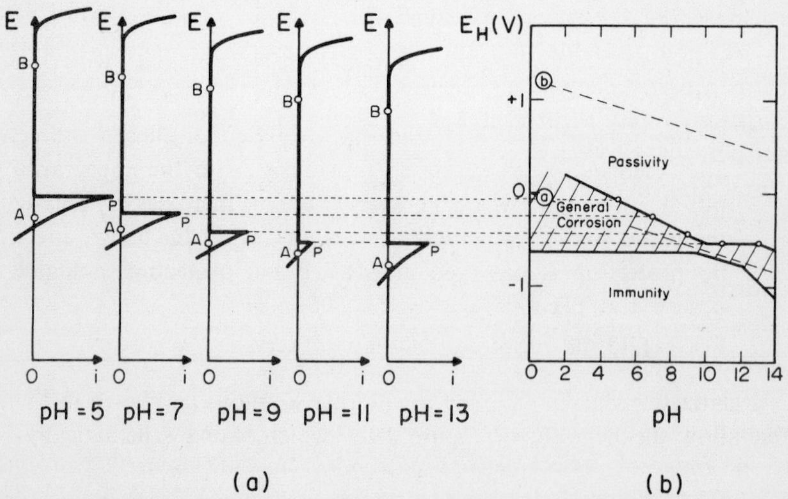

Figure 101. Behavior of iron in chloride-free solutions: (a) polarization curves in the presence of solutions, for pH 5 to 13; and (b) experimental conditions of immunity, general corrosion, and passivity.

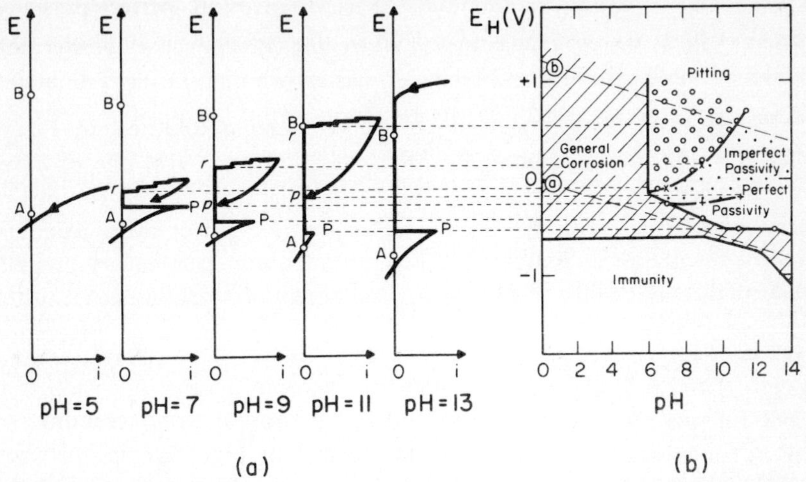

Figure 102. Behavior of iron in solutions containing chloride (355 ppm): (a) polarization curves in the presence of solutions, for pH 5 to 13; and (b) experimental conditions of immunity, general corrosion, perfect and imperfect passivity, and pitting. Protection potential against pitting.

to the region of perfect passivity, it should be pointed out that such cathodic protection should be impressed with greatest care since by decreasing the potential too low the area of general corrosion would be encountered and the material would again deteriorate. For example, in a solution of pH 9 with 355 ppm chloride and saturated with oxygen, nonpolarized iron exhibits a corrosion potential higher than $+0.2\ V_{SHE}$ and suffers pitting. This material may be protected cathodically by polarizing the metal to a lower potential between about -0.20 and -0.40 V to give perfect passivity, or protection may be achieved by lowering the potential to -0.62 V to achieve the immune state. If full protection is wished, the potential range between -0.40 and -0.62 V should be avoided for general corrosion would be possible.

When the existence of the above-mentioned protection potential was discovered in 1961 in research work performed in collaboration with C. Vanleugenhaghe *et al.*,[44] it was not possible to suggest a scientific explanation for the process of cathodic protection by lowering the potential into the range of perfect passivation. However, work recently performed at the Naval Research Laboratory in Washington, D.C. by B. F. Brown

[44] See *Corr. Sci.* **3**, 239–259 (1963).

et al.[45,46] concerning the stress corrosion cracking of various steels, and at CEBELCOR in Brussels by J. van Muylder and M. Pourbaix[47] concerning the crevice corrosion of iron and steels, has shown that, inside both active cracks and active crevices, the pH is about 2.7 to 3.8 and the electrode potential is about -0.3 to $-0.5\ V_{SHE}$. These pH values and electrode potentials are easily predicted by thermodynamic concepts primarily related to hydrolysis equilibria. In Figure 97, we have represented these conditions of pH and electrode potential, which, incidentally, should be very close to those conditions for pits, by a circle ⊙. We have also represented the possible values of potentials of the iron surface outside the pits or cracks (about $+0.2$ to -0.6 V, depending on the oxygen content) when exposed to water in the pH range 7.8 to 8.5 containing 10^{-2} g-atom Cl^-/liter. We have also drawn for two chloride levels, 10^{-2} and 10^{-1} g-atom, lines representing the rupture potential. In both cases, the protection potential lies between -0.2 and -0.3 V and, thus, is slightly higher than the potential inside the pits or cracks.

It appears that there is a great similarity in the behavior of copper and iron and steel as shown in Figures 96 and 97, relative to the processes of pitting and protection therefrom. In both cases, the solutions inside the pits are acid due to the hydrolysis of primary corrosion products and the resultant formation of stable oxides. In both cases, the growth of pits may be avoided by maintaining the potential of the metal outside the pits lower than a critical "protection potential." This potential is slightly higher than the potential inside the active pit. In both cases, a full understanding of the process of pit formation and pit growth can be obtained from a systematical electrochemical study of both the passivation phenomenon and the behavior of the active metal. There are a number of similarities between copper and iron alloys with respect to pitting. The values of pH and potential inside active pits and cracks for these materials together with the respective values of protection potentials are summarized in the Table XXVI.

Research work recently performed at the Naval Research Laboratory NRL (B. F. Brown *et al.*)[44,45] and at CEBELCOR (in collaboration with J. van Muylder)[46] has shown that a cathodic treatment of steels leads to a decrease of electrode potential and to an increase of pH inside cracks or

[45] B. F. Brown, C. T. Fujii, and E. P. Dahlberg, Methods for studying the solution chemistry within stress corrosion cracks, *J. Electrochem. Soc.* **116**, pp. 218–219 (1969).

[46] B. F. Brown, On the electrochemistry of stress corrosion cracking of high-strength steels, CEBELCOR Publication E. 76 (1969).

[47] See *Rapports Techniques CEBELCOR* RT. 166 (1969), RT. 179 (1970), RT. 199 (1971).

TABLE XXVI. The pH and Potentials Inside Active Pits and Cracks and the Protection Potentials Against Their Growth

	Pits (for copper and steels) and cracks (for steels)		Protection potentials
	pH	E (V_{SHE})	E (V_{SHE})
Copper	3.5	+0.27	+0.35 to +0.47
Iron and steel	3.8	−0.50 to −0.30	−0.40 to −0.20

pits, which become progressively neutral, and then alkaline. This explains why both cracks and pits become inactive when the "protection potential" is reached. However, as stated by Brown, these trends relate only to failure processes involving metal dissolution, and not hydrogen embrittlement. The latter may be enhanced by too great a degree of cathodic polarization of susceptible alloys, such as high strength steels.

Figure 103 shows schematically, after J. van Muylder and M. Pourbaix, the influence of cathodic polarization on the potential and pH conditions inside a pit or crevice for a carbon steel in contact with an aerated solution of 0.001 M NaCl and 0.001 M NaOH (external pH 10.0). Dotted lines marked 0 refer to nonpolarized steel; dotted lines marked 1 to 9 refer to increasing cathodic currents. Dotted line 3 shows that when the protection potential, $-0.40\ V_{SHE}$, is reached, the pH inside the pit, which was initially about 2.7 to 4.7, had increased to about 7; further polarization increased this pH to about 11.

Figure 104 (after B. F. Brown) shows the influence of cathodic polarization on the potential and pH conditions inside a *crack* for an AISI 4340 chromium steel in contact with a 3% NaCl solution. This figure also shows the relationship between these potential–pH conditions and the speed of crack propagation of several steels. It may be seen that the minimum speed occurs for a potential of about −0.5 to $-0.6\ V_{SHE}$ inside the crack.

Figures 103 and 104 show that, inside cracks as well as inside pits and crevices, the potential and pH conditions may be very different from those existing on the external passive surfaces. Even when the conditions on the external surfaces are high above the hydrogen line a, the conditions in the internal cavities are below this line, and this means that hydrogen evolution is possible inside these cavities. Thus, as stated by Brown, cracking

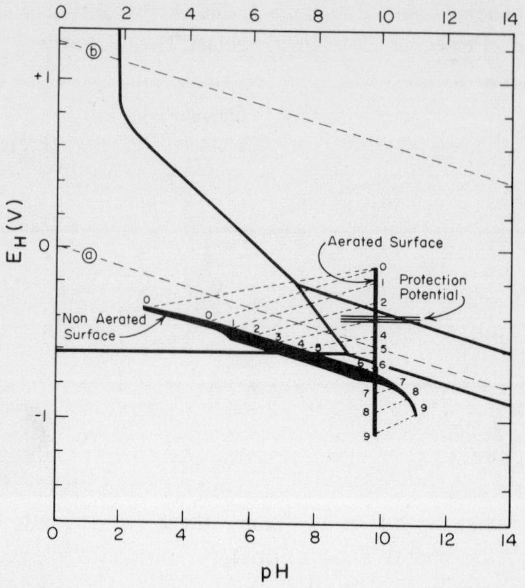

Figure 103. Influence of a cathodic polarization on the potential and pH conditions inside a pit (ordinary carbon steel with an aerated solution of 0.001 M NaOH and 0.001 M NaCl).

observed on some steels at potentials higher than the "protection potential" may be due, not only to dissolution of iron, but also to some hydrogen embrittlement; hydrogen embrittlement without dissolution of iron may also occur at potentials less than the protection potential.[48]

All that has been said here concerning pitting corrosion and stress corrosion cracking is also valid for crevice corrosion and for any other form of "occluded cell corrosion" where corrosion occurs under conditions of restricted diffusion: intergranular corrosion, filiform corrosion, tuberculation, corrosion inside recesses, corrosion under deposits including mill scale and barnacles, corrosion under paints, and corrosion at the waterline. Further conditions concerning this approach and its application to corrosion technology may be found in reports of CEBELCOR and of the University of Florida (*Rapports Techniques CEBELCOR* RT. 167, 179, 186, and 191), for copper, iron, aluminum, titanium, iron–chromium and copper–nickel alloys.

[48] See also M. Pourbaix, Electrochemical aspects of stress corrosion cracking, Proc. NATO Research Conference on Stress Corrosion Cracking, Ericeira, March 29–April 2, 1971, *Rapports Techniques CEBELCOR* RT. 199, **118** (1971) (in French).

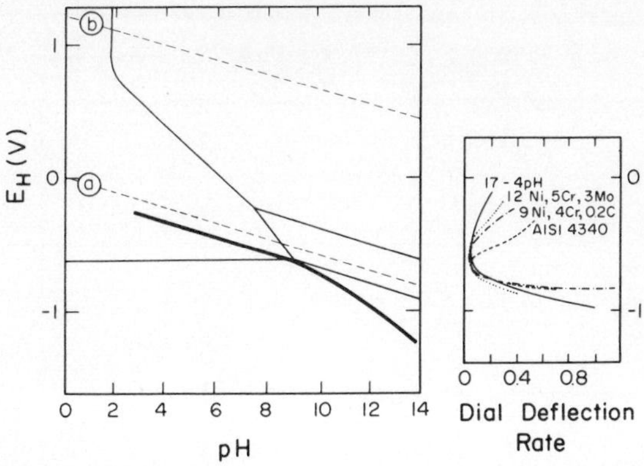

Figure 104. Influence of a cathodic polarization on the electrochemical characteristics and on the propagation rate of a stress corrosion crack (chromium steel AISI 4340 in contact with NaCl 3.5% solution) (after B. F. Brown).

The practical application of these considerations has not yet been systematically studied, but such ideas should be applicable to *all* familiar metals and alloys. They should be able to ensure perfect protection against pitting corrosion and crevice corrosion occurring in solutions containing chloride, notably sea water. They should help to solve, among others, the extremely important problem of stress corrosion cracking.

Definitive work on local electrochemical processes is an urgent necessity.

6.6.6.1.3. *Protection Procedures Using External Current and Sacrificial Anodes*

The electrode potential of iron may be lowered toward the state of immunity by passing an electric current through the material to be protected. The flow of electricity must be such that the reaction $\mathbf{M} \rightleftarrows M^+ + e^-$ is driven to the left; thus, electrons must be supplied *to* the metal. The amount of current to be supplied is determined by the E vs i relationship from established polarization curves. In Figure 68, full protection of the iron is achieved at point C where the forward reaction $\mathbf{M} \rightarrow M + e^-$ is completely stopped. However, when the potential is lowered to point A, the hydrogen ion reduction reaction is accelerated, $2H^+ + 2e^- \rightarrow \boldsymbol{H_2}$, and the solution becomes alkaline. The rate of the hydrogen reduction is obtained from the E vs i behavior of hydrogen ion reduction in Figure 68.

The source of electrons necessary to confer immunity may be either from an external source, together with an auxiliary electrode immersed in the same environment, or from a "sacrificial anode" in electrical contact

TABLE XXVII. Anodes Used as Groundbeds
(after U. S. Gerrard and G. G. Page in Shreir, Ref. 30)

Material	Maximum current density (A/dm^2) in soil	Maximum current density (A/dm^2) in water	Approximate consumption (kg/A-year)
Steel	0.05	0.05	6.8 to 9.1
Cast iron	0.05	0.05	4.5 to 6.8
Aluminum	—	0.2	4.1
Silicon iron	0.3	0.3 to 0.4	0.4 to 0.9
Graphite	0.1	0.2	0.9
Lead	—	1.1 to 2.2	—
Lead–platinum	—	110	—
Platinum	—	<110	—
Platinized titanium	—	<110	—
Platinized tantalum	—	<110	—

with the metal to be protected. The sacrificial anode operates by furnishing electrons to the protected metal from a relatively fast oxidation reaction.

Table XXVII lists various anodes used for protection in the external current technique of cathodic protection. Table XXVIII lists the important materials which are used as sacrificial anodes.

With regard to the quantity of current necessary to ensure protection, it has often been assumed empirically that it must be sufficient to lower the electrode potential of the steel by 0.3 V with respect to its free potential (where there is no external current). Also, it has been suggested that the current must be equal to that which produces a discontinuity in the current–potential polarization curve. Often, these two criteria are not satisfactory and lead to erroneous results. The criteria outlined above are preferable. Depending on the particular case, these consist of placing the metal in a state of immunity (generally below $-0.62\ V_{SHE}$ for ordinary steels) or in a state of perfect passivation (below the protection potential previously determined by preliminary electrochemical experiments). As has been shown by J. Horvàth and his collaborators,[49] the protection potential of

[49] J. Horvath, L. Häckl, and A. Rauscher, Investigations of the possibilities of corrosion inhibition in metal, Sulfur–water ternary systems on the basis of potential–pH equilibrium diagrams (43), *C. R. 2nd Symposium Corrosion Inhibitors, Ferrara, 1965*, pp. 477–505 (1966).

TABLE XXVIII. Composition and Properties of Materials Used as Sacrificial Anodes (after Gerrard and Page)

Anode	Composition %											Current efficiency (%)	Electrode potential (volts *vs* Cu–$CuSO_4$)	Consumption (kg per A-year)
	Mg	Al	Mn	Sn	Si	Cu	Ni	Fe	Pb	Cd	Sn			
Magnesium	Bal	5.3 to 5.7	0.15 min	2.5 to 3.5	0.3 max	0.05 max	0.003	0.003	—	—	—	50	−1.55	7.7
Zinc	—	0.1 to 0.3	—	Bal	—	0.005 max	—	0.0014	0.006	0.03 to 0.7	0.05 to 0.2	90	−1.1	12
Aluminum–zinc	—	Bal	—	3	0.10 max	0.02 max	—	0.17 max	—	—	—	50	−1.1	3–6

iron may be lower in sulfide media on account of the possible formation of nonprotective **FeS**.

The previous empirical criteria may be particularly dangerous in the case of protecting stainless steel in sea water. Thus, lowering the potential from the passive region may bring the potential to the active region where corrosion would be very rapid. For stainless steels, as well as for other passivable alloys (aluminum alloys, titanium alloys, nickel alloys, etc.), the setting up of a completely safe procedure for protection as well as the elaboration of new alloys, is greatly facilitated by experimental data as presented by Figure 102 (showing for iron the general conditions of immunity, general corrosion, perfect passivation, imperfect passivation, and pitting) as well as represented by Figure 103 (showing the influence of cathodic protection on the potential and pH conditions inside pits, crevices, cracks, and other recesses).

As a general rule, the amount of current and the possible sacrificial anodes used must be such that the whole surface of the structure is protected. The electrode potential must permanently be less than the "protection potential." Of course, it is of no avail to lower the potential further: this results in both an excessively high consumption of current or reactive metal and may even risk embrittling the steel by hydrogen absorption.

Protection by means of an external current[50] may be carried out, as outlined on page 281 and in Figure 105, by procedures of drainage, polarized drainage, and impressed current using anodes sometimes called "groundbeds." These groundbeds may be corrodible metals, mostly iron (old rails) which dissolves according to Faraday's law to the extent of 9.1 kg/A-year. In certain cases, aluminum alloy groundbeds are used to protect water supply installations.

Often uncorroded or mildly corroded conducting materials are used as groundbeds: silicon–iron, graphite, lead, and platinum both as massive or sheet material and as an electrolytic deposit on another material. Actually groundbeds of platinized titanium seem to be especially promising when properly constructed and employed.

Table XXVII lists a few of the characteristics of groundbeds taken from work published by J. S. Gerrard and G. G. Page.[51]

The *protection by sacrificial or reactive anodes*, although generally more costly than protection by means of an external current source, is preferred for installations of a complex nature or relatively small size where

[50] For example, see Shreir, Ref. 30, pp. 11–37.

[51] See J. S. Gerrard and G. G. Page, Ref. 30, pp. 11–36.

the corrosive medium has fairly low resistivity (not exceeding 4.000 ohm-cm) as well as when there is no readily available electric current.

Potentially, *magnesium* is by far the most useful anode material because of its extremely low corrosion potential (−1.55 V with respect to the copper sulfate electrode). However, from the viewpoint of material wastage for the amount of cathodic protection achieved, the cost of protection by magnesium is appreciably higher than that of zinc because of the low electrochemical efficiency of protection by magnesium (about 50% as compared with 90% of zinc). This results from the transitory formation of unstable monovalent Mg^{+} ions which then oxidize to bivalent ions, Mg^{++} (with hydrogen evolution). Consequently, *zinc* remains the material of choice in all cases where it is technically feasible. This implies that the metal has a considerable hydrogen overpotential in the presence of the corrosive medium and that anodic polarization produces general corrosion without significant elevation of the electrode potential. For a metal of suitable chemical composition, such conditions are obtained in the presence of sea water as well as in cold tap water; they are not obtained in tap water above a temperature of 60°C on account of the passivation which then affects zinc.

The material to be chosen for use as a reactive anode, depends on the local conditions and especially on the aggressive nature of the medium against the anode, on the conductivity of the electrolyte (soil, water, etc.), on the extent of the surface to be protected by one series on anodes, etc.

From the work of J. S. Gerrard and G. G. Page, Table XXVIII gives the principal characteristics of several metals and alloys used as sacrificial anodes for the protection of buried structures. When reactive anodes are used for the protection of buried structures, usually the anode is enclosed in a "backfill" (surrounding material) which ensures uniform corrosion of the anode. For magnesium, this "backfill" is composed generally of 75% gypsum, 20% bentonite, and 5% Na_2SO_4. For zinc, the bentonite may be replaced by argile. Usually, the quality of the given materials employed as the reactive anode or "backfill" may be controlled by electrochemical tests as outlined in Figure 85.

6.6.6.1.4. *Current Requirements for Cathodic Protection*

As stated earlier, the electric current necessary for protection is that which is required to lower the electrode potential of all points of the structure to its protection value. For structures in the presence of hard water, the local alkalization caused by the protection treatment leads to the formation on the protected surfaces of relatively adherent deposits which,

TABLE XXIX. The Order of Magnitude of Current Densities Necessary for the Cathodic Protection of Steel
(after H. H. Uhlig)

Surrounding environment	Current density (mA/cm^2)
Sulfuric acid pickling	4,000
Soils	0.1–5
Sea water in motion	1.5 (initial)–0.3 (final)
Warm water saturated with air	1.5
Circulating cold water	0.5

particularly when water containing magnesium is involved (sea water), make it possible to reduce appreciably the quantity of the protection current. From H. H. Uhlig's[52] work, Table XXIX gives some values of the current density for protection in the case of steel.

By employing a sufficient quantity of electric current it is generally perfectly feasible to cathodically protect bare structures. However, it is usually convenient to combine cathodic protection with the application of a coating thereby making it possible to reduce considerably the applied current density. Among the quantities necessary for such a coating, the most important is a good chemical resistance to alkaline solution since at all weak points of the coating where solution gains access to the metal alkalization is inevitably produced, owing to the consumption of hydrogen ions. In this particular case, such chemical resistance is more important than good electrical insulation or the absence of porosity, which may be remedied by employing a sufficient amount of current. H. H. Uhlig[52] has remarked that in the case of pipelines buried in the ground, the application of a coating makes it possible to extend from 30 to 8000 meters the length protected by a single magnesium anode; in the case of protection by an external current source, a single groundbed may protect up to 80,000 meters.

Note that cathodic protection cannot be effective on parts of a metallic structure in contact with electrolyte (waters) which, for example, are shielded from the action of the current by intervening insulating material. For instance, cathodic protection is without effect in "spiral corrosion" which sometimes occurs when using protective tapes which become insufficiently adherent to piping. Tapes for pipe-wrapping should, in order to be suitable

[52] H. H. Uhlig, Ref. 29, p. 189.

for cathodic protection, have "wetting" power; that is to say, they must fill all surface irregularities and exclude all air. The overlap seal must be permanent. Tapes with a compound on both sides, termed "biadhesive," ensure this and also act as a key for the outerwrap. Such a multilayer system has not only excellent anticorrosive qualities, but is also mechanically tough. On March 4, 1965, an important gas pipeline failed by corrosion in Natchitoches, Louisiana; this failure resulted in a fire which led to the death of 17 people. The responsible American authorities carried out a thorough inquiry which attributed the rupture to stress corrosion cracking, which, according to Townsend,[53] was caustic embrittlement stress corrosion cracking due to the combined effects of (1) high temperature resulting from adiabatic gas compression, (2) stress resulting primarily from internal gas pressure, and (3) a caustic solution at the steel surface resulting from the accumulation of alkaline products of cathodic protection. Near the broken part of the pipe, lack of adherence of the coating and many small pits were observed. As a result of this inquiry, the Department of Transportation of the USA has stipulated that extreme precautions be taken in the future for the choice of materials to be used as external coating of the pipes: these materials must have sufficient adhesion to the metal surface to prevent underfilm migration of moisture. Besides this, we consider it essential that the external surface of the pipes be free from any trace of corrosion when the coating is applied.

6.6.6.2. Applications of Cathodic Protection

The techniques of cathodic protection, which may be considered as "perfect" techniques, are mainly applicable to iron, copper, and often to aluminum and some alloys (brasses, "stainless steels"), in the instance where these metals or alloys are immersed in an aqueous solution or surrounded by wet material. Of course, it is necessary to place the "anodes" close to the surfaces to be protected thereby ensuring the flow of protective current to these surfaces.

Piping buried in the ground, bars within reinforced concrete, hulls of ships, reservoirs for water such as water tanks and central heating installations, timber-lined metal, sluicegates, and much industrial equipment may be protected. Cathodic protection is applicable only to immersed or embedded surfaces. It does not apply to structures subject to atmospheric corrosion or to parts of reservoirs above the water line.

[53] H. E. Townsend, Jr., Natural Gas Pipeline Stress Corrosion Cracking: A Failure Analysis, *Materials Protection and Performance,* **11**(10), 33–37 (1972).

6.6.6.2.1. *Cathodic Protection of Buried Piping*

Figure 105 illustrates schematically five procedures for the cathodic protection of buried piping; four of these involve the use of an external source of current. In case a (*drainage*), protection is obtained merely by connecting the piping to an electric rail whose potential is always sufficiently negative. The arrangement b (*polarized drainage*), provides for the case where, due to the traffic of electric vehicles (trains or trams) along the rail, the potential of the rail is not always sufficiently negative, in which case simple drainage may cause temporary corrosion during periods when the polarity of the rail is inverted; in this instance, a relay cuts the conduction path between the pipeline and the rail. In case c (*impressed current*), an external source with transformer and rectifier permanently taps off negative current from the pipeline to a rail at higher potential. In case d (*groundbed*), an external current source (a series of accumulators or rectified a.c.) taps off some negative current from the pipeline to a "groundbed" which, depending on the particular case, may be made up of iron or an incorrodible or mildly corrodible metal or alloy (graphite, silicon–iron). Generally, this is enclosed in a "backfill" (surrounding material) ensuring satisfactory electric conduction around the groundbed. In case e (*reactive anode*), the current necessary for protection is supplied by a metal (zinc, magnesium or special aluminum–zinc alloys) corroding at low potential in a suitable "backfill." The principles of these reactive or "sacrificial" anodes have been outlined in Figures 83 to 85 in Section 6.5.2.

In each of these five cases, control of the protection may be made by simple measurement of the electrode potential of the piping which at all

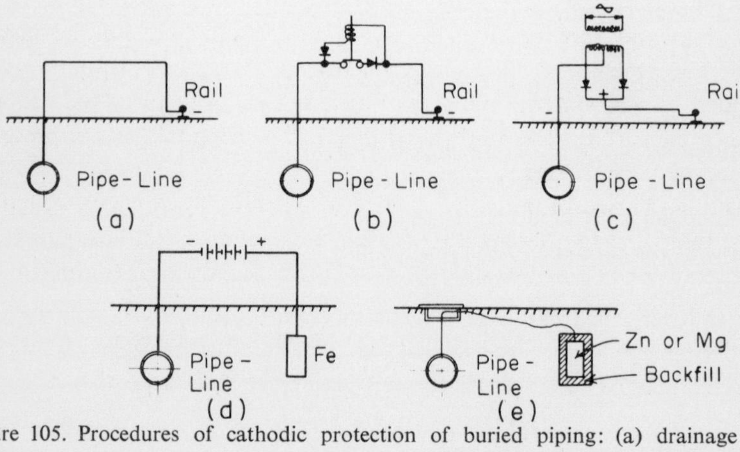

Figure 105. Procedures of cathodic protection of buried piping: (a) drainage; (b) polarized drainage; (c) impressed current; (d) groundbed; and (e) reactive anode.

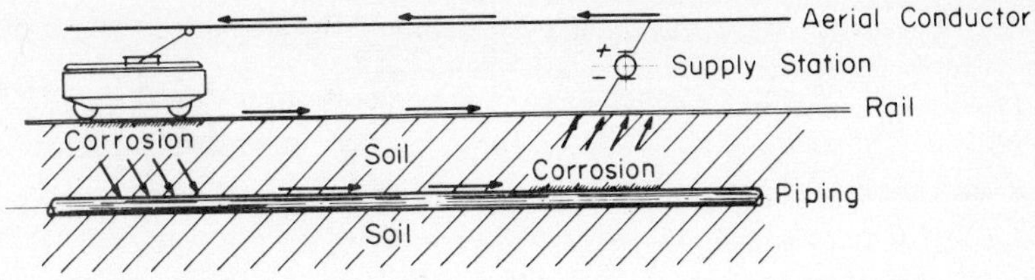

Figure 106. Corrosion by stray currents (after U. R. Evans).

points should be permanently below the protection potential. This is generally about −0.62 V with respect to the standard hydrogen electrode. Generally, cathodic protection should be continuously maintained. However, due to the effect of passivation that accounts for the alkalinity produced during the treatment, certain interruptions of the treatment are permissible. This is fortunate in cases where current generators are driven by wind vanes which, of course, cease to function in calm weather.

Figure 106 taken from a diagram of U. R. Evans,[54] illustrates the generation of stray currents affecting a pipeline buried near an electric rail. The current flowing in the line and supplied by the positive aerial wire, having passed through the motor of the vehicle, returns through the rail to the negative terminal at the generating station. Sometimes however, and particularly if the resistance of the rail is high (from defective welding) and if the ground is locally highly conducting (from the spreading of calcium chloride to melt ice and snow), a portion of the current may return through the ground and through nearby metallic structures. Corrosion then occurs at points where the current leaves both metal structures. The corrosion at the rail is produced in various places depending on the position of the vehicle, and so is of a fairly general character. Usually, this does not present a serious problem as in many cases the rail must be periodically replaced for other reasons. This is not the case for a buried pipeline in the situation illustrated by Figure 106 for in this instance corrosion is not general but localized to a region near the station. Here the remedy applied is a drainage technique, as illustrated in Figure 105a by a cable directly linking the bare portion of the pipeline to part of the rail near the negative terminal of the station.

However, such cathodic protection arrangements must only be used with extreme caution because nearby metallic structures which are not protected may be endangered, possibly leading to serious corrosion. Hence,

[54] U. R. Evans, Ref. 24, p. 264.

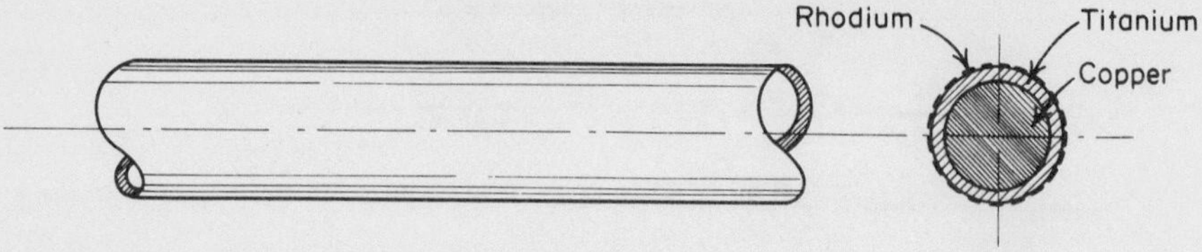

Figure 107. Incorrodible anode for cathodic protection of a structure of great length.

there are regulations to observe and this requires close cooperation between all users of the subsoil. In Belgium, particularly due to the pioneering work of R. de Brouwer and his assistant and successor A. Weiler, cooperation has been achieved between the organizations responsible for stray currents (railways, tramways), the users of buried structures (water, gas, and fuel pipes, electric cables) and the specialists of cathodic protection.

6.6.6.2.2. *Cathodic Protection of Immersed Structures. Internal Walls of Piping*

In the case of piping in which it is often difficult or impossible to place the conventional anodes, cathodic protection of the internal wall may be effected using incorrodible anodes made of bimetallic cables of which the interior is copper and the exterior is titanium, coated with platinum or rhodium (Figure 107). By combining the high electrical conductivity of copper and the corrosion resistance of titanium and of platinum or rhodium, it is possible to form anodes of great length from such cables, thereby opening a vast field of application. Figure 108 illustrates an arrangement for cathodic protection of a pipe by means of three such cables attached to the wall by insulating bands; Figure 109 shows a cable fitted with an insulating tress which is easily attached without the risk of a short circuit with the surface to be protected.

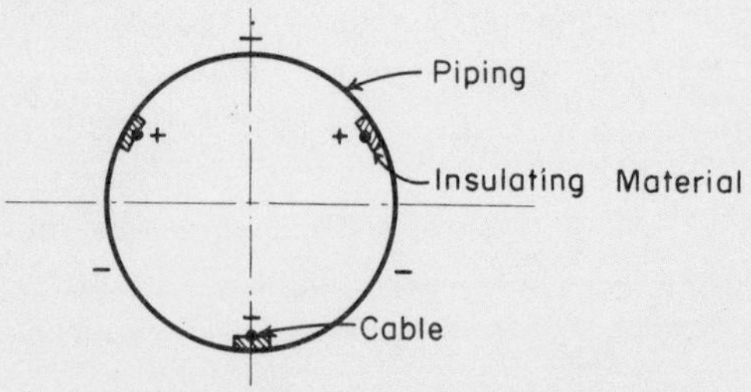

Figure 108. Protection arrangement of pressure piping by means of three bimetallic anodes (for hydroelectric power plant).

Figure 109. Anode with insulating tress.

Note, that cathodic protection by sacrificial anodes of magnesium, zinc, or aluminum–zinc alloys, or by incorrodible anodes of titanium or platinized titanium has many applications, some of which have already been mentioned above: protection of ships, floating docks, sluicegates, reservoirs of water such as water tanks, central heating installations, and gasometers as well as much industrial equipment. The function of galvanized steel is based, at least in part, on this principle of cathodic protection.

6.6.7. Protection by Corrosion Inhibitors[55]

6.6.7.1. General Remarks

A corrosion inhibitor is a substance which, when added in small quantities to a corroding medium, brings about an appreciable reduction of the corrosive action. For example, such inhibitors are: chromates, in many types of water; "etching inhibitors" (e.g., diorthotolylthiourea) which, when added in small quantities to a solution of sulfuric acid used for etching steel covered with oxide, considerably decreases the dissolution of the metal without modifying the rate of dissolution of the oxide; "vapor phase inhibitors" (e.g., "filming amines" and dicyclohexylammonium nitrite) which, being slightly volatile, are used for the protection of steel surfaces in contact with condensed vapor and mechanical parts sealed in virtually airtight containers.

According to a suggestion made, I think, by U. R. Evans, it is usual to divide corrosion inhibitors into two categories, *anodic inhibitors* and *cathodic inhibitors*, depending on whether their action causes a decrease of the anodic reaction (oxidation: generally termed the corrosion reaction) or of the cathodic reaction (reduction: hydrogen evolution reaction or the reduction of oxygen). In the present discussion, which is based on an electrochemical treatment of the problem, we consider successively *oxidizing inhibitors* which act by passivating the metal (that is, by raising the electrode

[55] A comprehensive account of corrosion inhibitors may be found in the proceedings of three *Symposiums Européens sur les Inhibiteurs de Corrosion*, organized by the Centre de Corrosion Aldo Dacco at Ferrara in September–October 1960 (Ref. 59), in September 1965 (Ref. 60), and in September 1970 (Ref. 61).

potential of the metal causing the formation of an oxide film) and *adsorption inhibitors* which act by adsorption of organic substances on certain parts of the metal surface.

6.6.7.2. Oxidizing Inhibitors

As may be seen from Figure 86, note that, contrary to the *cathodic protection* of iron which acts by lowering the electrode potential generally to the region of immunity, oxidizing inhibitors act by *anodic protection*, i.e., by raising the potential of the metal to the region of passivation. Obviously, as already outlined in Section 6.6.1.2, such protection may be induced sometimes in water containing oxygen, by simple alkalization, without increasing the pre-existing oxygen content; then oxygen dissolved in the water ceases to be activating and becomes passivating (Figure 74). Generally, oxidizing inhibitors are particularly effective if, in the circumstances under which they are used, their reduction gives rise to an insoluble derivative which is deposited at the bottom of fissures and other weak points of the $\mathbf{Fe_2O_3}$ film, causing the passivation of iron. In this way a highly protective film is produced, at least in the absence of active substances likely to damage the quality of this film. In Figure 57, where the white zones represent the theoretical conditions of stability of solid bodies (iron, oxides of iron, solid products formed by reduction of the oxidizing inhibitor), it can be seen that the protective efficiency of oxygen, hydrogen peroxide, and permanganate may be labile and transitory: in the case where there is an insufficient concentration, an extremely vigorous corrosion may appear in areas of low potential. Conversely, as has been shown experimentally by Cartledge, the hyperosmiates and the pertechnetates are extraordinarily protective. This seems due, at least partly, to the inclusion in the $\mathbf{Fe_2O_3}$ film of deposits of $\mathbf{OsO_2}$ and **Os**, of $\mathbf{TcO_2}$ and **Tc**, as illustrated in Figure 57.

The well-known effect of chromates is due, as has been verified experimentally by T. P. Hoar and U. R. Evans and in a general way by N. Brasher, to the inclusion of partially hydrated $\mathbf{Cr_2O_3}$ in the $\mathbf{Fe_2O_3}$ film. In highly acid media, in which $\mathbf{Cr(OH)_3}$ and $\mathbf{Cr_2O_3}$ are soluble, chromates are not passivating but activating. From Figure 57, the passivating action of molybdates and tungstates is much less than that of chromates while that of vanadates is fairly similar. Selenates, arsenates, and antimonates may be passivating in a weakly oxidizing medium or even a highly acid medium, due to deposits of elementary **Se**, **As**, or **Sb** on the metal. Concerning the latter two species, these are in agreement with experimental results obtained respectively by CEBELCOR (RT. 106) and by Piontelli and Fagnani[56] as well as by Burns and Bradley.[57]

As already indicated in Section 6.3.4, the action of oxidizing inhibitors is not entirely free from danger and may sometimes prove to be more harmful than advantageous, on one hand because, as Figure 72 shows, an insufficient increase of electrode potential produces a considerable increase in the rate of corrosion, and on the other hand because if the oxide film formed by passivation is imperfectly protective and has local weak points (which may exist in the presence of chlorides), the general corrosion which occurs in the absence of an oxidant may be replaced by extremely vigorous *local corrosion.* This restriction of the effectiveness of oxidizing inhibitors, which essentially results from the damaging action of chlorides on oxide films, probably does not apply in the case where passivation is due, not to the formation of an oxide film (oxygen, chromates), but to the deposition of an oxide-free element (selenium, arsenic, antimony, and sometimes osmium and technetium).

In Section 6.6.3, we have pointed out that the addition of orthophosphate to a corrosive medium enables the protective quality of an oxide film formed in a chloride-containing solution to be restored. Most probably, this is the basis of numerous procedures of *oxidizing phosphatization* which have been developed particularly since 1945. We will return to this topic in Section 6.6.7.4.6 when examining the protection of steel by painting.

6.6.7.3. Adsorption Inhibitors[58]

The action of the second group of inhibitors is based on the adsorption, at certain points of the metallic surface, of organic substances which, by more or less separating the metal from the action of aggressive ions in solution, sometimes cause a considerable reduction of the corrosion rate. In this instance, there is no question of passivation associated with an elevation of the electrode potential and formation of a protective oxide film. Inhibition by adsorption is generally produced without appreciable modification of the electrode potential which remains in the region of corrosion illustrated in Figure 86.

Some important work has already been carried out and more is in progress concerning the mechanism of adsorption inhibitors, adsorption

[56] R. Piontelli and L. Fagnani, *Korr. und Metallschutz*, **19**, 259 (1943).

[57] R. M. Burns and W. W. Bradley, *Protective Coatings for Metals* (2nd ed.), Reinhold, New York (1955), p. 274.

[58] H. H. Uhlig, Ref. 29, p. 189.

which seems to be associated in many cases to some values of electrode potential. Such matters have been extensively studied. The team of Norman Hackerman in Austin, Texas and at Rice University, and the Centre d'Étude sur la Corrosion Aldo Dacco, at Ferrara, which was founded by L. Cavallaro, have studied this. Three important European Symposia on Corrosion Inhibitors took place at the latter place in 1960, 1965, and 1970. Recently the secretariat of the "Corrosion Inhibitors" study group of the Federation Europeene de la Corrosion was entrusted to this centre. It may be advantageous to consult the proceedings of these three meetings (Refs. 59, 60, 61).

A second important group of inhibitors, acting by adsorption, includes "soluble oils" consisting of an aqueous emulsion of oil, soap, fatty acid, or alcohol the addition of which to water sometimes appreciably reduces the corrosive action. These inhibitors may prove to be dangerous when they do not have access to all of the metal surface to be protected and should only be used with caution.

6.6.7.4. Use of Corrosion Inhibitors

The use of corrosion inhibitors has developed considerably in many areas of application. In most cases, electrochemical methods of study and control have been invaluable. Below we briefly illustrate a few of these applications, some of which are still being developed.

6.6.7.4.1. *Drinking Water*

For the treatment of corrosive drinking water, the possibility of a passivating action due to oxygen has, as indicated in Section 6.6.4, been exploited in aeration treatments. Of course, in order for the passive film to be of good quality, generally the pH of the water is raised and it is saturated with $\mathbf{CaCO_3}$ (by filtration over $\mathbf{CaCO_3}$ or $\mathbf{MgO \cdot CaCO_3}$ or by addition of $Ca(OH)_2$ or NaOH). Sometimes sodium silicate is added. Recall that this treatment may not be applied to water of high chloride content.

6.6.7.4.2. *Central Heating Water*

As we have seen in Section 6.6.4.3, water flowing in *central heating installations* with steel radiators may often be rendered noncorrosive by careful treatment using di- or trisodium phosphate which induces satisfactory passivation of the steel. Carefully controlled treatments with mixed inhibitors should be used when other metals (copper, aluminum) are present. Adjustment of the treatment and control of its effectiveness may be carried out by electrochemical means, for example, by polarization curves

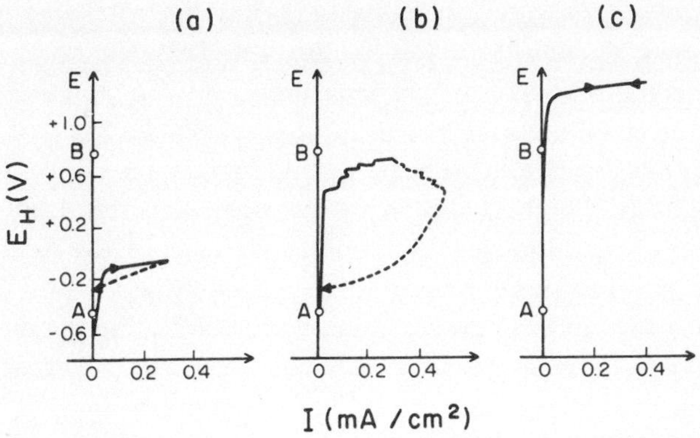

Figure 110. Influence of the addition of increasing quantities of disodium phosphate on the behavior of mild steel in the presence of oxygen-free mains water: (a) without phosphate; (b) with 3 g/liter $Na_2HPO_4 \cdot 12H_2O$; and (c) with 10 g/liter $Na_2HPO_4 \cdot 12H_2O$.

determined potentiokinetically. Figure 110 illustrates the influence of additions of increasing quantities of disodium phosphate on the behavior of radiator water in the presence of mild steel. The curve illustrated in Figure 110c indicates the circumstances under which steel radiators may be used with complete safety without risk of corrosion (*Rapports Techniques CEBELCOR* RT. 133, RT. 134).

6.6.7.4.3. *Antifrigerant Solutions*

Antifrigerant solutions (see U. R. Evans, Ref. 28, p. 154) which generally contain ethylene glycol and are used as antifreeze for automobile radiators, must be treated by the addition of corrosion inhibitors. Sometimes this treatment is awkward due to the need to simultaneously protect, at various temperatures, often markedly dissimilar metals and alloys: cast iron, steel, copper and copper-based metals, and aluminum. Oxidizing inhibitors and alkalizers which are effective for cast iron and steel produce a harmful effect on copper-base metals as well as aluminum. Probably, it is for this reason that more complex systems such as adsorption inhibitors, have been sought. The work of the Teddington school (of W. H. J. Vernon and F. Wormwell) has led to mixtures of sodium nitrate and sodium benzoate. Mixtures of triethanolamine phosphate and mercaptobenzothiazole phosphate are also used extensively.

6.6.7.4.4. *Pickling Baths*

Before applying a protective coating on steel, such as a paint, it is necessary to remove all rust and mill scale from the surface. Mostly, this is done mechanically (dry or moist sand-blasting, shot-peening, etc.) or chemically (see U. R. Evans, Ref. 28, pp. 156–160). Chemical cleaning of steel is generally carried out by immersion in a warm solution of sulfuric acid or in a cold solution of hydrochloric acid. Besides the desired dissolution of iron oxides, which constitute rust and mill scale, these solutions also corrode the metal with evolution of hydrogen and this leads to several complications. G. Batta (Ref. 4) seems to have been the first to observe that the addition of small quantities of formaldehyde to a sulfuric etching bath considerably reduce this corrosion without affecting the dissolution of the oxides (and possible calcium carbonate deposits), which is the object of the operation. The subject of pickling inhibitors, whose action depends essentially on adsorption phenomena, has expanded significantly. Actually, the most used inhibitors are generally organic molecules containing sulfur and/or nitrogen, such as diorthotolylthiourea and dihydrothioorthotoluidine.

6.6.7.4.5. *Boiler and Packaging Condensates; Volatile Inhibitors*

The action of volatile inhibitors already mentioned depends on the fact that when these substances are present in a gaseous phase in contact with a metallic surface on which water vapor condenses, they dissolve in this water and render it noncorrosive by the phenomena of adsorption. In this way "filming amines" (e.g., morpholine and cyclohexylamine) added to boiler water and carried by the steam ensure protection of condensers and the return circuits of condensed water (see U. R. Evans, Ref. 28, pp. 154 and 155). Other more volatile amines (e.g., dicyclohexylamine nitrite) are used successfully for the protection of mechanical parts in restricted or confined spaces and also for the impregnation of packaging materials. Hence, all condensation of water on these parts forms a noncorrosive solution probably because of the combined action of passivation by oxidation due to the nitrite radical and adsorption due to the amine radical.

6.6.7.4.6. *Treatment of Steel Surfaces Before Painting*

In Section 6.6.6.1.2, we discussed corrosion pits which frequently form on the surface of iron or ordinary steel and quickly lead to degradation by differential aeration corrosion. As stated on page 272 and shown in Figures 97 and 103, the solution inside active pits is acid, and if some chloride is present, this acid is hydrochloric acid, which is, as is well known, particularly aggressive. As indicated above, the only means of preventing such corrosion is to apply a correctly calculated and maintained cathodic protection to the metallic structure or to remove, by mechanical or chemical etching, all trace of rust present on the metal (Figure 78c)[59] (see U. R. Evans, Ref. 27, pp. 160 to 162). In the latter case, having largely opened up all the corrosion pits, perfect passivation of all the metal surface must be achieved by first an inhibitive undercoat (primer) on the rust-free surface, and then a covering layer which is, as far as possible, impermeable to moisture. The methods of providing an inhibitive undercoat are described below.

6.6.7.4.6.1. *Phosphatization*; "*Wash Primers*"

For a long time it has been the practice to treat steel surfaces before painting by phosphatization. This involved originally the action of solutions of phosphoric acid saturated with iron phosphate. These treatments have been progressively refined by using "accelerators" (nitrates, nitrites, chlorates, and nitrated organic compounds such as nitroguanidine), whose effectiveness is probably due to the beneficial action of an "oxidizing phosphatization" which we have discussed in Section 4.6.7.

During the years 1939–1945, research sponsored by the US Navy and carried out with a view to improving the protection of ships against corrosion led to the introduction of procedures for treating steel, zinc, and aluminum surfaces before painting, based on the use of phosphoric acid, vinyl resin, and zinc chromate.[60] Applied as a thin layer (0.01 to 0.03 mm) to well-cleaned metal, the products act both as an inhibitive wash and as an undercoat to which the final coat of paint more strongly adheres. Sensational results are often achieved in this way. These products may be used in the workshop to prevent corrosion during storage and assembly. During the last twenty years many variants have been developed under the name of *wash primers* and have led to important progress in corrosion protection.

The extraordinary efficiency of wash primers is, in our opinion (*Rapports Techniques CEBELCOR* RT. 19, RT. 66), due to the fact that they induce excellent passivation on iron. Consisting essentially of a mixture of phosphoric acid and chromate, these solutions have the effect of covering the metal with a protective film of ferric oxide in which the weak points are blocked by scarcely soluble chromic oxide, $\mathbf{Cr_2O_3}$ (*Atlas*, Refs. 33 and 8, p. 270)[61] (which leads to passivation of the *whole* surface), and of iron phosphate which, as the experimental electrochemical studies associated

[59] Recent work seems to show that rust may sometimes be rendered harmless by the application of kelates, acting as efficient "rust converters."

[60] See *Rapports Techniques CEBELCOR* RT. 16, RT. 17 (1953).

[61] D. M. Brasher and E. R. Stove, *Chem. and Ind.* (London), 171 (1952).

with Figure 95 have shown, improves the protective quality of these oxides in the presence of chlorides which endanger oxidizing inhibitors. At weak points of the oxide film, where iron tends to corrode, iron in solution is precipitated as insoluble phosphate which blocks the pores of the defective film and in this way prevents the local attack generally resulting from deficient passivation. In summary, wash primers maintain, under a thin layer of resin (e.g., vinyl), a film of iron and chromium oxides which, due to the presence of phosphate, have enhanced resistance to aggressive agents. Thus, they constitute an improvement over classical phosphoric etchants because, when properly used, their chromate content strongly enhances the protective action of phosphate.

6.6.7.4.6.2. *Inhibitive Primer Undercoats*

Undercoats most generally used contain either some red lead or zinc chromate. Mayne and Van Rooyen have attributed the protective action of paints based on lead and linseed oil to the formation of lead soaps such as lead linoleate which act as adsorption inhibitors. On the other hand, the action of undercoats based on zinc chromate is due principally to the passivating action of chromate acting as an oxidant. Therefore, these two groups of paints generally have the advantages and disadvantages which characterize these two groups of inhibitors (see Section 6.6.1). Zinc chromate paints are absolutely effective when properly used (on a perfectly cleaned metal) and in the presence of a mildly aggressive atmosphere (nonmarine); if these conditions are not obtained there may be failure. Paints based on red lead, which are less effective than chromate paints when used under ideal conditions, generally are superior to them when improperly used. For example, in the painting of ships, chromate paint is recommended when work can be carried out sheltered from sea spray, as in the interior of the ship or while in dry dock. Red lead paint is preferred for exterior work during a voyage.

6.6.7.4.6.3. *Zinc Plating*

Relatively recent work has shown that zinc plating, carried out preferably by hot dip or by spraying, or possibly by using good zinc-dust paints, often constitutes the best undercoat for the application of paint. Due to the cathodic protection that zinc confers on steel, a defect in the paint does not immediately lead to the formation of rust: the appearance of rust is delayed by sacrificial corrosion of zinc. Besides, in the case of zinc coatings heavy abrasion is not necessary before repainting, as is often the case for other coatings, because, unlike rust, the corrosion products of zinc are weakly adherent and do not cause the development of blisters under paint. Repainting operations are much less costly and difficult, provided they are carried out before the characteristic brown stains of rust appear; however, good adhesion of the paint to the zinc should be ensured.

6.6.8. Protection by Coatings

Since the present discussion is essentially concerned with electrochemical corrosion, we will only briefly discuss protection by coatings (paints, enamels). Usually, these coatings serve to protect the metal underneath from access of the corrosive medium, which is generally an aqueous solution. So the action of this coating generally is not electrochemical in nature, unless the chemistry of the undercoat interacts with the environment to reduce the corrosion of the underlying metal.

In every case when a coating is applied, particular care must be taken in the pretreatment of the surface. An excellent paint, if it is badly applied, is generally less effective than a well-applied paint of lesser quality. Mostly, the surface to be treated must be perfectly clean, free from rust and grease. In certain cases, the presence of rust formed in a clean rural environment may not be a source of great annoyance, especially if a fully reliable "rust converter" is used. This is not so for rust formed by urban, industrial, or marine atmospheres, containing sulfates and chlorides; in such cases the removal of rust is essential. It should not be forgotten that rust containing chloride or sulfate means the presence of chlorhydric acid or sulfuric acid in the underlying pits.

As we have said above, *zinc plating* is one of the best undercoats there is for the application of paint. Protection systems involving the application of high quality paint on a zinc-plated surface with possible use of a "wash primer" often yield remarkable results. As is well known, a simple zinc plating of the steel (by hot dip, by spraying or by painting with zinc dust) on its own often gives satisfactory results. Of course, because of the corrodability of zinc such protection is not permanent; in the presence of a given atmosphere the duration of protection is proportional to the thickness of the zinc coating.

Consequently, zinc plating is expected to develop considerably in the next few years as a technique for the protection of steel. Equally, an increase may be predicted in *aluminum plating* of steel. Immersion in a molten aluminum–silicon alloy or spraying with molten aluminum gives rise to coatings which are particularly resistant to corrosion in nonmarine atmospheres as well as to hot corrosion.

We also draw attention to cement paints and to the remarkable corrosion resistance of well-applied epoxy rosins and of plastic coatings which, for instance, may be used in the form of tapes wrapped around the piping to be protected. Cement paints may be applied onto a wet surface since water is one of the constituents; plastic tapes, if not *perfectly* waterproofed, may become particularly effective if perfectly inhibiting or passivating pastes are interposed between them and the metal.

6.6.9. Other Corrosion Problems

The strictly limited scope of this chapter dealing with some applications of electrochemical corrosion in the study of the corrosion of iron and unalloyed steels precluded treatment of many problems which are of considerable scientific and technical interest; such are, notably, many problems related to the corrosion of nonferrous metals, of alloy steels, and of other alloys, which also may be examined by the electrochemical principles and methods elucidated in the present lectures.

A general comparison between the electrochemical methods of evaluation of corrosion and the behavior of materials in service was the subject of a lecture given in Frankfurt in June 1964.[62] Evaluations of the development of electrochemical methods in the study of corrosion were given during several colloquia held in Brussels during the annual "Corrosion Weeks" of CEBELCOR, particularly since 1965,[63] as well as on the occasion of an ASTM conference held in Toronto, 1970.[64] Already existing electrochemical methods should be used more than is presently the case for studies related to the pitting and crevice corrosion of copper and of alloy steels; to the denickelification of cupronickels and to the dezincification of brasses;[65] to the stress corrosion cracking of high-strength steels, aluminum alloys, and titanium alloys;[66] and to the behavior of high-strength low-alloy "patinable" steels (HSLA "weathering steels").[67] Such steels, when exposed to certain atmospheric conditions, are covered with a rust which becomes

[62] *Rapports Techniques CEBELCOR* **98**, RT. 123 (1964).

[63] *Rapports Techniques CEBELCOR* **104**, RT. 139 (1966); **106**, RT. 144, 145 (1967); **108**, RT. 151 (1969); **111**, RT. 163, 165 (1969); **116**, RT. 187, 188 (1971).

[64] *Rapports Techniques CEBELCOR* **116**, RT. 186 (1971).

[65] *Rapports Techniques CEBELCOR* **109**, RT. 157 (1969); **112**, RT. 167 (1970); **114**, RT. 179 (1970); **116**, RT. 186 (1971); **117**, RT. 191 (1971).

[66] *Rapports Techniques CEBELCOR* **114**, RT. 179 (1970); **118**, RT. 199 (1971). Proc. NATO Scientific Conference on Stress Corrosion, Ericeira, March 29–April 2 (1971).

[67] *Rapports Techniques CEBELCOR* **107**, RT. 147 (1968); **109**, RT. 160 (1969).

gradually protective. These steels may thus be used without paint, and surely have an appropriate future. Recent electrochemical methods for the quick prediction of atmospheric corrosion behavior of metals and alloys may be of great assistance to the scientific and technical development of these low-alloy steels and of new alloys. The *Center for Applied Thermodynamics and Corrosion* has been recently created for promoting studies in these fields.[68]

When a metal surface is underroof and not subject to the action of rain, significant corrosion will occur only when there is condensation of water. In the case of surfaces clean and free from capillary interstices, this is only produced if the temperature of the metal is below the dew-point temperature, the value of which varies as a function of the water vapor pressure of the gas. Recall that, for a given temperature, the critical content of water vapor may be considerably reduced if capillary interstices or hygroscopic substances (e.g., calcium or magnesium chloride, *Rapports Techniques CEBELCOR* **115**, RT. 184, No. 9) are present on the surface. From the viewpoint of corrosion prevention, it is advantageous to avoid as far as possible capillary interstices (for instance resulting from riveted assemblies) and to ensure complete cleanness of surfaces.

According to F. L. Laque, there is generally no fear of corrosion in hangers unless the relative humidity exceeds 30%; this value may be raised to 45% if the steel surfaces are coated with oil or grease; a relative humidity of 20% is dangerous if the surfaces are contamined with magnesium chloride.

Slight heating or satisfactory drying of spaces (by means of calcium chloride or silica gel) often is sufficient to ensure protection of mechanical parts during storage.

In the presence of sulfurous gases, such as combustion gases of pyrites and exhaust gases of numerous fuel oil engines, the prevention of corrosion involves taking the necessary precautions to ensure that the temperature of surfaces remains, as far as possible, above the dew-point temperature of these gases. In the case of the combustion of appreciable quantities of sulfur, this temperature may reach 150° to 160°C on account of the formation of sulfuric acid solutions.

[68] *Rapports Techniques CEBELCOR* **117**, RT. 192 (1971).

Chapter 7

FURTHER APPLICATIONS OF ELECTROCHEMISTRY TO CORROSION STUDIES

7.1. INTRODUCTION

In the preceding chapters of these lectures, we have considered fundamental principles of electrochemical thermodynamics and kinetics and some of their applications to the study of corrosion phenomena in aqueous solutions at 25°C. These principles, which have been described, have an extremely broad applicability. The general concept of plotting both chemical and electrochemical equilibria as a function of thermodynamical variables provides a tool of immense value to scientists and engineers. It is the purpose of this chapter to describe some of the ways in which chemical and electrochemical data can be used in understanding corroding systems which are more complex than those described. It should also be mentioned in passing that the techniques and procedures described in this chapter can be applied broadly in many other areas of technology with substantial benefit.

The "potential–pH equilibrium diagrams," which have been drawn in Chapter 4 for binary systems which consist of a metal or metalloid in water, are strictly valid only for aqueous solutions containing substances which are unlikely to cause other solubilization reactions or to produce other solid compounds. When such additional substances are formed, then it is appropriate to construct more complex diagrams consisting of three, four, or five components. The team at CEBELCOR has prepared such diagrams for the following ternary and quinary systems:

$$Fe–CO_2–H_2O$$

$$Cu–Cl–H_2O$$

$$Cu–CO_2–H_2O$$

$$Cu–SO_3–H_2O$$

$$Cu–Cl–CO_2–SO_3–H_2O$$

Other diagrams for ternary systems have been published in 1961 by E. Mattson ($Cu–NH_3–H_2O$, $Zn–NH_3–H_2O$);[1] in 1962 by J. Horváth and co-workers for many metallic sulfides;[2] in 1963 by J. P. Brenet and co-workers ($Fe–S–H_2O$);[3] in 1966 by J. Niemiec and F. Letowski ($Cu–NH_3–H_2O$, $Co–NH_3–H_2O$, $Ni–NH_3–H_2O$, $Ag–NH_3–H_2O$).[4] Potential–pH diagrams for temperatures above 25°C and up to 300°C have been developed since 1966 by D. Lewis[5] and in 1968 by R. G. Robins.[6]

In addition to the use of potential–pH diagrams to describe corrosion processes in aqueous environments a substantial amount of work has been performed in connection with other electrolytes such as fused salts, organic solvents, and semiconducting oxides. With respect to fused salts, diagrams have been prepared; in 1960 by G. Delarue[7] and by C. Edeleanu and R. Littlewood[8] for molten chlorides; for carbonates in 1963 by R. Buvet and co-workers[9] and in 1965 by M.D. Ingram and G. J. Janz[10] and by M. Pour-

[1] E. Mattson, *Electrochim. Acta* **3**, 279–291 (1961).

[2] J. Horváth, *Acta Chem. Acad. Sci. Hungaricae* **33**, 221–235 (1962); **34**, 455–467 (1962); **38**, 151–165 (1963); *Proc. 2nd European Symposium Corrosion Inhibitors, Ferrara, 1965*, 477–505 (1966).

[3] J. Bouet and J. P. Brenet, *Corr. Sci.* **3**, 51–63 (1963).

[4] J. Niemiec and F. Letowski, *Ann. Soc. Chim. Polo.*, **40**, 1159 (1966); **41**, 1483–1490 (1967); *Freiberger Forschungshefte B* **128**, 19–24 (1968).

[5] D. Lewis, private communication.

[6] R. G. Robins, The application of potential–pH diagrams to the prediction of reactions in pressure hydrothermal processes, Warren Spring Lab. SBN 900790083 (1968).

[7] G. Delarue, *Bull. Soc. Chim. France* **266**, 1654–1659 (1960); Thesis No. 665, Paris (15-12-1960); *Silicates Industriels* **27**, 69–78 (1962).

[8] C. Edeleanu and R. Littlewood, *Electrochim. Acta* **3**, 195–207 (1960); R. Littlewood, Tube investments research laboratory report n° **138** (1961); *J. Electrochem. Soc.* **109**, 525–534 (1962).

[9] J. Dubois and R. Buvet, *Bull. Soc. Chim. France* 2522 (1963); N. Besson, S. Palous, R. Buvet, and J. Millet, *C. R. Acad. Sci. Paris* **260**, 6097–6100 (1965).

[10] M. D. Ingram and G. J. Janz, *Electrochim. Acta* **10**, 783–792 (1965).

baix;[11] and for sulfates in 1968 by A. Rahmel.[12] For organic solvents by J. Perichon and R. Buvet.[13]

Other diagrams of electrochemical equilibria showing the influence of temperature on the affinity and the electrode potential of chemical and electrochemical systems consisting of condensed and/or gaseous phases have been drawn in 1947 by M. Pourbaix and C. Rorive–Boute for some one-, two-, and three-component systems such as O, C–O, Zn–O, C–Zn–O.[14] A general bibliography concerning diagrams of electrochemical equilibria has been published recently.[15]

Examples of these more complex diagrams and their practical applications will be considered in the following sections.

7.2. POTENTIAL–pH EQUILIBRIUM DIAGRAMS FOR COMPLEX SYSTEMS CONTAINING COPPER AT 25°C

7.2.1. System Cu–Cl–H_2O

Two potential–pH equilibrium diagrams for the simple binary system, Cu–H_2O, at 25°C are shown in Figures 34 and 37. The simple potential–pH relation for copper has been described in considerable detail elsewhere.[16] These diagrams take into account the equilibrium conditions of 33 chemical and electrochemical reactions involving copper and its derivatives. The following 13 substances, for which the standard free enthalpies of formation are available, are the ones considered in the simple binary diagrams: six solid substances (**Cu, Cu_2O, Cu(OH), CuO, $Cu(OH)_2$**, and **$Cu_2O_3 \cdot nH_2O$**), six dissolved ions (Cu^+, Cu^{++}, CuO_2H^-, CuO_2^{--}, Cu^{+++}, and CuO_2^-), and one gas (***CuH***).

In order to extend the potential–pH diagrams from the simple binary ones to a diagram including chlorine, it is necessary to identify the possible

[11] M. Pourbaix, Observations on the presentation and the practical usefulness of diagrams of electrochemical equilibria in the presence of molten salts, *Rapports Techniques CEBELCOR* RT. 136 (1965) (in French).

[12] A. Rahmel, *Electrochim. Acta* **13**, 495–505 (1968).

[13] J. Perichon, C. Chevrot, and R. Buvet, *C. R. Acad. Sci.*, Séries C.

[14] M. Pourbaix and C. Rorive–Boute, *Discussions Faraday Soc. 1947*, **4**, 139–154, 223–228, 239–240 (1948); M. Pourbaix, *Electrochim. Acta* **13**, 1420–1423 (1968).

[15] *Rapports Techniques CEBELCOR* **116**, RT. 186 (1971).

[16] *Rapports Techniques CEBELCOR* **85**, RT. 100 (1962).

substances which it will be appropriate to consider in the copper–water–chlorine system. In addition, it is necessary then to prepare a set of chemical and electrochemical equations involving these compounds providing that suitable data for the standard free enthalpies are available. Such work has been undertaken in collaboration with J. van Muylder and N. de Zoubov,[17] in which the following additional substances have been considered: three solid substances (**CuCl**, $\mathbf{CuCl_2}$, and $\mathbf{3Cu(OH)_2 \cdot CuCl_2}$), and seven dissolved ions (Cl^-, $CuCl_2^-$, $CuCl_3^{--}$, $CuCl^+$, $CuCl_{2aq}$, $CuCl_3^-$, and $CuCl_4^{--}$). Other derivatives of chlorine such as dissolved HClO and ClO^-, $HClO_2$ and ClO_2^-, ClO_3^-, ClO_4^-, and gaseous and dissolved Cl_2 have not been considered in this complex diagram since the calculations become extremely complicated and the graphical representation is confusing. The inclusion of these additional ten substances increases the number of equilibrium formulas from 33 to 87. Since for a three-component system it is difficult to represent in a single diagram the effect of varying the chloride activity, it is necessary to draw several diagrams for several different chloride ion concentrations. Figure 111 represents four potential–pH diagrams drawn at four activities of chloride ion, namely: 10^{-3}, 10^{-2}, 10^{-1}, and 1 g-ion Cl^-/liter (i.e., 35, 355, 3550, and 35,500 ppm).

Listed below are examples of equilibrium equations for substances involving the two solid chlorides **CuCl** (nantokite) and $\mathbf{3Cu(OH)_2 \cdot CuCl_2}$ (paratacamite) besides **Cu** and its oxides, $\mathbf{Cu_2O}$ and **CuO**, previously considered in the binary potential–pH work.

The equations are as follows:

$$\mathbf{2CuCl} + H_2O = \mathbf{Cu_2O} + 2Cl^- + 2H^+$$
$$\log[Cl^-] = -5.66 + pH \qquad (51)$$

$$\mathbf{3Cu(OH)_2 \cdot CuCl_2} = \mathbf{4CuO} + 2Cl^- + 2H^+ + 2H_2O$$
$$\log[Cl^-] = -7.40 + pH \qquad (53)$$

$$\mathbf{Cu} + Cl^- = \mathbf{CuCl} + e^-$$
$$E = +0.137 - 0.0519 \log[Cl^-] \qquad (55)$$

$$\mathbf{2Cu_2O} + 4H_2O + 2Cl^- = \mathbf{3Cu(OH)_2 \cdot CuCl_2} + 2H^+ + 4e^-$$
$$E_0 = +0.451 - 0.0295\,pH - 0.0295 \log[Cl^-] \qquad (62)$$

As already explained in Section 6.6.6.1.2, the diagrams in Figure 111 show that **CuCl** is not stable in the presence of neutral water and thus

[17] *Rapports Techniques CEBELCOR* **85**, RT. 101 (1962).

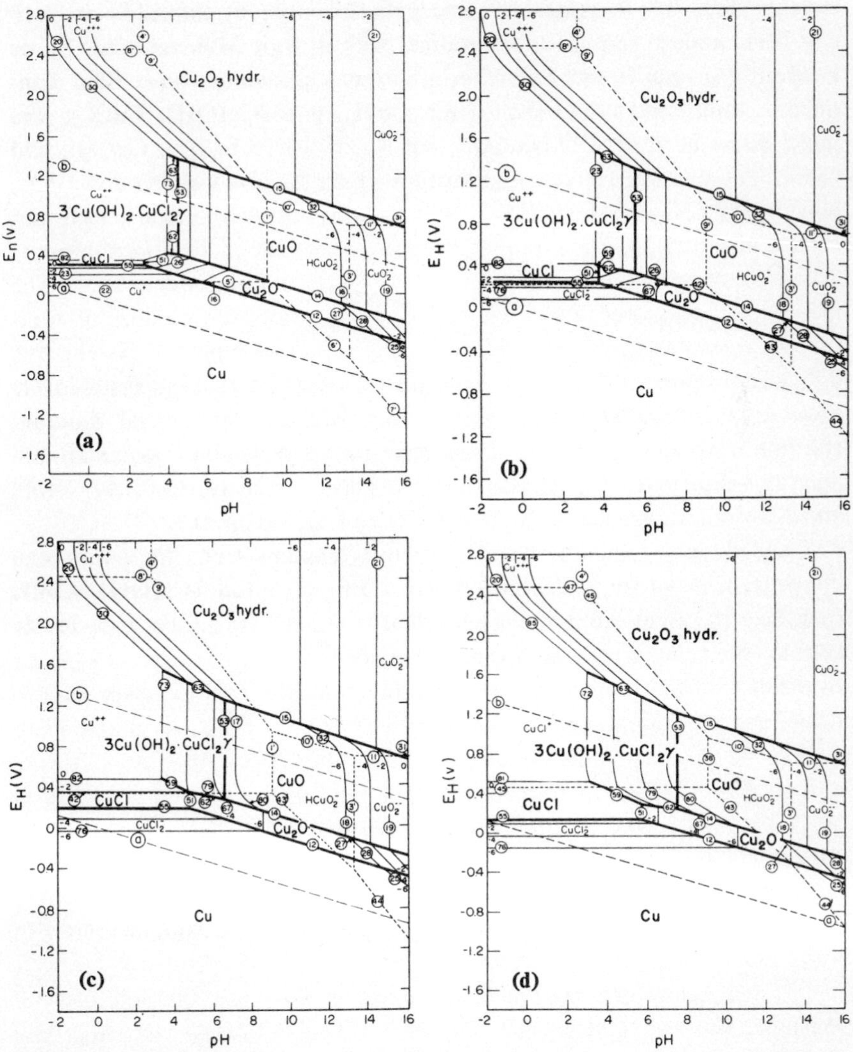

Figure 111. Potential–pH equilibrium diagrams for the ternary system Cu–Cl–H_2O at 25°C for (a) 10^{-3}, (b) 10^{-2}, (c) 10^{-1}, and (d) 1 g-ion Cl^-/liter (35, 355, 3550, and 35,500 ppm).

tends to be hydrolyzed with the formation of $\mathbf{Cu_2O}$ and HCl according to the reaction $2\mathbf{CuCl} + H_2O \rightarrow \mathbf{Cu_2O} + 2H^+ + 2Cl^-$. They also show that paratacamite, which is the Chilean copper mineral, is the stable form of copper derivatives in oxidized and acid rocks containing chloride.

The chloride content inside an active copper pit in Brussels tap water is about 355 ppm (instead of 22 ppm in the bulk of the water). The equilibrium conditions inside such a pit are indicated in Figure 111b by the triple point at $E = +270\ mV_{SHE}$, pH = 3.5, where lines 12, 51, and 55 cross. As previously discussed in Section 6.6.6.1.2.1, the knowledge of these equilibrium concentrations inside pits made it possible to establish remedies for pitting in copper.

7.2.2. System Cu–CO_2–H_2O

Equilibrium potential–pH diagrams for the Cu–CO_2–H_2O system have been drawn in collaboration with J. van Muylder and N. de Zoubov; the following additional substances have been considered: two solid substances (malachite, $\mathbf{CuCO_3 \cdot Cu(OH)_2}$; azurite, $\mathbf{2CuCO_3 \cdot Cu(OH)_2}$), and six dissolved substances (H_2CO_3, HCO_3^-, CO_3^{--}, $CuCO_{3\,aq}$, $Cu(CO_3)_2^{--}$, $CuCO_3(OH)_2^{--}$). Other derivatives, such as gaseous $\boldsymbol{CO_2}$, have not been considered in order to avoid unnecessarily complicated diagrams. By including the eight additional compounds obtained by adding CO_2 to the system, the number of equilibrium formulas increases from 33 to 113. As in the case of the ternary system including chloride, it is necessary here to draw separate diagrams for constant activity of CO_2. The four diagrams of Figure 112 show the influence of pH on the solubility on malachite, azurite, and tenorite ($\mathbf{CuO}$), for four activities of total dissolved CO_2 at values of 10^{-3}, 10^{-2}, 10^{-1}, and 1 mole/liter (i.e., 44, 440, 4400, 44,000 ppm). Thus, Figure 112 shows the range of conditions of pH and CO_2 activities where each of the three copper derivatives is the most stable. Figure 113 shows four potential–pH equilibrium diagrams relating to the same activities of total CO_2.

Listed below are examples of equations involving the solubility of malachite as Cu^{++}, $CuCO_3(aq)$, $Cu(CO_3)_2^{--}$, $CuCO_3(OH)_2^{--}$, $HCuO_2^-$, and CuO_2^{--}, in the presence of solutions containing H_2CO_3, HCO_3^-, and CO_3^{--}, and which were used for drawing Figure 112:

$$2Cu^{++} + H_2CO_3 + 2H_2O = \mathbf{CuCO_3 \cdot Cu(OH)_2} + 4H^+$$

$$\log[Cu^{++}] = +5.15 - 2pH - 0.07\log[H_2CO_3] \qquad (79)$$

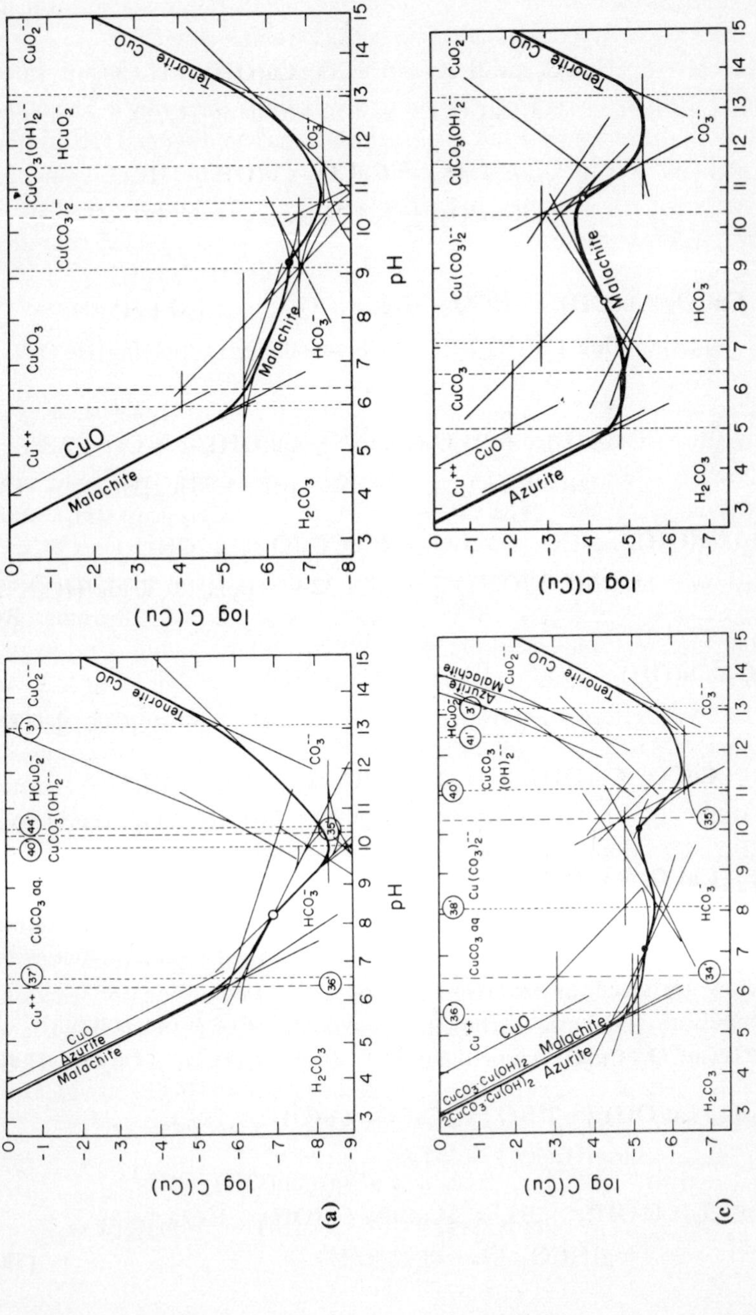

Fig. 112. Influence of pH on the solubility of tenorite, CuO; azurite, $2CuCO_3 \cdot Cu(OH)_2$; and malachite, $CuCO_3 \cdot Cu(OH)_2$ at 25°C for (a) 10^{-3}, (b) 10^{-2}, (c) 10^{-1}, and (d) 1 mole CO_2 per liter (44, 440, 4400, and 44,000 ppm).

$$2Cu^{++}+HCO_3^-+2H_2O = \mathbf{CuCO_3 \cdot Cu(OH)_2}+3H^+$$
$$\log[Cu^{++}] = +0.90-1.33pH-0.57\log[HCO_3^-] \quad (80)$$

$$2CuCO_{3aq}+2H_2O = \mathbf{CuCO_3 \cdot Cu(OH)_2}+H_2CO_3$$
$$\log[CuCO_3] = -4.50+0.50\log[H_2CO_3] \quad (83)$$

$$2CuCO_{3aq}+2H_2O = \mathbf{CuCO_3 \cdot Cu(OH)_2}+HCO_3+H^+$$
$$\log[CuCO_3] = -1.31-0.50pH+0.50\log[HCO_3^-] \quad (84)$$

$$\mathbf{CuCO_3 \cdot Cu(OH)_2}+3HCO_3^- = 2Cu(CO_3)_2^{--}+2H_2O+H^+$$
$$\log[Cu(CO_3)_2^{--}] = -8.44+0.50pH+1.50\log[HCO_3^-] \quad (87)$$

$$2Cu(CO_3)_2^{--}+2H_2O = \mathbf{CuCO_3 \cdot Cu(OH)_2}+3CO_3^{--}+2H^+$$
$$\log[Cu(CO_3)_2^{--}] = +7.08-pH+1.50\log[HCO_3^{--}] \quad (88)$$

$$\mathbf{CuCO_3 \cdot Cu(OH)_2}+HCO_3^-+2H_2O = 2CuCO_3(OH)_2^{--}+3H^+$$
$$\log[CuCO_3(OH)_2^{--}] = -21.32+1.50pH+0.50\log[HCO_3^-] \quad (91)$$

$$\mathbf{CuCO_3 \cdot Cu(OH)_2}+CO_3^{--}+2H_2O = 2CuCO_3(OH)_2^{--}+2H^+$$
$$\log[CuCO_3 \cdot (OH)_2^{--}] = -16.16+pH+0.50\log[CO_3^{--}] \quad (92)$$

$$\mathbf{CuCO_3 \cdot Cu(OH)_2}+2H_2O = 2HCuO_2^-+CO_3^{--}+4H^+$$
$$\log[HCuO_2^-] = -29.63+2pH-0.50\log[CO_3^{--}] \quad (94)$$

$$\mathbf{CuCO_3 \cdot Cu(OH)_2}+2H_2O = 2CuO_2^{--}+CO_3^{--}+6H^+$$
$$\log[CuO_2^{--}] = -42.78+3pH-0.50\log[CO_3^{--}] \quad (96)$$

Below are given, as examples, the equilibrium conditions of reactions involving both the basic carbonates, malachite ($\mathbf{CuCO_3 \cdot Cu(OH)_2}$), and azurite ($\mathbf{2CuCO_3 \cdot Cu(OH)_2}$) or the oxides, $\mathbf{Cu_2O}$, $\mathbf{CuO}$, and $\mathbf{Cu_2O_3 \cdot \mathit{n}H_2O}$.

$$2[\mathbf{2CuCO_3 \cdot Cu(OH)_2}]+2H_2O = \mathbf{3CuCO_3 \cdot Cu(OH)_2}+H_2CO_3$$
$$\log[H_2CO_3] = -1.85 \quad (57)$$

$$2[\mathbf{2CuCO_3 \cdot Cu(OH)_2}]+2H_2 = \mathbf{3CuCO_3 \cdot Cu(OH)_2}+HCO_3^-+H^+$$
$$\log[HCO_3^-] = -8.22+pH \quad (58)$$

$$\mathbf{CuCO_3 \cdot Cu(OH)_2} = \mathbf{2CuO} + H_2CO_3$$
$$\log[H_2CO_3] = -4.87 \quad (62)$$

$$\mathbf{CuCO_3 \cdot Cu(OH)_2} = \mathbf{2CuO} + HCO_3^- + H^+$$
$$\log[HCO_3^-] = -11.25 + \text{pH} \quad (63)$$

$$\mathbf{CuCO_3 \cdot Cu(OH)_2} = \mathbf{2CuO} + CO_3^{--} + 2H^+$$
$$\log[CO_3^{--}] = -21.58 + 2\text{pH} \quad (64)$$

$$\mathbf{Cu_2O} + \mathbf{H_2CO_3} + H_2O = \mathbf{CuCO_3 \cdot Cu(OH)_2} + 2H^+ + 2e^-$$
$$E = +0.525 - 0.0591\text{pH} - 0.0295\log[H_2CO_3] \quad (68)$$

$$\mathbf{Cu_2O} + HCO_3^- + H_2O = \mathbf{CuCO_3 \cdot Cu(OH)_2} + H^+ + 2e^-$$
$$E = +0.337 - 0.0295\text{pH} - 0.0295\log[HCO_3^-] \quad (69)$$

$$\mathbf{Cu_2O} + CO_3^{--} + H_2O = \mathbf{CuCO_3 \cdot Cu(OH)_2} + 2e^-$$
$$E = +0.031 - 0.0295\log[CO_3^{--}] \quad (70)$$

$$\mathbf{CuCO_3 \cdot Cu(OH)_2} + H_2O = \mathbf{Cu_2O_3} + H_2CO_3 + 2H^+ + 2e^-$$
$$E = +1.792 - 0.0591\text{pH} + 0.0295\log[H_2CO_3] \quad (74)$$

$$\mathbf{CuCO_3 \cdot Cu(OH)_2} + H_2O = \mathbf{Cu_2O_3} + HCO_3^- + 3H^+ + 2e^-$$
$$E = +1.980 - 0.0886\text{pH} + 0.0295\log[HCO_3^-] \quad (75)$$

$$\mathbf{CuCO_3 \cdot Cu(OH)_2} + H_2O = \mathbf{Cu_2O_3} + CO_3^{--} + 4H^+ + 2e^-$$
$$E = +2.286 - 0.1182\text{pH} + 0.0295\log[CO_3^{--}] \quad (76)$$

These equations as well as the solubility curves given in Figure 112 were used for drawing the equilibrium diagrams of Figure 113. This figure provides the basis for Figure 114, which shows for the four activities of dissolved CO_2 the theoretical conditions of immunity, corrosion, and passivation of Cu at 25°C.

According to Figure 114, small amounts of CO_2 (10^{-3} and 10^{-2} M) exert a passivating action in somewhat neutral and somewhat alkaline

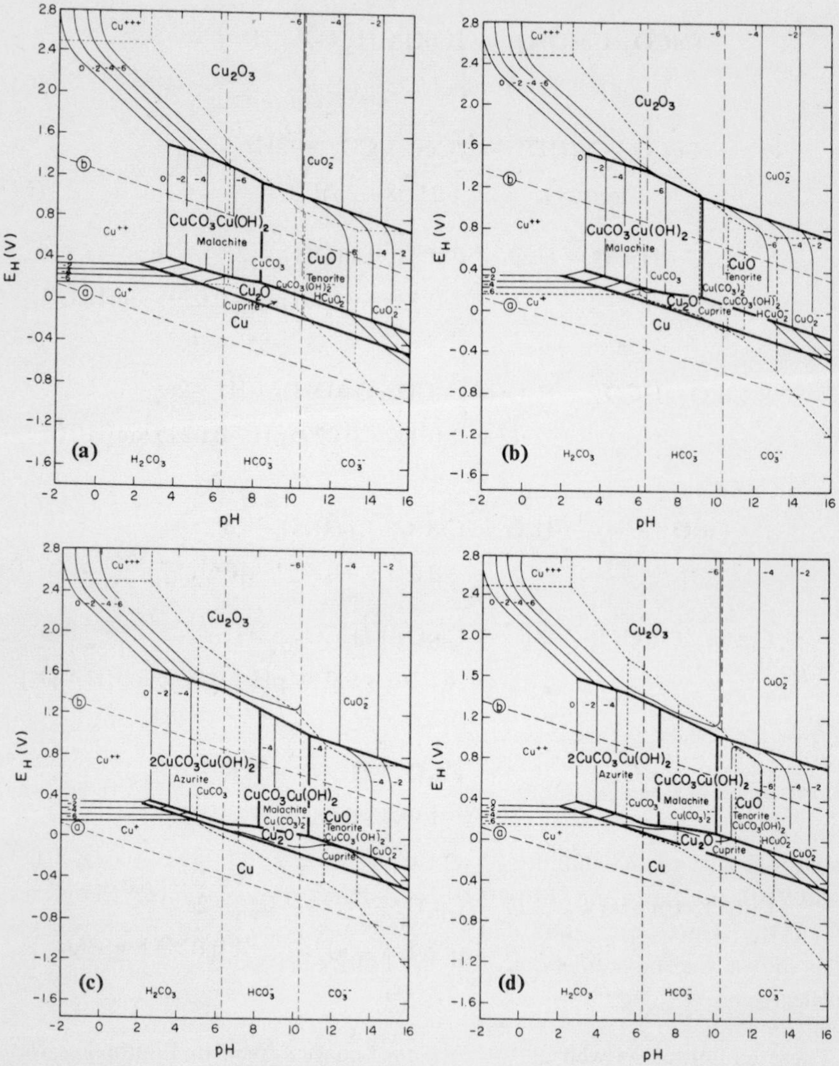

Figure 113. Potential–pH equilibrium diagrams for the ternary system Cu–CO_2–H_2O at 25°C for (a) 10^{-3}, (b) 10^{-2}, (c) 10^{-1}, and (d) 1 mole total dissolved CO_2 per liter (44, 440, 4400, and 44,000 ppm).

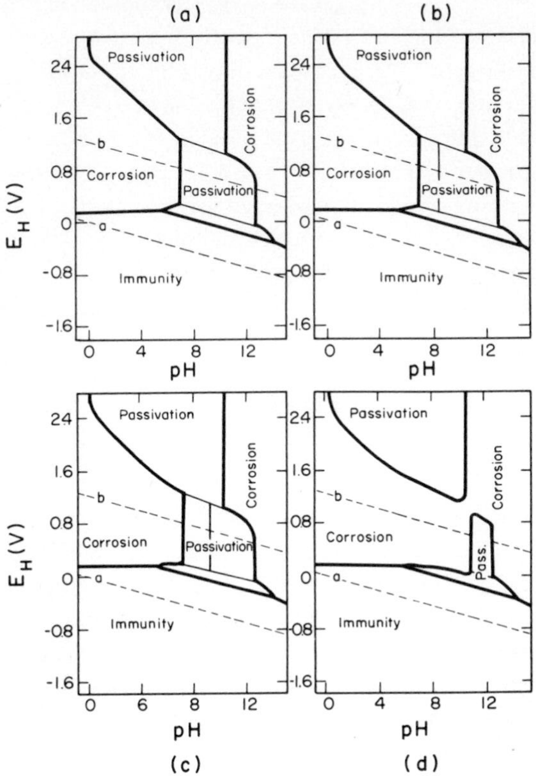

Figure 114. Influence of the amount in total dissolved CO_2 on the theoretical conditions of immunity, corrosion, and passivation of copper at 25°C: (a) CO_2-free solutions; (b) 10^{-3} mole CO_2 per liter; (c) 10^{-2} mole CO_2 per liter; and (d) 10^{-1} mole CO_2 per liter.

solutions due to the stability of malachite. Larger amounts of CO_2 (10^{-1} and 1 M) are activating owing to the stability of the soluble complexes $CuCO_{3aq}$, $Cu(CO_3)^{--}$, and $CuCO_3(OH)_2^{--}$. A brief series of experiments has shown that the predictions of Figure 114 are in accord with experimental observation.[18]

7.2.3. System Cu–Cl–CO_2–SO_3–H_2O

Diagrams such as those described above may be drawn for other ternary systems providing that one knows the values of the standard free

[18] *Rapports Techniques CEBELCOR* **101**, RT. 133 (1965). A more extensive report on this subject is under preparation.

enthalpy of formation for the species to be considered. This work has been done by J. van Muylder and N. de Zoubov for the system Cu–SO_3–H_2O. By sumperimposing the three ternary diagrams containing Cl, CO_2, and SO_3 at given activities in Cl, CO_2, and SO_3, it is possible to obtain diagrams for the quinary system Cu–Cl–SO_3–CO_2–H_2O. Figure 100 shows a diagram relating to solutions containing 22 ppm Cl, 229 ppm CO_2, and 46 ppm SO_3. These concentrations have been selected since they are the ones existing in Brussels tap water. As shown in Section 6.6.6.1.2.1, this diagram may be useful in predicting the conditions of equilibrium of copper derivatives when copper is in contact with either aerated or nonaerated Brussels water.

7.3. DIAGRAMS OF ELECTROCHEMICAL EQUILIBRIA FOR MOLTEN SALTS

It is possible to construct diagrams of thermodynamic equilibria for molten salts in a fashion analogous of that for the construction of the potential–pH diagrams for aqueous solutions. To recapitulate, the potential–pH equilibrium diagrams for aqueous solutions are based on the fact that liquid water can on the one hand dissociate into an acidic component, H^+, and an alkaline component, OH^-, in accordance with the homogeneous chemical reaction $H_2O = H^+ + OH^-$. On the other hand, water can dissociate into an oxidized component, $\boldsymbol{O}_2$, and a reduced component, $\boldsymbol{H}_2$, in accordance with the heterogeneous chemical reaction $2H_2O = 2\boldsymbol{H}_2 + \boldsymbol{O}_2$. The latter reaction can be expressed as two electrochemical reactions as follows:

Reduction $\quad 4H^+ + 4e^- = 2\boldsymbol{H}_2 \quad$ (a)

Oxidation $\quad 2H_2O = \boldsymbol{O}_2 + 4H^+ + 4e^- \quad$ (b)

$$2H_2O = 2\boldsymbol{H}_2 + \boldsymbol{O}_2$$

The conditions of the equilibrium of these two electrochemical reactions can be expressed using the following two electrochemical equations which depend on both pH and potential at 25°C:

$$E_a = 0.000 - 0.0591\text{pH} - 0.0295 \log p_{H_2}$$

$$E_b = +1.228 - 0.0591\text{pH} + 0.0147 \log p_{O_2}$$

In these equations, the electrode potential is in terms of volts relative to the

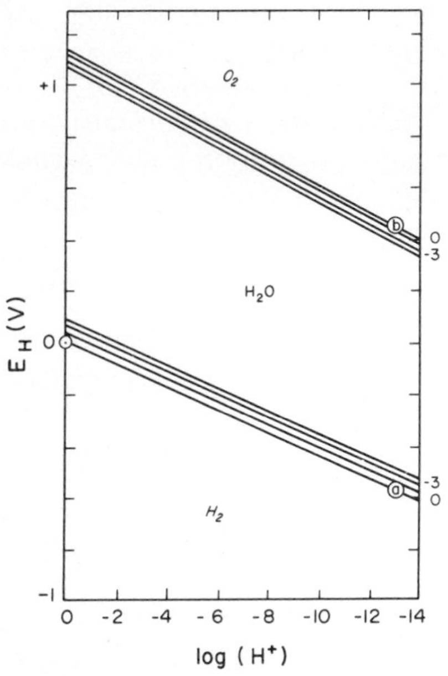

Figure 115. Potential–pH equilibrium diagram for water at 25°C (77°F).

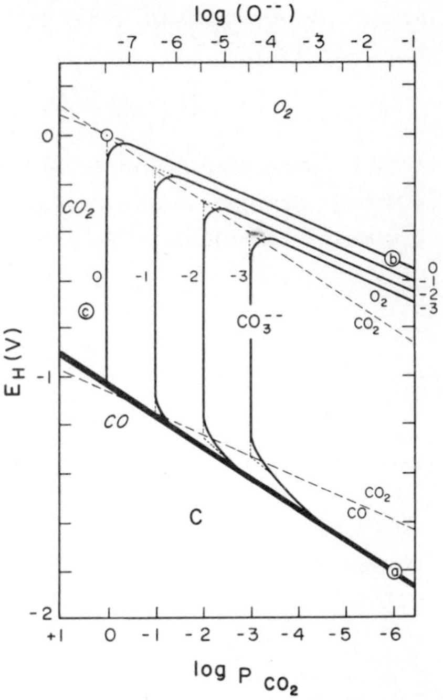

Figure 116. Potential–log p_{CO_2} equilibrium diagram for molten carbonate at 600°C (1112°F).

standard hydrogen electrode; the partial pressures, p_{H_2} and p_{O_2}, are in atmospheres, and the activity of the water forming the solvents is considered to be constant and equal to that of pure water. These two equations make possible the graphic representations shown in Figure 115. This figure shows for the temperature of 25°C the conditions of electrochemical equilibrium of water as affected by the pH, electrode potential, and partial pressure of the prospective gases.

Using the procedure described above for considering the behavior of aqueous solutions, we now consider the case of molten carbonate solutions. The carbonate ion, CO_3^{--}, present in molten carbonates can also dissociate into an acidic component $\boldsymbol{CO_2}$ and an alkaline component, O^{--}, in accordance with the chemical reaction

$$CO_3^{--} = \boldsymbol{CO_2} + O^{--}$$

Further, it is well known that $\boldsymbol{CO_2}$ can be resolved into an oxidized component, O_2 and reduced components, C and CO, in accordance with the chemical reactions

$$CO_2 = C + O_2 \text{ or } 2CO_2 = 2CO + O_2$$

Both of the above-mentioned chemical reactions can be expressed as two separate electrochemical reactions containing oxidation and reduction processes as follows:

Reduction $$CO_2 + 4e^- = C + 2O^{--} \quad \text{(a)}$$

Oxidation $$2O^{--} = O_2 + 4e^- \quad \text{(b)}$$

$$CO_2 = C + O_2$$

or

Reduction $$2CO_2 + 4e^- = 2CO + 2O^{--} \quad \text{(c)}$$

Oxidation $$2O^{--} = O_2 + 4e^- \quad \text{(d)}$$

$$2CO_2 = 2CO + O_2$$

As stated elsewhere,[19] the equilibrium conditions of these various chemical and electrochemical reactions make possible the graphical representation shown in Figure 116. This figure, drawn for a temperature of 600°C, shows the area of stability of the fused carbonate, CO_3^{--}, as a function of the logarithms of the partial pressure of CO_2 (and of the activity in dissolved O^{--} ions) as the x axis and the electrode potential, E_{SOE}, *versus the standard oxygen electrode* (under an oxygen pressure of 1 atm) as the y axis. The various lines are drawn for total pressures of O_2+CO_2+CO which correspond to 10^{-3}, 10^{-2}, 10^{-1}, and 10^{-0} atm. Figure 119 demonstrates particularly clearly the conditions under which the carbonate melt oxidizes in gaseous oxygen, can be acidified in gaseous CO_2, and can be reduced to gaseous CO or to solid carbon. It also shows the conditions where, in the presence of a carbonate melt, oxygen can be reduced, and CO and carbon can be oxidized. This knowledge is quite useful in fuel-cell technology.

It was previously described in aqueous considerations that the stability of the metal, relative to an environment, can be estimated by simply superimposing equilibrium diagrams. For example, it was possible to determine the stability of silver in the presence of water, simply by superimposing their

[19] See footnote 11 on page 299.

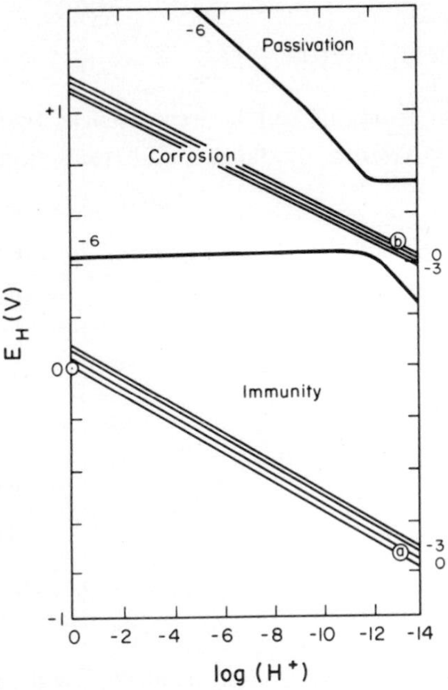

Figure 117. Theoretical conditions of corrosion, immunity, and passivation for silver in the presence of aqueous solutions at 25°C.

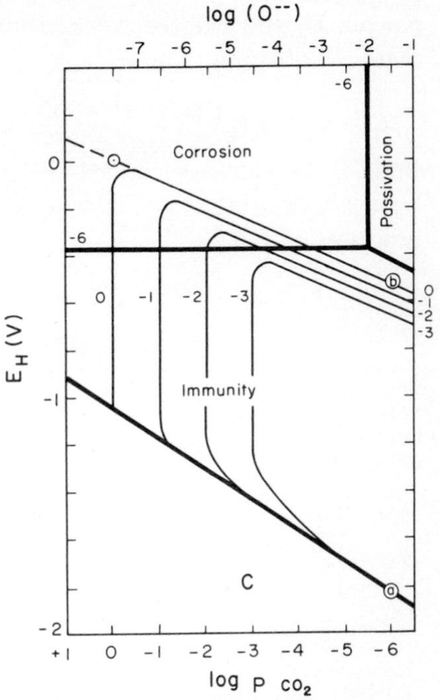

Figure 118. Theoretical conditions of corrosion, immunity, and passivation for silver in the presence of molten carbonate at 600°C (1112°F).

respective diagrams. This is done in order to obtain Figure 117 by the use of Figure 115 and we obtain the general conditions of corrosion, immunity, and passivation for silver in aqueous solutions at 25°C. In this same way, superimposing the stability conditions of silver, its oxide, $\mathbf{Ag_2O}$, and the ion, Ag^+, in fused carbonate on Figure 116 yields Figure 118 which represents the corrosion, immunity, and passivation conditions for silver in the presence of carbonate melts at 600°C.

The use of these graphical techniques and especially their various possible superpositions can be of great value in investigating the many processes which involve fused salts.[20] In particular, the investigation of operating conditions for fuel cells, corrosion, fusion electrolysis, etc. may be greatly aided by such considerations.

[20] M. Pourbaix, Diagrams of electrochemical equilibria and experimental applications, *Rapports Techniques CEBELCOR* **102**, RT. 137 (1966) (in French).

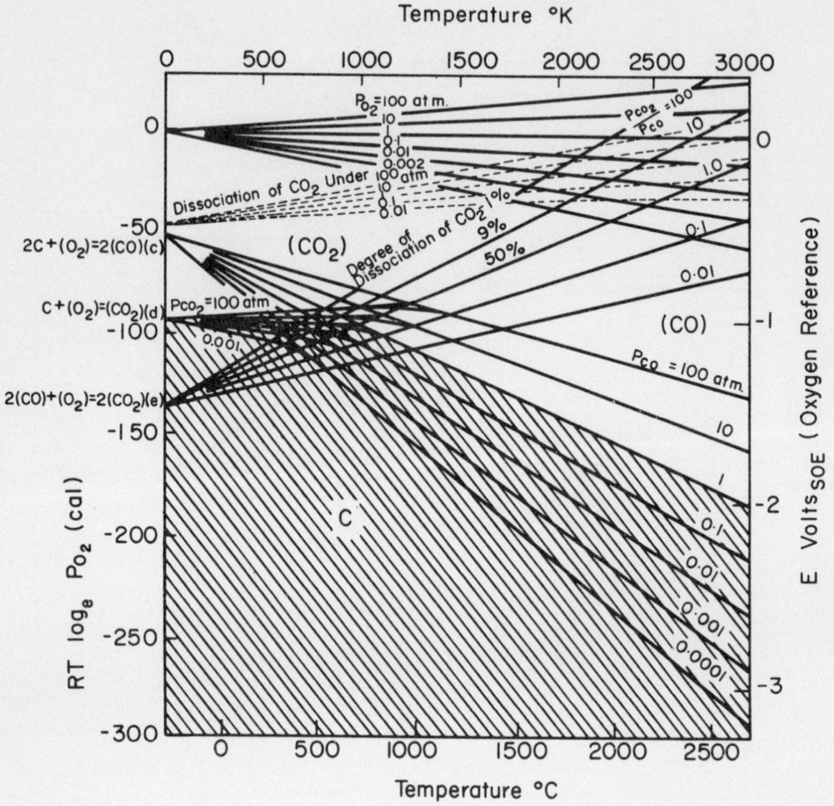

Figure 119. Equilibria carbon–oxygen.

7.4. THE INFLUENCE OF TEMPERATURE ON ELECTROCHEMICAL EQUILIBRIA. ELECTRODE POTENTIAL–TEMPERATURE DIAGRAMS

In a very early publication[21] and on the occasion of discussions with C. L. Goodeve and H. J. T. Ellingham, we described a method for studying complicated equilibria in which graphs are used with temperature, T, as the x axis and the function $RT \ln p_{O_2}$ as the y axis for the case of oxidation equilibria. Such diagrams, as said by H. J. T. Ellingham when they were

[21] M. Pourbaix and C. Rorive–Boute, Graphical study of metallurgical equilibria, *Discussions Faraday Soc.* **4**, pp. 139–154, 223–227, 239–240 (1948).

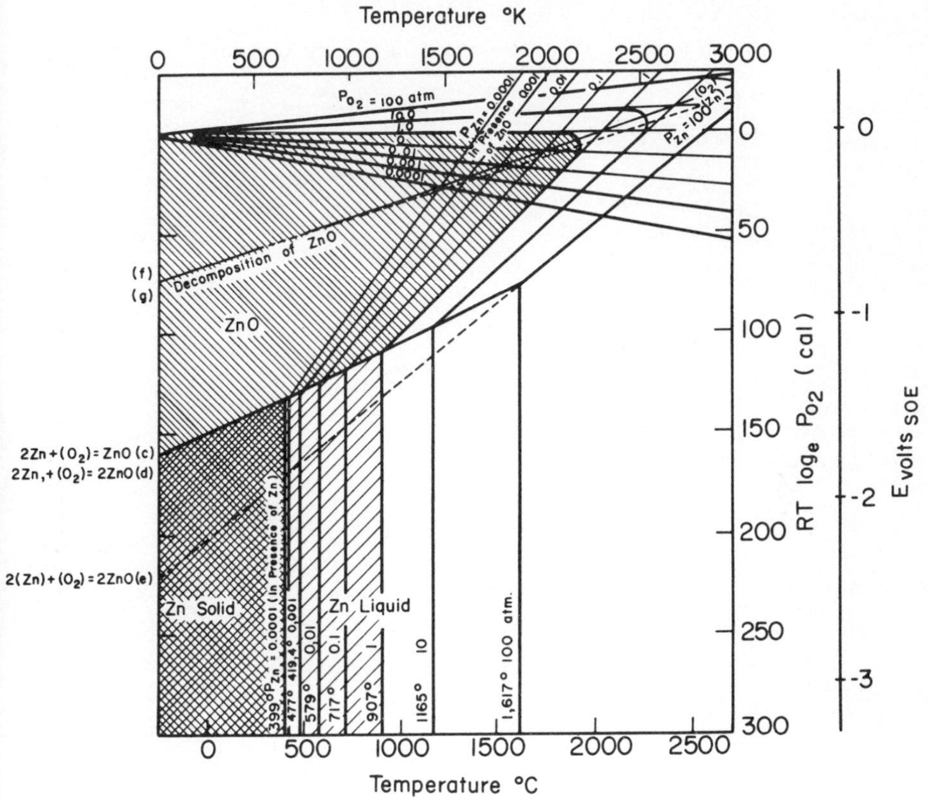

Figure 120. Equilibria zinc–oxygen.

presented,[22] allow for the direct reading of the influence of temperature on the "driving force" (or affinity) expressed by the free energy, $-\Delta G$, as measured per mole of the reaction of diatomic oxygen, O_2, or by an electromotive force. Figures 119, 120, and 121 show three such graphs in which the ordinate scale is set for both the free energy (or free enthalpy), $-\Delta G$, and the electromotive force. This electromotive force in reality measures the *equilibrium electrode potential*, E_0, of each of the reactions plotted, as taken as if the reaction takes place electrochemically in the presence of an electrolyte. These potentials refer numerically to a standard oxygen electrode which would function reversibly under an oxygen pressure of 1 atm in the presence of the same electrolyte at the same temperature. For example, as shown by Figure 119 relating to the carbon–oxygen system,

[22] H. J. T. Ellingham, *Discussions Faraday Soc.* **4**, 227–228 (1948).

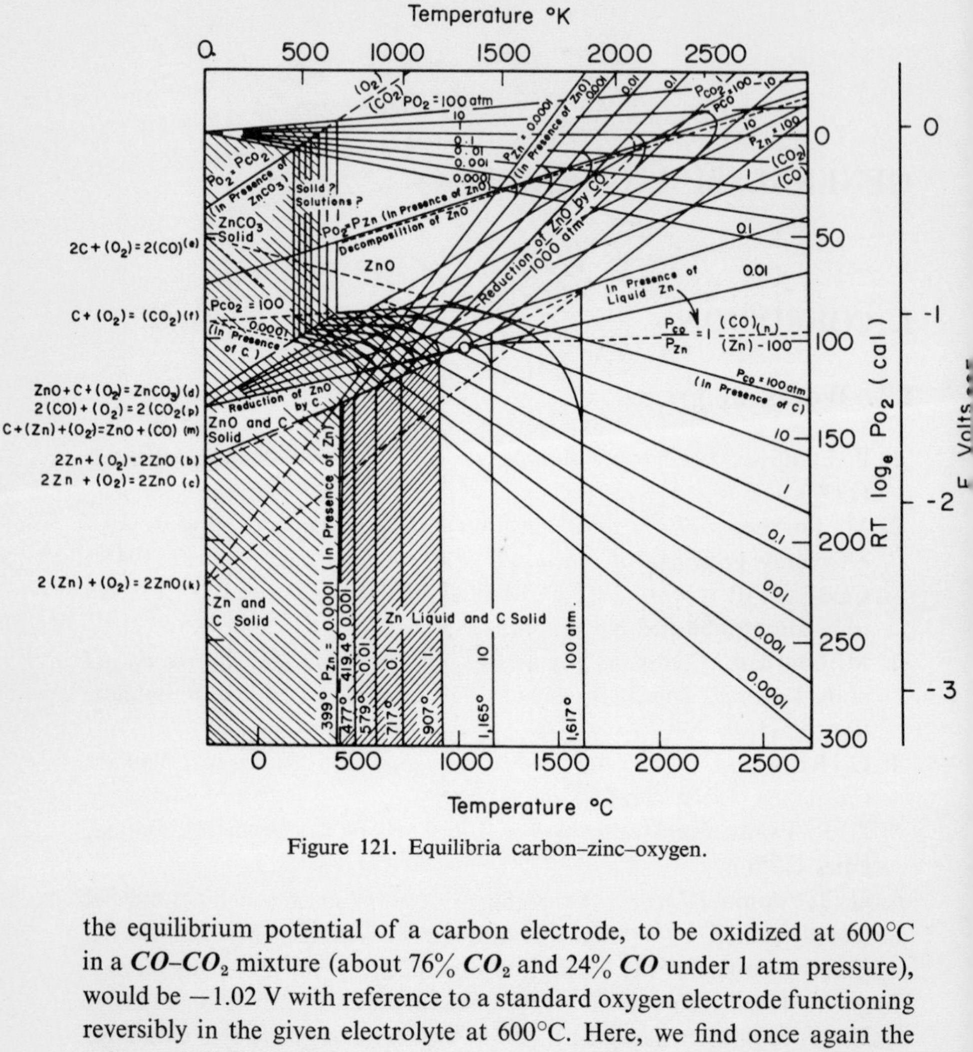

Figure 121. Equilibria carbon–zinc–oxygen.

the equilibrium potential of a carbon electrode, to be oxidized at 600°C in a ***CO–CO***$_2$ mixture (about 76% ***CO***$_2$ and 24% ***CO*** under 1 atm pressure), would be −1.02 V with reference to a standard oxygen electrode functioning reversibly in the given electrolyte at 600°C. Here, we find once again the value −1.02 V which is, as may be seen in Figure 116, the equilibrium electrode potential of carbon in a carbonate melt at a temperature of 600°C. Similar equilibrium diagrams may be drawn in the presence of gases other than oxygen such as chlorine, sulfur, and hydrogen.

The procedure used in the above illustration may be applied to all metals and nonmetals in the presence of all electrolytes such as water, organic solvents, fused salts, and semiconducting oxides. Figures 120 and 121 show such diagrams for the systems zinc–oxygen and carbon–zinc–oxygen (*Rapports Techniques CEBELCOR* **115**, RT. 181, RT. 182 (1970)).

GENERAL BIBLIOGRAPHY[1]

1. CORROSION

1.1. Works in French

1. P. Erculisse, *Les bases théoriques de la corrosion*, A.B.E.M., Brussels (1937).
2. M. Pourbaix, *Thermodynamique des solutions aqueuses diluées. Représentation graphique du rôle du pH et du potentiel*, Thesis, Delft (1945); CEBELCOR reprint (1963), CEBELCOR Publication F. 227. (English translation published by Arnold; see Ref. 22.)
3. M. Pourbaix, *Corrosion, passivité et passivation du fer. Le rôle du pH et du potentiel*, Thesis, Brussels (1945), extract; CEBELCOR Publication F. 21.
4. G. Batta and E. Leclerc, *Chimie des matériaux techniques*, Vaillant Carmanne, Liège (1946).
5. U. R. Evans, *Précis de corrosion* (translated by G. Dechaux), Dunod, Paris (1951).
6. G. Y. Akimov, *Théorie et méthodes d'essais de la corrosion des métaux* (supplemented by N. D. Tomashov, translated by S. Medvedieff), Dunod, Paris (1957).
7. A. J. Maurin, *Manuel d'anticorrosion*, Eyrolles, Paris (1961).
8. M. Pourbaix *et al.*, *Atlas d'équilibres électrochimiques*, Gauthier-Villars, Paris, and CEBELCOR, Brussels (1963); CEBELCOR Publication F. 226. (English translation published by Pergamon Press; see Ref. 34.)
9. J. Bénard *et al.*, *L'oxydation des métaux, I. Processus fondamentaux*; *II. Monographies*, Gauthier-Villars, Paris (1964).

[1] The publications in this list may be consulted in the laboratories of CEBELCOR.

10. H. Coriou and L. Grall, *Problèmes de corrosion aqueuse de matériaux de structure dans les constructions nucléaires*, Rapport C.E.A., R. 2.600, Centre d'Études Nucléaires de Saclay (1964).
11. *Les progrès dans la transformation de l'acier*, C. R. 2° Congrès Acier Communauté Européenne Charbon–Acier (CECA), Luxembourg, October 1965 (1967). (See also Ref. 39.)
12. Publications of CEBELCOR in French (F) (mainly *Rapports Techniques*, R.T.). List on request.
13. Reviews:
    Corrosion et Anticorrosion, Paris,
    Chimie des Peintures, Brussels,
    Galvano, Paris,
    Navires, Ports et Chantiers, Paris,
    Peintures, Pigments et Vernis, Paris.

1.2. Works in German

14. *Werkstoff-Tabellen*, Dechema, Frankfurt.
15. F. Ritter, *Korrosionstabellen metallischer Werkstoffe*, Springer, Vienna (1958).
16. W. Wiederholt, *Taschenbuch des Metallschutzes*, Wissenschaftliche Verlagsgesellschaft, Stuttgart (1960).
17. F. Tödt, *Korrosion und Korrosionsschutz*, De Gruyter, Berlin (1961).
18. K. J. Vetter, *Electrochemische Kinetik*, Springer, Berlin (1961) (English translation by Academic Press, see Ref. 40).
19. H. Kaesche, *Die Korrosion der Metalle*, Springer, Berlin (1966).
20. Reviews:
    Metalloberfläche, Munich,
    Korrosion und Metallschutz, Berlin,
    Werkstoffe und Korrosion, Weinheim-Bergst.

1.3. Works in English

21. H. H. Uhlig *et al.*, *Corrosion Handbook*, Wiley, New York, and Chapman and Hall, London (1948).
22. M. Pourbaix, *Thermodynamics of Dilute Aqueous Solutions, with Applications to Electrochemistry and Corrosion*, Thesis, Delft (1945) (Translation of Ref. 2 by J. N. Agar), Arnold, London (1949).
23. R. Rabald, *Corrosion Guide*, Elsevier, Amsterdam (1951).
24. U. R. Evans, *The Corrosion and Oxidation of Metals: Scientific Principles and Practical Applications*, Arnold, London (1960).

25. M. G. Fontana, *Course 14: Corrosion*, American Society for Metals (1960).
26. E. C. Potter, *Electrochemistry, Principles and Applications*, Cleaver-Hume, London (1961).
27. F. L. Laque and H. R. Copson, *Corrosion Resistance of Metals and Alloys*, American Chemical Society, Reinhold, New York (1963).
28. U. R. Evans, *An Introduction to Metallic Corrosion*, Arnold, London (1963).
29. H. H. Uhlig, *Corrosion and Corrosion Control: An Introduction to Metallic Corrosion Science and Engineering*, Wiley, New York and London (1963).
30. L. L. Shreir *et al.*, *Corrosion: I. Corrosion of Metals and Alloys; II. Corrosion Control*, Newnes, London (1963).
31. F. Champion, *Corrosion Testing Procedures*, Chapman and Hall, London (1964).
32. J. M. West, *Electrodeposition and Corrosion Processes*, D. van Nostrand, New York (1965).
33. K. Hauffe, *Oxidation of Metals*, Plenum Press, New York (1965).
34. M. Pourbaix *et al.*, *Atlas of Electrochemical Equilibria in Aqueous Solutions*, Pergamon Press, Oxford, and CEBELCOR, Brussels (1966) (Translation of Ref. 8 by J. A. Franklin).
35. J. C. Scully, *The Fundamentals of Corrosion*, Pergamon Press, Oxford (1966).
36. N. Tomashov, *Theory of the Corrosion and Protection of Metals*, MacMillan, New York (1966) (Translation of Ref. 55).
37. N. Tomashov and G. Chernova, *Passivity and Protection of Metals Against Corrosion*, Plenum Press, New York (1967) (Translation of Ref. 58).
38. M. G. Fontana and N. D. Greene, *Corrosion Engineering*, McGraw-Hill, New York (1967).
39. *Progress in Steel Processing*, Proc. 2nd Steel Congress, European Coal and Steel Community, Luxembourg, October 1965 (1967). (See also Ref. 11.)
40. K. J. Vetter, *Electrochemical Kinetics—Theoretical and Experimental Aspects*, Academic Press, New York (1967) (Translation of Ref. 18).
41. U. R. Evans, *The Corrosion and Oxidation of Metals: First Supplementary Volume*, Arnold, London (1968).
42. M. G. Fontana and R. W. Staehle, eds., *Advances in Corrosion Science and Technology*, Plenum Press, New York (Vol. 1, 1970; Vol. 2, 1972).
43. *NACE Corrosion Course*, NACE, Houston (1971).

44. W. H. Ailor, ed., *Handbook on Corrosion Testing and Evaluation*, Wiley, New York (1971).
45. W. E. Berry, *Corrosion in Nuclear Applications*, Wiley, New York (1971).
46. J. C. Scully, ed., *The Theory of Stress-Corrosion Cracking in Alloys*, Proc. NATO Ericeira Conference 1971, NATO, Brussels (1971).
47. G. Wranglén, *An Introduction to Corrosion and Protection of Metals*, Institut för Metallskydd, Stockholm (1972).
48. Publications of CEBELCOR in English (E) (mainly *Rapports Techniques*, R.T.). List on request.
49. Reviews:
    Anticorrosion, Methods and Materials, London (formerly *Corrosion Technology*),
    Corrosion, Houston,
    Corrosion Abstracts, Houston,
    Corrosion Abstracts, Stockholm,
    Corrosion, Prevention and Control, London,
    Corrosion Technology, London,
    Materials Protection, Houston,
    Material Research Standards, Philadelphia.

1.4. Works in Italian

50. R. Piontelli, *Elementi di Teoria della Corrosione a umido dei Materiali Metallici*, Longanesi, Milan (1961).
51. G. Guzzoni and G. Storage, *Corrosione dei Metalli e loro Protezione*, Hoepli, Milan (1964).
52. G. Bianchi and F. Mazza, *Corrosione e Protezione dei Metalli*, Tamburini, Milan (1968).
53. P. Gallone, *Principi dei Processi Elettrochimici*, Tamburini, Milan (1970).
54. Review:
    Metallurgia Italiana, Milan.

1.5. Works in Russian

55. N. Tomashov, *Teoriya korrozii i zashchity metallov* (Theory of the Corrosion and Protection of Metals), Izd. AN SSSR, Moscow (1959) (English translation published by Macmillan, see Ref. 36).
56. A. V. Shreider, *Oksidirovanie alyuminya i ego splavov* (Oxidation of Aluminum and Its Alloys), Scientific Publishing House for Ferrous and Nonferrous Metallurgy, Moscow (1960).

57. F. N. Tavadze and C. N. Mandygaladze, *Korroziya i zashchita metallov v naturalnikh lechebnikh vodakh* (Corrosion and Protection of Metals in Medicinal Waters), Izd. AN SSR, Moscow (1963).
58. N. Tomashov and G. Chernova, *Passivnost': zashchita metallov ot korrozii.* (Passivity and Protection of Metals Against Corrosion), Izd. AN SSR, Moscow (1965) (English translation published by Plenum Press, see Ref. 37).

1.6. Works in Various Languages

59. *C. R. 1er Symposium Européen sur les Inhibiteurs de Corrosion*, Ferrara, September-October 1960, University of Ferrara (1961).
60. *C. R. 2e Symposium Européen sur les Inhibiteurs de Corrosion*, Ferrara, September 1965, University of Ferrara (1966).
61. *C. R. 3e Symposium Européen sur les Inhibiteurs de Corrosion*, Ferrara, September 1970, University of Ferrara (1971).
62. *Steel in the Chemical Industry*, Proc. 3rd Steel Congress, European Coal and Steel Community, Luxemburg, July 1968 (1972).
63. Reviews:
    Corrosion Science, Oxford,
    Electrochimica Metallorum, Milan.

2. GENERAL THERMODYNAMICS AND PHYSICAL CHEMISTRY

65. T. de Donder and P. van Rysselberghe, *L'affinité*, Gauthier-Villars, Liège (1936).
66. I. Prigogine, *Thermodynamique des phénomènes irréversibles*, Desoer, Liège (1945).
67. F. E. C. Scheffer, *Toepassingen van de Thermodynamika op chemische Processen*, Waltman, Delft (1945).
68. I. Prigogine and R. Defay, *Thermodynamique chimique*, Desoer, Liège (1950). (English translation by Longmans Green, see Ref. 71).
69. W. I. Latimer, *Oxidation Potentials* (2nd ed.), Prentice Hall, New York (1952).
70. R. Defay, *Eléments de chimie physique*, Presses Universitaires de Bruxelles, Brussels (1954).

71. I. Prigogine and R. Defay, *Chemical Thermodynamics*, Longmans Green, London (1954) (Translation of Ref. 68 by D. H. Everett).
72. P. van Rysselberghe, *Electrochemical Affinity*, Hermann, Paris (1955).
73. A. I. Beljajew, E. A. Shemtschushina, and L. A. Firsanowa, *Physikalische Chemie Geschmolzener Salze*, Deutscher Verlag für Grundstoffchemie, Leipzig (1964).
74. G. L. Clark and G. G. Hawley, *The Encyclopedia of Chemistry* (2nd ed.), Reinhold, New York (1966).
75. R. Defay, I. Prigogine, A. Bellemans, and D. H. Everett, *Surface Tension and Adsorption*, Longmans Green, London (1966).

3. ELECTROCHEMISTRY

76. Proceedings of the Meetings of C.I.T.C.E. (*Comité International de Thermodynamique et de Cinétique Electrochimiques*)
    Second Meeting: Milan 1950, Tamburini, Milan,
    Third Meeting: Berne 1951, Manfredi, Milan,
    Sixth Meeting: Poitiers 1954, Butterworths, London,
    Seventh Meeting: Lindau 1955, Butterworths, London,
    Eighth Meeting: Madrid 1956, Butterworths, London,
    Ninth Meeting: Paris 1957, Butterworths, London.
77. G. Kortüm and J. O'M. Bockris, *Textbook of Electrochemistry*, Elsevier, The Hague (1951).
78. P. Delahay, *New Instrumental Methods in Electrochemistry. Theory, Instrumentation, and Applications to Analytical and Physical Chemistry*, Interscience, New York and London (1954).
79. V. Gaertner, *Electrochimie pratique. Principles et technologie*, Eyrolles and Gauthier-Villars, Paris (1955).
80. J. O'M. Bockris, *Modern Aspects of Electrochemistry*, No. 1, Butterworths, London, and Plenum Press, New York (1959).
81. G. Milazzo, *Electrochemistry. Theoretical Principles and Practical Applications*, Elsevier, Amsterdam (1963).
82. C. Hampel *et al.*, *The Encyclopedia of Electrochemistry*, Reinhold, New York (1964).
83. A. de Bethune and N. A. Suendenman Loud, *Standard Aqueous Electrode Potentials and Temperature Coefficients at 25°C*, Hampel, Skokie (Illinois).
84. L. I. Antropov, *Teoreticheskaya elektrokhimia* (Theoretical Electrochemistry), Vyssh. Shkola, Moscow (1964).

85. A. Vanhaute, *Industriele Elektrochemie*, Waltman, Delft (1966).
86. J. O'M. Bockris and A. K. N. Reddy, *Modern Electrochemistry* (2 volumes), Plenum Press, New York (1970).

4. ANALYTICAL CHEMISTRY

87. G. Charlot, *Théorie et méthode nouvelle d'analyse qualitative*, Masson, Paris (1949).
88. G. Charlot and R. Gauguin, *Les méthodes d'analyse des reactions en solution*, Masson, Paris (1951).
89. G. Charlot and D. Bézier, *Méthodes électrochimiques d'analyse*, Masson, Paris (1954).

5. GEOLOGY

90. R. M. Garrels, *Mineral Equilibria at Low Temperature and Pressure*, Harper's Geoscience Series, New York (1960).
91. R. M. Garrels and C. L. Christ, *Solutions, Minerals and Equilibria*, Harper and Row, New York (1965).

6. METALLURGY

92. A. G. Guy, *Physical Metallurgy for Engineers*, Addison-Wesley, Reading, Massachusetts (1962).
93. J. A. Kirkaldy and R. G. Ward, *Aspects of Modern Ferrous Metallurgy*, University of Toronto (1964).
94. A. D. Merriman, *A Concise Encyclopedia of Metallurgy*, Elsevier, Amsterdam (1965).
95. J. Mackowiak, *Physical Chemistry for Metallurgists*, George Allen & Unwin Ltd., London (1966).
96. A. R. Poster, *Handbook of Metal Powders*, Reinhold, New York (1966).

AUTHOR INDEX

SUBJECT INDEX

Italic numbers refer to section headings in the text.